HANDBOOK OF

Spices, Seasonings, and Flavorings

SECOND EDITION

HANDBOOK OF

Spices, Seasonings, and Flavorings

SECOND EDITION

Susheela Raghavan

CRC Press
Taylor & Francis Group
Boca Raton London New York

CRC Press is an imprint of the
Taylor & Francis Group, an **informa** business

CRC Press
Taylor & Francis Group
6000 Broken Sound Parkway NW, Suite 300
Boca Raton, FL 33487-2742

First issued in paperback 2019

© 2007 by Taylor & Francis Group, LLC
CRC Press is an imprint of Taylor & Francis Group, an Informa business

No claim to original U.S. Government works

ISBN-13: 978-0-8493-2842-8 (hbk)
ISBN-13: 978-0-367-39009-9 (pbk)

Library of Congress Cataloging-in-Publication Data

Raghavan, Susheela.
    Handbook of spices, seasonings, and flavorings / Susheela Raghavan. -- 2nd ed.
        p. cm.
    Includes bibliographical references and index.
    ISBN 0-8493-2842-X
    1. Spices--Handbooks. 2. Cookery (Spices) I. Title.

TX406.R34 2006
641.3'383--dc22   2006046571

Visit the Taylor & Francis Web site at
http://www.taylorandfrancis.com

and the CRC Press Web site at
http://www.crcpress.com

Dedication

To my parents . . .
Pathmavathy Kumaran and Kattary Raghavan

I dedicate this book to my ma and cha who planted the seed of taste within me, from which my thirst for knowledge of spices and flavors grew. It was this exposure that enabled me to truly appreciate and enjoy the many diverse foods and flavors from around the world.

Table of Contents

Preface

My gastronomic heritage began while I was growing up in Malaysia. Watching my late grandma (whom we called Periama) grinding the soaked rice–lentil mixture for Sunday's breakfast, picking kari leaves for my late ma's aromatic crab curry, and listening to my late cha's (father's) food adventures during meals, all created in me a passion for food, spices, and cultures. For mom, cooking was a creative process—every day there had to be something new and different on the table. She never hurried her cooking and never settled for less than the best in her choice of spices and ingredients. I observed and learned her pride in creating the ultimate flavor and absorbed her approach to freshness, flavor, and healthy eating. For cha, food had no boundaries. It surpassed all cultures and religions. He taught me to explore and try all foods and flavors available in Malaysia, whether Malay, Chinese, Indian, Indonesian, Thai, or Western. His adventure for unique flavors and his cultural experiences from his travels around the world provided me the beginnings to understand the many global foods. Mom's endless search for the ultimate flavor profiles and her zest for cooking, and cha's enthusiasm and appreciation of all ethnic cuisines gave me an appetite for adventurous eating and a curiosity about new flavors.

Their passion for new foods instilled in me an appreciation for all cultures and their cooking, and their spirit ultimately influenced my career as a food developer. My search for unique flavors began at home, in Malaysia, with mom's fragrant fish sambals, sister-in law's (Santha's) delectable acar, or the many visits to the hawkers to taste their variety of noodle preparations. To me every meal was a discovery! I have traveled all over the globe to experience the exotic floating markets of Thailand, to cook aromatic fish curries on the stone charcoal ovens of Kerala, India, to prepare a variety of smoked chilies for moles in Oaxaca, Mexico, to dine in the country pubs of England, to savor the freshly prepared grilled sardines of Portugal, and to taste the perfect chili in Texas, United States.

All of this I continue to do, in search of new and unique flavors, and to study their cultural origins. I am not alone in my search. Food professionals in the United States, and around the world, are continually looking to use innovative spices and flavorings because of the growing global demand for variety, exotic flavors, and authenticity. The increasing pursuit of fresh and all-natural foods with no preservatives (or using natural preservatives and natural colors), no MSG or HPP, and using foods to prevent ailments and to attain a healthier lifestyle, continue to grow. I wrote this book as a guide for food developers to understand and gather a vast technical and culinary knowledge of spices, seasonings, and flavorings and to continue to meet the consumers' growing demands for new and exotic flavors, and to follow a healthy approach to their lifestyles.

Today's food development continues to grow in the direction of a "techno-culinary" trend—connecting science or technology with the culinary arts. Cultural influences, food trends, and nutrition are incorporated into the development of foods. A food developer needs to combine technical knowledge, creative talent, and an understanding of the cultural aspects of the cuisine and its preparation, in order to develop products that will be a success in the marketplace. This book incorporates technical information about spices—forms, varieties, properties, applications, medical properties, regulations, labeling, and quality specifications—with trends, spice history, and the culture behind their cuisines. I designed the format of this book so that it becomes a significant tool for the many professionals who develop and market foods. It will be a handy guide in today's multicultural and acculturated society. The product developer needs to understand the creative use of spices and flavorings in addition to technical know-how. The chef needs to add technical information to balance his/her creativity. The flavorist needs to learn the origins and varieties of spices and how they are prepared and used in ethnic cuisines, to understand their differing flavor profiles. Nutritionists need to be aware of the eating habits for a diverse ethnic population and the spices used in their cooking to develop menus that are appropriate for them. Food marketers and sales professionals can use the cultural and technical information to keep abreast of food trends. With the information I provide in this book, they can work as a team to develop and market successful products.

First and foremost, I have tried to make this book a comprehensive guide to spices. Spices are the building blocks of flavors and set apart one cuisine from another. They define flavors and cuisines, and are important tools for providing consistency and color. They create the desired taste, characterize cuisines, and differentiate one recipe from another. Understanding spices in their fullest capacities is the cornerstone of successful product or seasoning development. Today's consumers are becoming more sophisticated about the use of spices. Thus, our knowledge of spices and their technology, their use and compatibility with other ingredients, and how they are prepared, becomes of paramount importance. This book contains detailed descriptions of each spice, arranged alphabetically. While many reference books on spices include alphabetized descriptions, I believe that the similarity between this book and others ends there.

Consistent with my desire to create a truly comprehensive and global reference on spices, this book goes beyond a dry technical description of spices. It describes each spice's varieties, forms, and chemical components that typify its flavor and color. It includes a description of spice properties, both chemical and sensory, and the culinary information that will aid in product development. This book also explains how each spice is used in many applications around the world, lists the popular global spice blends that contain the spice, describes each spice's folklore and traditional medicinal usage and, most importantly, for purchasing, provides translations of each spice's name in global languages.

In researching this book, I became aware that there were no comprehensive guides to many of the spices that could help food developers create new products based on the popular ethnic cuisines from around the world. Consequently, I have attempted to provide detailed descriptions of many global varieties of spices or closely related ones, including chile peppers, mints, and black peppers. In addition, this book goes beyond other spice books by describing other important ingredients found among the world's cuisines that are essential in providing flavors, textures, colors, and nutritional value to foods. It describes how these ingredients, which I call flavorings (including wrappers, rhizomes, flowers, fruits, or seafood), are commonly used with spices to create authentic or new flavors.

The chapter on seasonings describes major regional ethnic cuisines and their characteristic flavor profiles and seasonings with some popular examples. In addition, I provide many regional variations of a seasoning or spice blend, including sofrito, adobo, curry, or hot sauce. This chapter is not intended as an in-depth study of each region's flavor profile, but it is written to provide a general understanding of some typical flavor profiles of each of these regions as a strong foundation for product development. The final chapter discusses seven examples of spice blend or seasoning formulations to give you some understanding of and to provide assistance on creating ethnic seasoning formulations for marinades, sauces, dips, salad dressings, or curries.

In writing this book, I have tried to create a complete modern reference book on spices, seasonings, and flavorings. I have included traditionally popular spices and flavorings, as well as those that are emerging in the United States to create authentic or fusion products. It is designed to meet the challenges and demands of today's dynamic market. My ultimate aim, however, was to share with the reader some of my passion and enthusiasm for spices, and how learning to understand and use spices has given me a sense of creative adventure in my everyday life and work.

A note on terminology in this book: throughout history, the various parts of plants have been cultivated and used for their aromatic, fragrant, pungent, or other desirable qualities. This book uses the term "spice" to refer to all of the edible parts of a plant used for flavoring foods—including the root, stem, bark, seeds, rhizome, and the leafy plant parts usually referred to as herbs in European and North American cuisines. There are several reasons for this usage. In the case of herbs, it avoids the shifting definitions of what an herb is, which have varied greatly over time. In addition, not all herbs are used in seasoning foods; many are not edible and do not function as spices, but as healthy tools. Moreover, many traditional cultures today do not separate these leafy spices into a distinct herb category. This book attempts to discuss and define spices from a global perspective; therefore, a global approach to defining flavoring ingredients is most appropriate. Finally, as with roots, stems, seeds, and flowers and other plant parts, the purpose of these leafy plant parts is to "spice up" food or beverage products, and their collective grouping is the most logical.

Susheela Raghavan

Acknowledgments

I wish to thank my daughter, Geeta, for encouraging me to write this book and showing patience for my absence during this period. She also wrote the beautiful dedication to my parents. With this second edition, she has added more flavor, with her own food and culture discoveries around the world. I am grateful to my husband, Bob Roach, who helped me immensely with editing and organization. He was also my best critic throughout. I wish to thank my family members around the globe, who gave me information and moral support during this time. And I wish to remember my parents' spirit in this undertaking, whether simply creating a meal together, sitting down at a meal or having a conversation at a meal. My mom was always there, patiently giving her support and providing me with information on the many healing properties of various spices that she and her mom had grown up with. And I gratefully thank my father, *cha*, for encouraging me to experience and appreciate all foods and cultures. I want to also give credit to my late sister Prem, who patiently read the first edition carefully to provide her insight and suggestions. I am thankful to my brother Sathee who helped in translating many spices into local dialects.

I also wish to express my thanks to many colleagues in the United States and around the world who encouraged me to write this book and who provided help in translating the names of spices into global names. In addition, I would like to say that I have developed a greater understanding of and insight into many of the local and global flavors during my travels around the United States and the world, whether for work or for pleasure. I continue to do so.

Author

Susheela Raghavan is founder and president of Horizons Consulting LLC., dba Taste of Malacca, a New Rochelle, New York, supplier of innovative and gourmet spice blends for retail, wholesale, and food service. Her consulting services include spotting trends and developing ethnic and new American products for the U.S. and global markets. Susheela has over twenty-five years of product development, corporate and research experience in the food, spice, and flavor industries, and in specialty food service establishments. Her scientific grounding, corporate experience, academic standing, culinary skills, and a keen study of world regional culinary culture, combined with a creative, independent spirit, bring uniqueness to her products and services. Susheela focuses on understanding authentic flavors and adapting them to emerging palates and healthy lifestyles.

In addition, Susheela actively contributes to the development of her field. She has authored numerous articles on trends, ethnic cuisines and flavors, and spices and seasonings, in *Food Technology, Food Product Design, Fine Cooking*, and *Restaurant USA*. She has also served significantly in the academic realm, as adjunct professor at New York University's Department of Nutrition, Food Studies and Public Health for seven years, guest lecturer at The Institute of Culinary Education and the French Culinary Institute, and as a scientific lecturer for the U.S. Institute of Food Technologists. In addition, she conducts training programs and symposiums and is a speaker for many professional organizations and academic institutions in the United States and abroad.

Susheela's keen interest and study of global cultures and their cuisines, spices, flavor origination, and cross-cultural cooking, have a foundation in the vast international experience she possesses. Born and raised in Malaysia, she has lived and worked in North America, Southeast Asia, South Asia, Latin America, Europe, and the Caribbean.

1 Spices in History

Today's search for unique and authentic spices is not new. In ancient times, spices were status symbols in Europe and throughout the Mediterranean for the wealthy who ate them. Spices had an enormous trade value, not only as flavorings for food, but as medicines, preservatives, and perfumes.

A brief tour of the history of spices—and modern trends, which is covered in Chapter 2—will serve as a good introduction to the use of spices in today's global cuisines.

A "SPICY" TALE: A SHORT HISTORY OF THE SPICE TRADE

The history of spices is entwined with exploration, adventure, religious missions, commerce, and conquest. Treasured like gold and precious stones, spices have had enormous commercial value in ancient and medieval times. Most spices and flavorings had origins in the tropics or subtropics. They were much sought after in the West, and the quest for spices tremendously changed the course of history.

The East is the birthplace of most popular spices and flavorings. India, Southeast Asia, and China have given us anise, basil, cardamom, cinnamon, clove, garlic, ginger, mace, mustard, nutmeg, onion, peppers, star anise, tamarind, and turmeric. Other spices, such as bay leaf, coriander, cumin, dill, fennel, fenugreek, rosemary, sage, sesame, and thyme came from the Middle East, North Africa, and other parts of the Mediterranean. The colder regions of Europe have provided us with juniper and horseradish, while the Americas gave us allspice, annatto, chile peppers, chocolate, epazote, and sassafras.

Ancient civilizations, such as Indians, Middle Easterners, Chinese, Aztecs, and Incas, have used spices since time immemorial. As with modern civilizations, these cultures spiced foods to enhance their palatability and to create different flavor profiles. In addition, spices were used to preserve meats or fish, to disguise tainted foods and counteract disagreeable odors, and even to create cosmetics and perfumes.

Early civilizations understood that spices had medicinal value and used them as antidotes for poisons, to help cure diseases, and to prevent ailments. During medieval times, spices such as cinnamon, garlic, and oregano were used as germicides to battle the spread of the plague.

Many cultures also believed that spices had magical properties, and they were used in religious functions and on ceremonial occasions.

EARLY USE OF SPICES IN THE AMERICAS

While stories of most spices begin in the East, a number of the more popular spices and flavorings in use today are native to the Western Hemisphere. Since the dawn of time, Native American Indians—Aztecs, Mayans, and Incas—flavored their food and drinks with spices and offered them to their gods in religious ceremonies. Chile peppers, sweet peppers, allspice, chocolate, and vanilla originated in the New World before being introduced to Europe and Asia.

Chile peppers, an essential spicing in today's cuisines, grew wild in the Andes and were used as early as 10,000 years ago. From South America, chile peppers were carried to Central America, Mexico, North America, the Caribbean, and throughout the world. Archaeological excavations in Mexico reveal chile pepper remains dating back to 7000 BC.

Anthropologists have been unable to define with certainty when chile peppers were first domesticated. It appears that Native Americans began domesticating chilies between 5200 and 3300 BC. By the time the Spanish arrived in Mexico in the sixteenth century, the Aztecs were growing dozens of pod types. Today, all domesticated cultivars are derived from five domesticated species of chile peppers, and none differ substantially from those domesticated by Native Americans.

In Pre-Columbian Americas, dried chile peppers were used in trade in what is now the Southwestern United States and regions of Mexico. Atole (a corn, cacao bean, sugar, and chile pepper drink) and posole (a corn and chile pepper stew) were some of the foods flavored with chile peppers, enjoyed by the Aztecs and Mayans.

Vanilla planifolia, a climbing, tropical orchid, grew wild in the hothouse jungles of Central America and northern South America. When the fruit pod of the vanilla orchid fell to the jungle floor before it was ripe, it would ferment and give off a marvelous aroma, which the Aztecs must have noticed. Called *tlixochitl* or "black pod" by the Aztecs, vanilla pods were harvested from wild climbing vines found in the jungles of southwest Mexico. Later, the Aztecs domesticated this exotic plant and cultivated its vines. In addition to flavoring, vanilla was used as a nerve stimulant, and was reputed to be an aphrodisiac. The Aztecs blended the smooth vanilla flavor with chocolate, chilies, corn kernels, and honey to create "royal" drinks reserved for the elite of society. By legend, the great Aztec emperor Montezuma presented this flavored chocolate drink, served in golden goblets, to the Spanish conquistador Cortez.

In Mexico and Guatemala, Mayans, Toltecs, and Aztecs took the seeds of cocoa pods, roasted them, crushed them into powder on stones, and whisked the powder with boiling water to create tchacahoua (in Mayan) or tchocoatl (in Aztec). This drink, often mixed with chile peppers, honey, or ground maize, was considered sacred food.

Allspice is the fruit of an evergreen-type tree that grew wild in southern Mexico, Central America, and on several Caribbean islands, including Jamaica and Cuba. The Mayan Indians used allspice berries to help preserve or embalm the bodies of their leaders. The fruit of the unripe allspice berry, which looks like a large black peppercorn, was sought by early Spanish explorers. Thus, they called these berries pimienta, or pepper, from which we get the name pimento.

THE ASIAN SPICE EMPORIUM

Many of the spices that are popular today are indigenous to India, where they have been savored for thousands of years. The Harappa civilization, one of the first cultures of the Indus valley in northern India, ground saffron and other spices on stones around 3200 BC.

One of the earliest written records regarding spices appears in the religious scriptures of the Aryan people of north India who had driven the earlier civilizations further south. The Vedas, written in Sanskrit between 1700 and 800 BC., refer to mustard (*baja*), turmeric (*haridra*), long pepper (*pippali*), and sour citrus (*jambira*).

The Sanskrit language, itself, however, contains words for spices that reflect the well-established use of spices by the most ancient peoples in India. For example, the Sanskrit word for tamarind (*chincha*) has aboriginal origins. *Haridra* or turmeric comes from the Munda, a pre-Aryan people who lived through much of North India. The Vedas refer to a community called Nishadas, which translates literally into "turmeric eaters."

The Aryans looked down on certain spices. Vedic literature describes garlic, leeks, mushrooms, and onions as native foods despised by the Aryans. Some scholars explain that this aversion arose from the common practice at the time of fertilizing these crops with a manure of human waste.

Later Vedic writings establish that early North Indians were engaged in a far-reaching spice trade. The Vedas report the Aryans using black pepper (*maricha*) brought from South India and asafoetida (*hingu*) from Afghanistan. In the Buddhist era (800 to 350 BC), we see the introduction in North India of ginger, cumin, and cloves from other parts of Asia.

The origins of ginger have been obscured by its wide domestication. It is native to Southeast Asia, but wild forms are found in India. Cumin appears in Vedaic writing around 300 BC and appears to be native to the Middle East. The Sanskrit term for cumin, *jeeraka*, comes from the Persian language. Clove originated in the Moluccas Islands in eastern Indonesia. It first appeared in the Ramayana, an Indian epic written between 350 BC and AD 1. Clove may have originally come to India through Malaysia because the Sanskrit word for clove, *lavanga*, appears to be derived from the Malay words for clove bud, *bunga lavanga*.

The Dravidians were the predominant civilization of South India. They used tamarind, black pepper, lemon, cardamom, cinnamon, clove, turmeric, and pomegranate to flavor their foods. Pepper plants, cardamom, and cinnamon grew wild in the south of India, particularly in the states of Kerala and Karnataka. Mysore, in Karnataka, was popular for its cardamom, and Kerala was known for its black pepper. While cinnamon also grew wild in South India, the best cinnamon came from Sri Lanka, off the coast of South India. In addition to flavoring foods, spices played a significant role in the religious and cultural lives of early Indian people. The colors yellow and orange were considered auspicious and festive because of their connection to the sun. Consequently, saffron and turmeric were used in religious ceremonies and in the important personal occasions in everyday life, such as childbirths, marriages, and funerals.

It was a common practice of the Aryans in the north and the Dravidians in the south to mark their forehead with *kumkum*, as a sign of religious respect and auspiciousness. Saffron was generally used by Aryans and the Dravidians used turmeric, made alkaline with slaked lime, to make *kumkum*. Arghya, consisting of water mixed with saffron, flowers, and sandalwood powder, was presented to deities in worship. Akshatas, or rice colored with saffron, was presented to God Vighneswara (presently called Lord Ganesha) in the *Puniah Vachna* ceremony, when praying for the removal of an obstacle in life. Saffron was also used to color other religious articles.

Turmeric had erotic significance for Indians and played an important role in wedding ceremonies of Hindus and Muslims. During the nuptial bath called *Nalangu*, the heads of the bride and groom were rubbed with sesame oil, and the exposed parts of their bodies were smeared with turmeric. In some communities, sweets made of nutmeg and saffron were also given to the newlyweds as aphrodisiacs, while perfumes of saffron, white sandalwood, cardamom, nutmeg, and mace were poured on the sacrificial wedding fire. The exquisite golden complexion of Naga women of North India was reputedly obtained through their constant use of turmeric.

Turmeric was widely available and since it was considered auspicious, it was also widely used in everyday life whenever good luck was desired. For example, garments dyed or marked on the corners with turmeric were considered lucky and possessed with protective powers.

Spices were also commonly used to cure disease and promote health in India. The sacred *Ayurvedic* texts, which were formulated before 1000 BC and dealt with matters of health and medicine, make frequent reference to the use of spices. For example, the Ayurvedic system of medicine suggests that cloves and cardamom wrapped in betel-nut leaves be chewed after meals to aid digestion. In about 500 BC the physician Susruta the Second described over 700 drugs derived from spices, including cinnamon, cardamom, ginger, turmeric, and peppers. Vapors of white mustard were used to fumigate the rooms of surgery patients, and sesame poultice was applied to wounds as an antiseptic.

Spices were also used in China for thousands of years. Confucius, who lived from 551 to 479 BC, mentions ginger in his analects. The use of cassia was noted in the *Eligies of Ch'u* in the fourth century BC, and the name of the South China state of Kweilin, founded in 216 BC, translates literally into "cassia forest." It is also reported that the Chinese officials ate cloves in the third century BC to sweeten their breaths when they addressed the Emperor.

THE FIRST SPICE TRADERS

Trade and travel have always been part of Indian culture. Some sources indicate that as early as 3000 BC Indian explorers and traders took sea trips from the Malabar Coast in South India to the Persian Gulf and the fertile valleys of the Tigris and Euphrates rivers. From at least as early as 600 BC until almost AD 1400, Hindu and Buddhist missionaries and traders from India colonized countries and converted peoples throughout Asia. During this time period, Indian culture was considered the height of civilization. Powerful Hindu/Buddhist kingdoms, influenced by India, arose in Sri Lanka and throughout Southeast Asia, including Malaysia, Thailand, Kampuchea, and Indonesia. Hindu Brahmins were sought by the courts of regional leaders as teachers and bureaucrats.

These early Indian merchants and colonizers understood the value of spices and traded in spices with China and throughout the Malay Archipelago, Indonesia, and the Spice Islands (or Moluccas). The Indian spice merchants brought back ginger from China, cinnamon from Sri Lanka, nutmeg from the Spice Islands, and cumin, cloves, and coriander from throughout Asia.

SPICE USE IN THE WEST

In the West, precious Eastern spices were collected and treated like jewels, given as gifts, or used for ransom or for currency when purchasing cows, goats, or sheep. From the Bible, we know that King Solomon counted spices among the valuables in his treasury. Ancient Egyptian rulers used spices such as sesame, fenugreek, cinnamon, anise, cardamom, saffron, caraway, and mustard for embalming, as body ointments, and as fumigants in their homes. Many of the spices described in ancient western texts and writings are not grown in the West but are native to India and other eastern countries. The great desire for these spices became a driving force in transcontinental trade between the East and the West.

Traders from the West seeking wealth in the spice trade came to India and other destinations in the Far East, such as China and the Spice Islands (Indonesia) for at least 3000 years. Arabs, Greeks, Romans, and other Europeans came to India's Malabar Coast, which they called the "spice emporium," for cloves, pepper, pippali, zedoary, nutmeg, and turmeric. They carried their precious cargoes to Africa, the Mediterranean, and Europe.

Spices came to Europe and the West from the Far East by land and sea. Spices were taken on long caravan trips from the Far East by the Silk Route that traveled from China through northwest India, Afghanistan, and Turkestan. They were also taken by the Incense Route that went through southern Arabia to Egypt and other parts of the Middle East.

Arab and Phoenician traders were the first to bring eastern spices to the Middle East and Europe. From at least as early as 950 BC, the Arabs were the dominant middlemen in the spice trade between India and the West. They braved rough sea trips to the Malabar Coast of Kerala, India, and brought back spices, such as black pepper, cinnamon, ginger, and cardamom. They traveled through the Persian Gulf, using the Tigris and Euphrates valleys to Babylon or went around the coast of Arabia and up the Red Sea to the Middle East and Africa. Over land, the Ishmaelites, who were Arab merchants, took long caravan routes to India, Burma, and the Spice Islands.

The Phoenicians were the dominant traders of the Mediterranean. Renowned as fearless seamen, they traded with the ancient people of Greece, Italy, Spain, Portugal, France, and Africa. They brought fish sauce (called garum) and spice essences from Ethiopia and cassia and cinnamon from the Arabs. In turn, the Phoenicians traded them around the Mediterranean, with many of these spices being taken to North Africa, making it the focal point for spice trade between the Far East and the West.

The Phoenicians introduced Asian spices, such as nutmeg, coriander, cumin, and cloves, throughout the region. With the wealth obtained by the Phoenicians from this trade, they built their great colonial city of Carthage on the North African coast. They remained in control of the Mediterranean spice trade until Alexander the Great conquered Egypt and established Alexandria in 332 BC.

GREEK AND ROMAN SPICE TRADERS

For centuries, Arab merchants sought to protect their spice trade by hiding their true sources. The Arabs told stories of a mythical land in Africa as the source of spices. They also spun tales of gods and creatures protecting the spices from harvest by human hands.

In the first century AD, a Greek sailor discovered the secret of the monsoon winds to and from India that hastened the trip and broke the Arab monopoly. Early Tamil poems from that time tell about the *Yavanas* or Greeks who spoke a strange language and traveled in well-built ships. The Greeks were quickly followed by Romans who established trading posts and warehouses in South India. In his writings, Ptolemy from Alexandria listed eleven ports and thirty walled towns along the coast of India. A Roman warehouse was excavated in Tamil-Nadu at the town of Arikamedu. Artifacts discovered there date the ruins to the first or second century AD. Coins from the Roman kings Augustus, Tiberius, Nero, and Caligula have been found at thirty sites, mostly in South India, and they bear witness to the extensive trade that took place. The Romans paid gold, silver, and wine for spices from India and, during the time of the Roman Empire, dominated the European trade for Asian spices.

During this period, the Romans sailed from Egypt to India to bring back spices such as black pepper and turmeric for food, wine, cosmetics, and medicine. The Romans became the first Europeans to cook with spices and used them lavishly. Black pepper was the most popular and most expensive spice during this period. Cumin and coriander were used for preserving meats and sausages. Fish were preserved with salt and leafy spices such as dill, mint, and savory, and flavored with pepper, cumin, and mint.

The Romans also carried spices overland using the Silk Road that passed from Xian in China, around the Himalayas in North India and across Persia and then by ship over the Caspian Sea to the Mediterranean. They brought back cumin, ginger, cloves, nutmeg, and cassia, and Constantinople, the Roman Empire's eastern capital, became known as the "spice city."

The Romans also traded in spices with their colonies. The Romans brought to Northern Europe temperate spices such as garlic, parsley, dill, mint, sage, thyme, and savory, as well as the exotic spices of Asia.

THE ARAB CONQUEST

The Arabs regained their monopoly on spices with their conquest of Alexandria in AD 641 and their subsequent expansion into northern Africa and southern Spain. With the growth of Islam, the Arabs again took control of the spice trade.

Arab influence also expanded beyond the lands they conquered because of the spread of Islam, which replaced Hinduism throughout Southeast Asia, including the Spice Islands, and influenced north and central India. The followers of the Prophet Mohammed traveled from Mecca, in Saudi Arabia, carrying Islam to the Far East and bringing back spices to North Africa, Turkestan, and Spain. The Arabs continued to dominate the spice trade in Asia until the late fifteenth century.

In sub-Saharan Africa, Arab traders supplanted the Indians, who first brought spices to East Africa in the third century BC. The Arabs established a clove-trading center on the island of Zanzibar.

Because the Arabs controlled trade routes in the Indian Ocean and throughout North Africa and the Middle East, the spice trade with Europe dramatically decreased during the seventh century AD. Without access to Asian spices, Europeans grew temperate spices such as mint, fennel, lovage, rosemary, sage, dill, poppy, and celery.

SPICE USE IN THE MIDDLE AGES

Trade with the Far East, especially with India and China, was reopened with the crusades in the eleventh century AD. During this time, Genoa and Venice became important trading ports. In AD 1271, a Venetian trader named Marco Polo captured Europe's imagination with tales of exotic lands and exotic spices. Marco Polo had traveled with his father and uncle to China and eventually to many other countries of Asia and the Near East, including India, Indonesia, Turkey, and Egypt.

Venice took control of the European trade in spices, buying the products of the Far East from Arabian middlemen. Once again, Europeans enjoyed spices such as ginger and galangal (referred to as mild ginger) from China, cloves from the Spice Islands, and cinnamon and pepper from the Malabar Coast of India. Venetians provided salt and a good deal of gold and silver in exchange.

As a consequence of this increased trade, the consumption of spices grew dramatically. In the Middle Ages, Arabic and Asian luxury goods became indispensable to the European upper classes. Asian spices, pepper in particular, became the most important luxury items. Indeed, the use of spices took on an almost ceremonial function. At dinner parties of the refined upper classes, spices were passed around on gold or silver trays from which guests helped themselves.

In European cooking, spices were used in astonishing quantities by today's standards. Food was buried in pepper and other spices. Spices were also served in beverages, such as powerfully spiced wines. The more excessive a dinner host's use of spices, the higher was his guests' perception of his social rank.

As the Middle Ages drew to a close, the middle and upper classes expanded, and the European appetite for spices grew even larger. Pepper sauce became a staple of the middle class diet. Old overland transportation routes and numerous middlemen limited the supply of spices. Increased tariffs on this precious cargo also drove up its cost. As a consequence, the price of pepper increased 30-fold during the fifteenth century.

The ensuing crisis led to the age of exploration, conquest, and the discovery of new trade routes by Europeans.

THE AGE OF EUROPEAN CONQUEST

Realizing the value of the spice trade, Europeans sought to discover new routes to Asia and to conquer the countries where spices grew. One by one, European nations took control over trade routes to Asia and the spice producing regions—first the Portuguese, next the Spanish, then the Dutch, and lastly the English.

Around the end of the fifteenth century, the Portuguese, led by Vasco da Gama, were the first of these Europeans to reach Calicut, India. This ended the Arab and Venetian monopoly on spices. The Portuguese eventually took control of the Indian and Far East spice trade. They paid gold and silver to the local Indian rulers for spices.

The Portuguese established trading ports at Goa, India, and Sri Lanka and moved farther east to Malacca, Malaysia, and the Spice Islands (the Moluccas) bringing back pepper, cloves, nutmeg, and mace. Until the late sixteenth century, they dominated the spice trade to Europe.

Competing with the Portuguese for the lucrative spice trade were the Spanish. A Spanish explorer, Ferdinand Magellan, sailed to the Spice Islands in 1519 looking for spices. The Spanish also sought a quicker western trade route to the spices of Asia and its greatest prize, black pepper, called pimienta in Spanish. Christopher Columbus began his great exploration looking for pepper. What he found instead were the Americas and the chile pepper, used abundantly by the Native Americans. By 1529, Spanish colonizers learned that the Aztecs had developed dozens of pod-type chile peppers. They called the fiery new plant pimienta picante to reflect its stronger taste profile. Today, its name chili or chile pepper is derived from the Nahuatl language, "chilli" meaning red.

The Spanish found that chile peppers were natural colonizers, readily transportable, and remained viable for several years. As a consequence, chile peppers were brought back from the Americas and quickly spread to other parts of the world. Chile peppers were known in Spain in 1493, Italy in 1526, Germany in 1543, and the Balkans in 1569. As *paprka* (paprika), they revolutionized Hungarian cooking. The Spanish also brought back nuts, beans, allspice, and other ingredients.

The Portuguese brought chile peppers from the New World to their colonies around the world, including Africa, Arabia, and Asia, where they grew rapidly. Unlike the rare and exotic spices brought to Europe, which were expensive and unattainable except to the upper classes, chile peppers grew easily in the tropical climates of Asia and were readily available to the common people. Chile peppers were so widely used and grown in Asia that by the mid-sixteenth century, European colonizers in India were not sure whether the "Calcutta pepper" was native to India or came from the New World. They also brought corn, potatoes, beans, and tomatoes from the New World and peanuts from Africa to the Spice Islands.

In the early seventeenth century, the Dutch took over from the Portuguese and became masters over Java, Malacca, and southern India. The Dutch controlled the Spice Islands (which had a brief British rule from 1811 to 1816) until the Japanese came during World War II. During the period when the Dutch colonized South Africa, they introduced chile peppers from the Americas and brought in Malay slaves from the Spice Islands. The Malays introduced many spices such as aniseed, turmeric, cumin, cardamom, coriander, mustard seed, garlic, tamarind, ginger, and fennel. These ingredients are now essential to Cape Malay cuisine, which combines Malay cooking with Dutch, English, Indian, and indigenous African flavors.

During the eighteenth century, the French introduced spices such as clove and nutmeg to Mauritius and the French territories in Vietnam and the Caribbean. During this era, the British took control of the Indian and the East Indies spice trade. The British defeated the Moguls of North India, who had established themselves as rulers from the eighth to the sixteenth century. Then, spices such as mustard, poppy seed, sesame, coriander, and cumin from the northern regions of India were available for export to Britain and its colonies. The British established Bombay and Calicut as the spice trade centers in India, and Penang as the major eastern port in peninsular Malaya.

In the early nineteenth century, Chinese traders and merchants (mainly from southern China), encouraged by a liberal immigration policy instituted by British colonial powers, traveled to peninsular Malaya and Singapore and married local Malay women. The descendants came to be called *Peranakan* or Straits Chinese (*Baba* for men and *Nonya* for women). Nonya cuisine evolved in these regions.

SPICES IN AMERICA

In the eighteenth century, North Americans from Boston and other northern coastal towns sailed to the Far East, to places like Malabar in India and Sumatra in the East Indies, and brought back pepper, cloves, cassia, ginger, and cinnamon. New York City, Baltimore, and San Francisco became major ports for the spice trade to the United States.

Today, the United States is the biggest spice importer and the largest consumer of spices in the world. As is shown in the next chapter on trends, North Americans continue to "explore" unique spices and are demanding flavorings from around the globe. This will continue the trend for authentic cuisines as well as "cross-over" flavors in North America.

2 Trends in the World of Spices Today

TRENDS IN FOODS AND SPICES

Throughout the ages, the opening of trade routes and changing immigration patterns have affected the way the world eats. Today, we are facing a new revolution in eating patterns and the way we use spices. North American palates continue to become more daring and adventurous. They seek variety and something new. They want foods with more intense flavors and hotter and spicier profiles. North Americans also want foods that are fresh, light, and healthy, and that have a perception of "natural," and that are convenient to prepare. At the same time, they want to indulge.

As in the age of colonial adventure, we are seeking new routes to find foods that provide the tastes we demand. As a result, our interest in tastes and flavors from faraway places is increasing. Cuisines once considered unusual from other countries are becoming commonplace. The foods and ingredients of the world—Southeast Asia, India, Latin America, and the Mediterranean—are more available than ever. Ingredients once considered "exotic" are infiltrating traditional North American foods through cross-cultural cooking and regional American fare, such as French-Thai, Indian-Mexican, Pan-Pacific, New Californian, or Floribbean. New and diverse cooking styles and ingredients are not simply part of a passing fad. To the contrary, basic changes in who North Americans are, and their lifestyles, are driving these trends (Figure 1).

The United States has become an increasingly diverse country. The U.S. Census Bureau reports that between 1990 and 2000 the Asian and Latino populations have grown substantially. They are the fastest growing ethnic groups and are becoming a greater part of our social fabric. Asian-Americans, increased by almost 55%, while Latino Americans increased by about 61%. Latinos are now the largest minority group in the United States, surpassing African Americans. In 2000, Latino Americans, Asian Americans, and African Americans made up nearly 30% of the U.S. population (Figure 1(b)). By 2010, these three groups will comprise more than one-third of the U.S. population. By 2010, the Latino population in the United States is expected to grow 96%, while Asian-Americans will grow by 110% according to the Census Bureau. By the middle of the twenty-first century (Figure 1(d)), they will make up half of the U.S. population.

These statistics do not tell the whole story. Ethnic groups have become more diverse as well. There are not simply more Asian Americans, but more Americans of Indian, Korean, Thai, Chinese, Japanese, Filipino, or Vietnamese descent. Likewise, the growing Latino population includes people of many different ancestries, including all of the regions in South America, Central America, Mexico, and the

11

U.S. Population – 1990

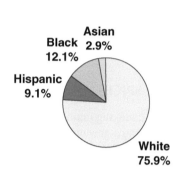

U.S. Population – 2000

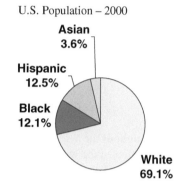

U.S. Population – 2010

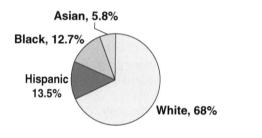

U.S. Population – 2040

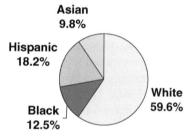

FIGURE 1 U.S. population (a) 1990, (b) 2000, (c) 2010, and (d) 2040. *Source*: U.S. Census Board.

Caribbean. The South American population has been growing in numbers, with Colombians, Ecuadoreans, Peruvians, and Argentinians constituting the majority of them.

The increased presence of ethnic groups in our communities is increasing our exposure to many different cultures, foods, and ingredients. Asian and Latino cuisines, which have more diverse and spicier flavor profiles, have become a greater part of our social fabric. Ginger, cilantro, and cinnamon have become mainstream items while newcomers include lemongrass, chipotle, star anise, wasabi, epazote, and kari leaf. Consumers are sampling these cuisines and their exciting new ingredients in a variety of ways. They are discovering fish sauces, flower essences, wrappers, and fermented soybeans in restaurants that feature ethnic and fusion menus. Ethnic grocery stores and bodegas carry specialized items such as tamarind, kokum, banana leaf, pandan leaf, and galangal. Consumers can also find authentic ethnic ingredients, such as nigella, ajowan, rocotos, or black cumin, as well as prepared ethnic foods in gourmet, health, and natural food stores, and even on the Internet.

Global travel is also driving the new ethnic cuisine revolution. Nearly 60 million Americans travel overseas each year. Most sample the cuisines of the countries they visit and when they return home, they crave for those flavors. Television media is introducing a whole range of ethnic cuisines, spices, and flavors to the public who want 'real' flavors or to add an interesting twist to everyday cooking. Media attention to ethnic cooking, growth of cookbooks, proliferation of restaurants, celebrity chefs, and ethnic convenience meals are all creating major changes in our eating habits.

Cooking schools in the United States, as well as overseas, are helping Americans to learn about authentic ethnic ingredients and cooking styles. The Internet is also making global commerce, communication, and the exchange of ideas an instantaneous affair.

What do these trends portend for the consumption of spices and their use in prepared foods? The answer is already available. Americans are buying an increasing volume and variety of Asian, Latin American, Caribbean, and Mediterranean spices.

The American Spice Trade Association's (ASTA) 2000 Spice Statistics Report says that within the last twenty years there has been a significant increase in the consumption of spices, with overall spice consumption being doubled. It reports that the hottest trend is our taste for hot spices such as mustard seeds, black and white peppers, ginger, and red pepper that have shown a 72% increase in sales volume since the late 1980s. The major spice supplying nations, reported by ASTA, in the year 2000 (with major spices), are the United States (dehydrated garlic and onion, paprika, chilies, and mustard seed), Canada (mustard seed, coriander, and caraway), India (sesame seed, black pepper, red pepper, turmeric, celery seed, cumin, and fennel), Indonesia (black pepper, cinnamon, white pepper, nutmeg, and vanilla), Peoples' Republic of China (garlic, red pepper, ginger, and sesame seed), Mexico (sesame seed, red pepper, and oregano), and Guatemala (sesame seed, cardamom, and allspice), which together provide 84% of the U.S. consumption of spices.

The demand for spices will increase, not only in total volume, but also in variety. Thus, we can expect increased sales volumes for familiar spices, such as garlic, onion, allspice, ginger, cumin seed, and mustard, while new demands for emerging spices, such as fennel seeds, star anise, Thai basil, guajillo, and cardamom, will grow as well.

There is also be an evolution in the nature of prepared foods. Prepared foods are already being presented with influences from Asian and Latino styles. Smaller portions of entrees, with thinner cuts of meat that are marinated, seasoned or "sauced up" rather than being dry are becoming popular. Entrees are being perked up with a variety of seasoned side dishes and condiments. Asian and Latino concepts of one-dish/bowl meals, using pasta or rice, with multidimensional flavors and textures, are becoming more popular because of their taste, convenience, and economics.

Authentic preparation techniques for spices and other ingredients are adding new flavor and texture dimensions to foods. These include dry roasting, "*tarkaring*" (frying in oil), and "*tumising*," (slow stir frying) which make spices or spice pastes more fragrant, less bitter and with more flavor intensity.

The demand for "healthy" ingredients and natural ways of preventing illnesses or diseases are also contributing to the increasing use of spices in the United States. For example, Indian vegetarian foods that use abundant spices are becoming more popular because they provide taste and nutrition. Chinese vegetarian cooking that showcases wonderful textures and flavors will emerge in mainstream cooking. Consumers' growing interest in using spices for their therapeutic properties will continue to rise.

While many Americans enjoy traditional foods, at the same time they expect their foods to be seasoned and well balanced with variety and exciting tastes. Ethnic foods with their multidimensional flavor and texture profiles will provide this, especially cuisines that include a variety of spices, seasonings, and condiments. Latino bodegas and Asian supermarkets and restaurants continue to grow to meet the demands of the ethnic consumers.

Many young Americans who have grown up with ethnic foods will continue to demand these flavors. These acculturated Americans enjoy the fusion or cross cultural flavors. They are also environmentally conscious, so ethnic style vegetarian and natural and organic ingredients appeal to them. Aging Baby Boomers' growing focus on a healthier lifestyle is also bringing meatless cuisine to nonvegetarians. Bland tasteless, texture-less, boiled vegetables are giving way to gourmet, ethnic style vegetarian meals. Interest in more seasoned and textured vegetables and an all-natural, environmentally conscious food trend is spearheading the 'true' and regional flavors of Asia and the Mediterranean.

Meals with fresh aromas and textures are becoming a significant point of purchase. Apart from leafy greens, grains, pickled vegetables or legumes to provide fresh appeal, fragrant whole spices add to their enhancement. Bowl meals, soups, and freshly made spring rolls with fragrant leafy spices and Thai, Vietnamese, Indian, and Japanese sauces and condiments are emerging favorites.

Mainstream cooking will evolve further with consumers' better understanding of authentic cuisines, their spices and flavorings that provide the new tastes and texture sensations. Ginger was an exotic spice ten years ago, but today it is added to everything — from sodas, teas, beverages, and candies to sauces and soups.

Recognizing these fast moving trends, we, as product developers, need to join the bandwagon to create authentic products or add excitement to mainstream taste buds. But before we can do this, we must first understand spices — the basic building blocks of flavors in ethnic foods — and how they are prepared and blended.

Some consumers seek excitement and adventure with meals while others want something familiar but with new flavor twists and new ways of preparing or serving meals. How do we meet these consumers' needs for familiarity, tradition, and novelty at meal tables? First, we need to have an in-depth understanding of spices and other flavorings to effectively utilize them to create new products. By effectively connecting spices and other flavorings, we can create authenticity or fusion products. Spices are great tools for "safely" providing authenticity or new flair to traditional foods. They can add a comforting new dimension to a traditional product or create a totally new and unique product. We can also satisfy niche markets by focusing on regional profiles within the United States and developing products that reflect the flavors of their ethnic mix.

UNDERSTANDING AND EFFECTIVELY MEETING THE GROWING DEMAND FOR AUTHENTICITY

Authentic ethnic flavors and their preparation techniques and presentation styles are becoming a regular part of our meals. For most Americans, certain ethnic foods are so common that they are not thought of as ethnic anymore, such as pizzas, spaghetti, tacos, burritos, sushi, egg rolls, or stir-fries. When exotic ingredients and dishes are presented in a more familiar setting, and consumers enjoy these foods, then they want to experience authentic foods and flavors. Many chain restaurants continue to add Asian and Hispanic flairs to their menus to attract mainstream Americans.

More upscale and innovative dishes with authentic cooking of Canton, Peking, Shanghai, and Szechwan are emerging from the standard Chinese American take-out fare. Noodles, bowl meals, and stir-fries will continue to appeal to many, but with more intense and exciting flavors. Lighter, subtle Cantonese style sauces are supplanting heavy, starchy types. More innovative bowl meals and dim sum dishes will appear at American restaurants. Many foods that "taste and look like" chicken, pork, or shrimp in Chinese vegetarian cooking will emerge, using mushrooms, tofu, and a variety of vegetables. Braised five-spice lentils, oven-roasted gingered potatoes, sesame-scented bean curd, and pan caramelized Chinese cabbage are appearing in upscale menus. Simplicity, freshness, taste, and presentation are paramount to Japanese cooking, which will continue to appeal to the adventurous young and mainstream. More authentic flavorings such as horseradish and wasabi condiments, soba and rice bowls, vegetarian sushi, tempura, yakitori (meat on skewers), and light miso soups are becoming more common on mainstream and fusion menus.

Youth and adults alike enjoy Tex-Mex cuisine, but where there is a growing Hispanic population, the "real" Mexican and the other Latino fares are gaining momentum. Regional Mexican cuisine and a host of other Latino foods and flavors from Central and South America are appearing on menus. While tacos, salsa, burritos, and nachos have become staples of American dining, many small ethnic eateries are exposing us to authentic Mexican foods—a variety of fillings for soft tacos (carnitas, pollo asado, chorizos, etc.), chili-based sauces (salsa verde, chipotle, recados, etc.), and bebidas/drinks (horchata, tamarindo, Jamaica, mamey, etc.).

Consumers are developing a craving for the true "el sabor Latino," so Venezuelan arepas, Brazilian chimmichurri, Peruvian papas rellenas, Cuban mojos, and Puerto Rican ropa viejas are found in upscale menus. South Americans are becoming a large and integral part of North America's Latino population especially in Florida, New York, California, and New Jersey. Colombians, Ecuadorians, and Peruvians constitute a major portion of this growing U.S. Hispanic category. South America's culture and "foodways" are as diverse as its geography and people, inlcuding Ecuadorian ceviche, Bolivian humitas, Paraguayan yerbe mate, Argentinean churrasco, Brazilian caipirinhas, or Bolivian corn pudding.

In many large cities, such as Miami, Chicago, and New York, Latino flairs are added to local flavors in fine dining restaurants, such as sofrito marinated chicken, arepas with tofu and portobella mushrooms, roasted corn and pepper salsa, mango mint mojo ,or coconut crusted dulce de leche ice cream.

A penchant for in-depth flavors is driving consumers to become more familiar with cuisines that use abundant spices, including South Asian and Southeast Asian cuisines, which have always focused on fresh ingredients and intensely flavored meals. Thai food has become very popular, even crossing over to mainstream menus. Thai cuisine is a unique combination of royal Thai elaborate presentations, Indian influenced curries, Chinese stir fries, and bowl meals, with ethnic Thai ingredients such as galangal, lemongrass, fish sauce, and aromatic basils. More authentic, regional flavors are emerging, such as Mussaman curry with coconut milk and spices from the coastal south; pad Thai with fish sauce and dried shrimp from the central region; ground spicy beef (or *laab)* with mint and basils from the Northeast, and spicy *som tam* (grated green papaya salad with garlic, shallots, and chilies) from the north.

Other Southeast Asian cuisines including Vietnamese, Malaysian, and Indonesian will become increasingly popular. Vietnamese cooking has fresh fragrant aromas and crispy crunchy textures involving simple preparation methods, with well-marinated grilled meats, al dente veggies, light fresh salads, aromatic rice and noodle dishes, and fresh spring rolls, all served with light and flavorful condiments. Malaysian cuisine is an emerging favorite, with the melting pot of flavors of its diverse ethnic mix. Chinese style noodle bowl meals and stir-fries with Malay spicy sambals, Indian curries, Nonya pungent laksas, and Portuguese influenced fish dishes add to its unique flavors.

Indian meals are popular with youth and many baby boomers. The concept of yellow, turmeric-based curry (kari) is changing. Consumers are beginning to understand that numerous curry themes based on regional preferences and cooking styles exist. Because Thai curry pastes are in vogue, a variety of curry mixtures from the Indian subcontinent are emerging. The variety offered by Indian vegetarian meals, with their numerous spices, vegetables, condiments, and breads, and preparation styles is making them trendy. Small vegetarian Indian eateries serving authentic meals in a *thali* style setting, flatbreads such as dosas and chappatis, with accompanying condiments or lunch boxes, are fast becoming popular in major cities and around campuses. Naan and parathas, tandoori chicken, chicken tikka masala, samosas, and vegetarian dishes will soon become mainstream items. Fusion-style grab-and-go sandwiches, bowl meals or wraps with Indian fillings (alu gobi, grilled paneer, tandoori chicken) accompanied by an assortment of condiments, are also a growing trend.

SPICES AND FLAVORINGS OF POPULAR, AUTHENTIC ETHNIC CUISINES

As discussed, today's consumers are looking beyond generic Chinese, Mexican, or Italian foods and seek the authentic newcomers—Thai, Indian, Vietnamese, or

Malaysian; Szechwan, Cantonese, and Peking; or Tuscan, Umbrian, and Ligurian; or Oaxacan, Puebla, or Michiocan.

To create this authenticity, we need to look at the actual spices and flavorings that give rise to these distinct flavor differences (including their history, origins, varieties, forms, properties, and uses), how they are prepared or cooked, and when and how they are incorporated or presented in a meal.

AUTHENTIC ETHNIC SPICES

Certain spices have a mass appeal for many ethnic groups, such as garlic, black pepper, ginger, or coriander leaf. However, within each ethnic group or region, preferences for spiciness, sweetness, or heat occur. Consequently, the type of spice and the amount of spice vary as well. Ethnic cuisines use locally available spices to bring out their characterizing flavor profiles. Hence, we have Cajun, Thai, Keralan, Tuscan, or Cantonese.

In addition, there are many varieties of each spice. Different ethnic groups often prefer a specific variety. For example, with cinnamon, Mexicans and South Asians use the more delicate canela, also called Ceylon cinnamon, while the Chinese and most Southeast Asians use the harsher and sweeter cassia. In the United States, cassia is commonly used, but Ceylon cinnamon is becoming better known. In the past, substitutions have been used for an "exotic" ethnic spice that was not easily available. Today, consumers want the real thing and we need to understand the many varieties of each spice and their properties for authenticating a product. Let's look at some of the typical spices and flavorings used in some of the popular and emerging cuisines.

SPICES, FLAVORINGS, AND SEASONINGS OF POPULAR AUTHENTIC ETHNIC CUISINES

Southeast Asian (Burmese, Cambodian, Filipino, Indonesian, Laotian, Malaysian, Singaporean, Thai, and Vietnamese): black peppercorn, cardamom, cayenne pepper, cinnamon, clove, coriander (leaf, seed, and root), cumin, fennel, fenugreek, mace, nutmeg, saffron, turmeric, shallot, Chinese chive, mustard (seed, powder, and oil), galangal, garlic, ginger, kaffir lime, basils (Thai, lemon, and holy), kari leaf, lemongrass, spearmint, sesame (seed and oil), star anise, tamarind, bay leaf, zedoary, white pepper, bird pepper, annatto, candlenuts, banana leaf, coconut milk, soy sauce, lily bud, shrimp paste, fish sauce, mushroom, palm sugar, pandan leaf, peanuts, laksa leaf, lime, kalamansi, jasmine essence, ginger flower; five-spice, adobo, bumbu, curry blends, garam masala, blends for bean pastes, oyster sauce, sambals, hoisin, plum sauce, and rendangs.

South Asian (Indian, Sri Lankan, Pakistani, Afghan, Bangladeshi, Nepali, and Tibetian): turmeric, ajowan, anise, bay leaf, cardamom, cayenne pepper, cinnamon, clove, nutmeg, mace, caraway, coriander (seed and leaf), asafoetida, black cumin, cumin, kari leaf, fennel, fenugreek, onion, garlic, ginger, mustard (seed and oil), long pepper, nigella, tamarind, dill, poppy seeds, rose petal, saffron, black pepper, paprika, mints, pomegranate, screw-pine leaf, kokum, almonds, cashew nuts, black, amchur, ghee, coconut milk, sesame (seed and oil): curry blends, garam masala, chat masala, panchphoron, kala masala, sambar podi, blends for pickles, chutneys, tandoori, and recheado.

East Asian (Chinese, Japanese, Korean, and Taiwanese): anise, black pepper-corn, cayenne pepper, onion, shallot, garlic, ginger, ginseng, sansho, cinnamon, clove, fennel, sesame seed, shiso, soy sauce, star anise, szechwan pepper, horserad-ish, wasabi, parsley, sesame (seed and paste), licorice, coriander (seed and leaf), lotus (leaf, root, and seed), scallion, chives, daikon, kudzu, five-spice, mirin, miso, soybean (whole and paste), dried shrimps, mushrooms, wine, orange peel, vinegar, bonito, ponzu, shichimi togatashi, blends for hoisin, oyster sauce, kochu'jang, bean pastes, kimchi, and plum sauce.

Caribbean: allspice, annatto, black pepper, chives, coriander (seed and root), cilantro, garlic, ginger, clove, paprika, cumin, cardamom, nutmeg, mace, cinnamon, Scotch Bonnet, cayenne pepper, mustard (powder and pastes), onion, bay leaf, thyme, parsley, cilantro, roselle, tamarind, turmeric, amchur, dried mango, papaya, coconut milk, salted codfish, cassareep; garam masala, curry blends, jerk blends, blend for rouille, Creole sauce, and hot sauces.

Latin American (Mexican, Cuban, Puerto Rican, Dominican Republic, Central American, and South American): annatto, bay leaf, canela, black pepper, chepil, poblano, chipotle, aji, ancho, chiltepin, serrano, rocoto, guajillo, habanero, jalapeno, cilantro, parsley, coriander (seed and root), cilantro, cumin, nutmeg, mace, clove, boldo leaf, epazote, garlic, tamarind, vanilla, chocolate, roselle, oregano, almond, pepian, olive, tomatillas, cassareep, aji dulce, naranga agria, lime juice, dende oil, ginger, bacalhau; adobos, sofritos, blends for salsas, moles, recados, chimichurris, pebre, molho malagueta, ajilimojilli, and xni pec.

Mediterranean (Northern African, Middle Eastern Portuguese, Spanish, South-ern French, Italian, Greek, and Cypriot): ajowan, anise, asafetida, bay leaf, garlic, ginger, pepper (black and white), caraway, cardamom, cayenne pepper, chervil, celery seed, cinnamon, clove, fennel, fenugreek, nutmeg, mace, coriander (seed and leaf), cumin, dill, juniper, long pepper, marigold, marjoram, meloukhia, mints, paprika, parsley, poppy seed, rosemary, rose petal, saffron, savory, sesame (seed and paste), sorrel, sumac, tarragon, sage, lemon verbena, lovage, lemon balm, oregano, sweet basil, myrtle, thyme, turmeric, vanilla, chocolate, capers, bell peppers, olive (fruit and oil), pomegranate, fig, mushrooms, balsamic vinegar, caper, anchovy, pine nut, cheese, aioli, bacalao, and blends for harissa, berbere, zhoug, sofrito, and ras-el-hanout.

Middle Eastern (Israeli, Iranian, Arabic regions, Greek, Turkish, Armenian, and Cypriot): bay leaf, garlic, cassia, mint, fenugreek, dill, caraway, coriander, cumin, cayenne, anise, clove, nutmeg, paprika, sesame seeds, sumac, Allepo peppers, parsley, saffron, rose essence, and blends for zhug, za'atar, tahini, baharat, muham-mara, and talia.

African: anise, clove, coriander (seed and leaf), cubeb pepper, cumin, egusi seeds, fenugreek, sesame paste, onions, garlic, ginger, grains of paradise/melegueta peppers, saffron, turmeric, licorice, mace, mint, nutmeg, piri-piri, utazi leaf, tama-rind, vanilla, ground nuts, guedge, palm oil, pepper, allspice, rose petal, cayenne;

garam masala, curry powders; and blends for berbere, harissa, ras-el hanout, atjars, blatjang, and piri-piri.

North American: sassafras, red cayenne pepper, tabasco, jalapeno, pepper (black and white), allspice, cinnamon, bay leaf, oregano, sweet basil, cumin, cilantro, mustard, garlic, onion, chive, thyme, mints, nutmeg, rosemary, sage, chili powder, olives, cheeses, and mushrooms.

AUTHENTIC ETHNIC FLAVORINGS

As the trend for authenticity and variety continues, "newer" spices such as ajowan seeds, nigella, turmeric leaf, or shiso will emerge. Consumers are also discovering unique flavor properties of parts of plants other than seeds or leaves, such as stalks, stems, roots, and flowers of spices. These are used in traditional spice blends of many global cultures, such as Latin American, Asian, and Caribbean. In addition, there are many other ingredients that provide "spicing" or flavoring to foods, such as oyster sauce, pomegranate, olives, mushrooms, candlenut, and dried fish. These are being used with increasing frequency in the United States in pastes, purees, or extracts and as natural and healthy ways to add flavor, color, or texture to foods and beverages. Some of the emerging global spices and flavorings are examined in Table 1.

TABLE 1
Emerging Global Spices and Flavoring

Seeds/fruits/bark	Ajowan, black cumin, canela, green cardamom, asafoetida, fennel seed, fenugreek, star anise, sumac, annatto, mustard seed
Leaves/stalks/stems	Thai basil, cilantro, kari leaf, kaffir lime leaf, epazote, shiso, lemongrass, recao leaf
Chile peppers	Chipotle, aji, ancho, bird pepper, cayenne, guajillo, habanero, poblano, rocoto, cachucha
Fruits	Citrus, cranberry, guava, mango, persimmon, green papaya, kalamansi, pomegranate, olive, kokum
Vegetables	Mushrooms, tomatoes, squash, taro, yams
Nuts	Almond, candlenut, cashew, peanut, pine nut, pistachio, gingko biloba
Seeds	Pumpkin, sesame, sunflower, lotus
Food wrappers	Corn husk, lotus leaf, pandan leaf, banana leaf, nori, hoba leaf, hoja santa
Flowers	Jasmine, orange blossom, rose, squash blossom, violet, roselle, lavender, ginger flower
Roots	Coriander, ginseng, lotus, turmeric, galangal, wasabi, licorice, daikon
Fish and shrimp	Blacan, nam pla, nuoc mam, trasi, guedge, dashi (dried, fermented, and smoked)
Legumes	Miso, oyster sauce, soybean pastes (fermented, pickled, and salted), dals, beans
Other	Balsamic vinegar, coconut milk, ghee, dende oil, palm sugar, chocolate, vanilla, coffee, tea

AUTHENTIC PREPARATION AND COOKING TECHNIQUES

Many spices are common to a number of ethnic cuisines, yet the same spice gives different flavors. Why? The way each spice is prepared and combined with other ingredients gives varying flavor profiles. Dry roasting of coriander seeds or chile peppers, or "popping" of mustard seeds in hot oil produces distinct flavors and colors. Raw, bitter notes are removed, certain flavor notes are intensified, and sometimes colors are modified. The timing in which the spices are added to the cooking process is also crucial in creating a well-balanced flavor profile.

In authentic ethnic cooking, corn husks and banana and pandan leaves are used to wrap chicken, fish, rice, and meat, that are then steamed, smoked, grilled, or barbecued. These wrappers provide unique flavors, moistness, color, texture, and visual appeal to foods. Though commercially we may not be able to present food in these wrappers, we can simulate the flavors they create. For example, in a chicken dish, you can add a grilled flavor with pandan notes to simulate the flavor of chicken wrapped and grilled in pandan leaf. In addition, a pandan leaf for serving the chicken could be packaged as part of a meal kit that can be a presentation tool to create authenticity or excitement at the meal table.

Therefore, to meet the growing demand for authentic or new flavors, we will need to understand not only true ethnic ingredients, but also their preparation and cooking techniques.

ETHNIC PRESENTATION STYLES

Ingredients and preparation techniques affect flavors and the way foods are presented and served. Globally, food is eaten from plates, bowls, banana leaves or "*thalis*" using different utensils: chopsticks, forks, knives, fingers, spoons or fingers. To fully perceive the different sensations of flavors and textures, it is important to know not only how the food is seasoned, but also how it is served and eaten in traditional cultures.

Presentation of meals achieves visual appeal and balance, to harmonize the differing tastes for achieving flavor and for maintaining health. Ethnic cuisines such as Japanese, Thai, or Caribbean create great eye appeal using contrasting colors, shapes, or sizes with ingredients. Spices, fruits, vegetables, wrappers, or flowers are sculpted or designed to create exciting appearances.

Serving containers are also an integral part of meal presentation. Various-sized brass baltis (wok-like utensils) provide an authentic and familiar touch to many Indian and Pakistani cuisines. An authentic edge can be added to meal solutions by including authentic utensils in the package, such as chopsticks, chutney dip bowls, banana leaves, corn husks, comals, tagines, skewers or clay pots.

In presenting authentic ethnic foods, side dishes play a prominent role. In traditional North American presentation, the focus is on a main course of meat or fish, with side dishes of rice or pasta and vegetables. In contrast, for many ethnic groups, the main dish is starch based, such as white rice, with side dishes of intensely seasoned chicken, fish, or vegetables.

Curries, stews, pickled vegetables, and other ethnic side dishes that are hot, sour, sweet, or crunchy provide variety and enhance the flavor and texture of the main entree. Likewise, condiments are essential in many ethnic meals. Every ethnic group has its typical condiments to achieve each individual's desired taste. The Chinese have their sweet and sour sauces or plum sauces, Mexicans have salsas, Japanese have teriyakis, Tunisians have harrissa, French have rouille, Indians have pickles or chutneys, Indonesians have sambals, and Koreans have kimchis. As with side dishes, when condiments are eaten with a meal, totally new sensations are produced.

In many Asian cuisines, the concept of a "main entrée" and side dishes blend together in the form of a "one pot" or "one dish" meal such as stews, soups, sauced or souped noodles, or fried rice, all of which are also served with condiments. Bowl meals provide varying tastes and textures simultaneously.

FUSION AND REGIONAL AMERICAN FLAVORINGS

While changing demographics are increasing the demand for authentic ethnic ingredients and foods in general, one of the most important trends in a diverse country such as the United States is the growth of fusion (or cross-cultural) and regional American foods. These new foods are emerging from the mixing of ingredients and preparation styles taken from the popular traditional American and ethnic cuisines.

Fusion food is not something new. The transformation of cuisines through fusing flavors, ingredients, and preparation techniques has happened around the world whenever two or more cultures have lived together. Mediterranean, Southeast Asian, and Caribbean cuisines are three of the better-known types of fusion foods that combine flavors of many cultures and regions.

In the United States today, fusion flavors have evolved from two or more ethnic groups, such as Japanese and Cuban, Italian and Thai, or French and Indian. These "new" cuisines have great appeal because they are the product of many cultures. Fusion foods also include newer versions of traditional American products with flavors derived from ethnic ingredients, such as ginger roast chicken, sofrito mashed potatoes, or chipotle potato chips.

Within the category of fusion foods, there are also regional American fusion flavors, such as Creole, Tex-Mex, New England, or Southern. These "traditional" fusion foods evolved when the cuisines of earlier immigrants to the United States were combined. Many of these regional foods are changing in flavors to meet the newer groups of immigrants and changing consumers' taste buds. Regionalization of cuisines will continue in the United States. More than half of the U.S population lives around large cities, such as New York, Miami, Chicago, Los Angeles, or Washington. The populations in these regions are becoming increasingly diverse and, thus, will give rise to unique fusion flavors. Newly arrived immigrants from Latin America, Asia, or the Caribbean are producing the "newer" fusion and regional American fare, such as Floribbean, Nuevo York, Texnamese, Nuevo Latino, Pan Californian, or Pan Asian. Because fusion and regional cooking will increasingly affect our eating patterns, we will examine them in greater detail.

Fusion Flavors

Ethnic ingredients and their preparation styles blend well with North American cooking techniques, such as roasting, broiling, or baking. Consumers are becoming more innovative and are creating exciting fusion foods using varied ingredients. It is not uncommon nowadays to see ethnic foods with traditional American ingredients. Ethnic groups have traditionally used wrappers, such as tortillas, pita breads, and other flatbreads as meals or snacks. They are popular as fusion lunch entrees with salad ingredients and topping sauces, grilled chicken, chili con carne, and other traditional North American fillings.

Fusion sauces can be used as condiments to perk up traditional grilled, steamed, or baked foods that are prepared with moderate seasonings. Alternatively, mainstream sauces, such as tartar sauce, steak sauce, or ketchup, can be made into a "new" sauce with ethnic spices and flavorings such as wasabi, chipotle, or pomegranate.

To create great fusion products, flavors need not be compromised. Creative food product designers can develop foods for a wide market by "toning down" authentic ethnic ingredients or fusing them with more traditional American flavors. By adding, deleting, or toning down chile peppers, fish sauces, fermented bean pastes, and other flavor-intense ingredients, we can create appealing fusion foods or new flavor twists for traditional North American comfort foods such as roast chicken, meatloaf, or mashed potatoes.

For mainstream consumers seeking an ethnic flair or something new, lemongrass, adobo, or cilantro can be added to chicken, rice, or pizza. For the adventurous consumer, tandoori chicken, poblano pasta, or sambal fish can be developed. Fusion foods can also be created to meet the needs of ethnic consumers. Ethnic comfort foods such as stir-fried noodles, pilafs, curries, rice and beans, or pastas can also be mainstreamed to meet the tastes of second- and third-generation ethnic populations who have adapted to a new way of living in the United States.

Regional American Flavors

In the United States, there is a long history of immigration. Often, various immigrant groups have been attracted to particular regions of the United States. Italians moved to the Northeast, Japanese to the West Coast, Scandinavians settled in the Midwest, and Cubans emigrated to Florida. As these various groups settled in a region, they mixed with the local population and influenced the local foods.

When food developers create products for the different regions of the United States, they need to understand each region's specific preferences. This is also true when developing products for the ethnic market. For example, the market for Latin American foods varies around the United States. In Miami, Hispanics are predominately Cuban, while Puerto Ricans and Dominicans dominate New York, with Mexicans the majority in Los Angeles. Therefore, when we create Hispanic-style foods for these niche markets, they must be directed to the tastes of those who live in these regions.

While the tastes of the new immigrants influence the foods of a region, their own tastes become less distinct. Bagels, chow mein, tacos, chilis and other "Americanized" ethnic foods emerged from the influence of the local population. These products were modified from their authentic profiles in order to suit regional consumers' palates. Thus, traditional tastes fade while new tastes grow.

New immigrant groups continue in the same path as those who arrived earlier. The interactions of these new ethnic groups with the existing population, the availability of a variety of ingredients, and the adventurous spirit of cooks in the United States have given rise to some creative regional American cuisines. The flavors that are emerging from these regions reflect the cultural groups that are locally dominant.

In Miami, there is a large population of Cubans, Central Americans, Haitians, and African-Americans from the South, and the cooking there reflects the influence of these groups. Yuca, black beans, habaneros, or mango are combined with key lime, lima beans, or potatoes to create a new Floribbean cuisine. In New York City, new fusion foods such as Italian-Thai, French-Indian, and Japanese-Latino reflect the global diversity of its population.

For the future, the forces that have created the present demand for authentic ethnic and fusion flavors will continue to grow in the twenty-first century. The population trends that have changed the tastes of Americans today will continue to have an ever-increasing effect on the foods we eat tomorrow. The result will be a further evolution of unique regional flavors in North America.

In the future, the "new" foods will continually evolve that will have different or more sophisticated flavor profiles, colors, and textures. In regional cooking, fusion themes will be taken to new levels of visual appeal and creativity, such as Malaysian-Tuscan, Chicago-Thai, Miami-Indian, or Oaxacan-Japanese. Pasta sauces, ketchup, salsas, or curries will develop clearly defined regional U.S. flavor profiles.

As consumers become more exposed to and knowledgeable about new and "exotic" ingredients, preparation techniques, and presentation of meals, food and beverage designers need to understand these factors to create these new products to challenge their taste buds.

NATURAL AND ORGANIC SPICE TRENDS

Today's consumers are demanding natural and organic foods with label-friendly ingredients, for maintaining a healthier lifestyle, preventing ailments, and concern for the environment. Sustainable methods of growing crops and manufacturing products without the addition of chemicals and pesticides, MSG, hydrolysed plant protein (HPP), salt, sugar, or chemical preservatives are the trend today. The organic food consumption is growing at a steady rate in the United States. The current annual growth rate for organic foods is 20% compared with the 2% to 3% growth rate for conventional foods. The younger generation, baby boomers as well as the Asian and Hispanic Americans, are the biggest consumers of natural and organic foods and beverages.

For these consumers who prefer natural and organic ingredients, the demand for these spices is growing rapidly. The sale of organic spices is growing 35% annually, compared to 16% for nonorganic spices. Since organic spices are produced under ecofarming and sustainable methods free from chemical contaminants and pesticide residues, the demand for these products should steadily increase. They are also free from genetically modified ingredients. Steam sterilization and ozone are used to kill microbes. Leafy spices, such as cilantro or basil, are fumigated with carbon dioxide or freeze-dried.

Thus chemicals, pesticides, or fumigants are not used during spices' growth, during their storage, or to eliminate microbial growth. Organic spices contain none of the fillers (sucrose, starch, or dextrose), synthetic anti-caking agents, artificial colors and flavors, or preservatives that may be found in conventional spices. Many spice-producing countries are already in the organic spice-production trade, including the United States, India, Guatemala, Turkey, Indonesia, and Egypt.

3 Forms, Functions, and Applications of Spices

INTRODUCTION

Spices are the building blocks of flavor in food applications. Food developers who wish to use these building blocks effectively to create successful products must understand spices completely. The word "spice" came from the Latin word "species," meaning specific kind. The name reflects the fact that all plant parts have been cultivated for their aromatic, fragrant, pungent, or any other desirable properties including the seed (e.g., aniseed, caraway, and coriander), leaf (cilantro, kari, bay, and mint), berry (allspice, juniper, and black pepper), bark (cinnamon), kernel (nutmeg), aril (mace), stem (chives), stalk (lemongrass), rhizome (ginger, turmeric, and galangal), root (lovage and horseradish), flower (saffron), bulb (garlic and onion), fruit (star anise, cardamom, and chile pepper), and flower bud (clove).

For people throughout the world, spices stimulate the appetite, add flavor and texture to food, and create visual appeal in meals. Called rempah (Malaysian/Indonesian), beharat (Arabic), besamim (Hebrew), epices (French), kruen tet (Thai), masala (Hindi), specie (Italian), especerias (Spanish), sheng liu (Mandarin), specerjien (Dutch), krooder (Norwegian), or kimem (Ethiopian), spices have been savored and sought around the world from the earliest times because of their diverse functions.

Nowadays, food professionals continually search for "new" and unique spice flavorings because of the growing global demand for authentic ethnic and cross-cultural cuisines. Consumers are also seeking natural foods and natural preservatives for healthier lifestyles and natural ways of preventing ailments. So, spices are also being sought for their medicinal value, as antioxidants, and as antimicrobials.

This chapter describes the different forms in which spices are sold and their composition, the primary and secondary functions of spices in applications, the techniques for preparing spices, the methods for applying spices in product development, and the methods for assuring proper quality control in spices.

SPICE FORMS AND COMPOSITION

Spices are available in many forms: fresh, dried, or frozen; whole, ground, crushed, pureed, as pastes, extracts, or infusions. Each form has its respective qualities and drawbacks. The form chosen by the food product designer will depend on the specific application, processing parameters, and shelf life.

FRESH WHOLE SPICES

Consumers and chefs frequently use fresh spices to give a fresh taste to foods. This fresh taste of ginger, cilantro, sweet basil, or chile peppers is due to the overall flavor, aroma, and texture. The "fresh" taste consumers seek from spices comes initially from their aroma. This aroma is due to the volatile component of the spice. It can be lost during harvesting, storing, processing, or handling. For some spices, the fresh forms have different flavor profiles than the dried forms, examples being ginger, galangal, cilantro, or basil. Fresh ginger has more pungency than dried ginger because of higher levels of pungent-producing gingerol which, during storage, degrades to the milder shogoals that are found in larger amounts in dried ginger.

Fresh ingredients, especially whole spices, when freshly ground, give prepared foods a fresh taste. Fresh spices provide crunchy, crispy textures and colorful appeal. Fresh whole spices also become very aromatic when they are roasted or fried in oil, and their aroma transfers to the application. This is especially true of whole or cracked seeds and leaves, such as bay leaf, kari leaf, or mustard seeds.

Whole spices provide aroma and, most importantly, texture and visual effect. Certain spices have a strong aroma when fresh, such as basil, kari leaf, ginger, or mint, due to their highly volatile (essential) oils. The essential oils disappear quickly at high temperatures, especially if the spices are processed in an aqueous system, but they can also be lost at room temperatures or when the spices are cut or bruised.

While uneven distribution of whole spices in a product can be problematic, this effect is sometimes desired to achieve nuttiness or a sensation of "bite" into a whole spice, such as whole sesame seeds on a breadstick or ajowan seeds on Indian naan bread. In this regard, whole spices can become the major flavor characterizing a product. Also, whole spices, especially the leafy spices, provide great visual appeal as garnishes.

Flavor is intact in a whole spice and is more slowly released than with the ground spice, especially when subjected to preparation techniques such as frying or roasting, during which time the whole spice slightly cracks open.

In a whole spice, the chemical components that provide the flavors vary in concentration throughout the spice. In chile peppers, the greatest concentrations of pungent compounds are found in the inner portions, such as the veins and seeds.

In many whole spices, cooking or processing changes the spices' chemical compounds and their proportions to varying degrees, often giving rise to different flavor profiles. For example, smoking, grilling, or drying certain chile peppers significantly changes their flavor and color. When jalapeno is smoked and dried, it changes its flavor and color completely, giving it a new identity, called chipotle.

Spices that do not have a strong aroma, such as bay leaf, chile pepper, or sesame seed, develop intense flavors after roasting or boiling. Mustard seed, star anise, and fagara (Szechwan pepper) are generally dry roasted to intensify their flavors for meat, fish, and poultry dishes.

Many spices, such as lemongrass, spearmint, basil, and chile peppers, are blended fresh and are used in making sauces and condiments with water, oil, wine, or vinegar. The fresh pureed or paste forms have intense flavors and need to be mixed well before application in sauces, soups, or gravies. Since the paste form usually contains oil, it can become rancid in a shorter period of time.

Consumers want to use "fresh" spices, but usually their flavors, colors, and textures are lost during storage and prolonged processing. Preliminary preparations, such as grinding, roasting, or flaking of whole spices, are generally carried out before adding the spices to processed foods.

Consistency is also more difficult to achieve in fresh spices because their origin, age, and storage conditions cause flavor variations. Therefore, dry spices and spice extractives are, by necessity, the forms most often used to formulate foods or beverages. Fresh whole spices are not frequently found in processed foods, but are generally used in restaurants, in home cooking, and in other smaller-scale applications.

The goal for a food designer is to develop products that will have the "fresh" quality desired by consumers but that have spice-sensory attributes that can withstand processing, freezing, and storage conditions.

DRIED SPICES

Spices are often used in their dried forms because they are not subject to seasonal availability, are easier to process, have longer shelf life, and have lower cost. These dried forms are most frequently used for processed products or for wholesale usage. Dried spices come whole, finely or coarsely ground, cracked, and as various-sized particulates. Spices are ground by milling them to various-sized particulates. This grinding also generates rapid air movement and heat that dissipate some of the volatile oils and even change some natural flavor notes through oxidation.

Depending on its form, the same dried spice will deliver different flavor percep-tions in the finished product. Ground spices have better dispersibility in food products than fresh whole spices. Some volatile oils are released through grinding, which partially breaks down the cellular matrix of the spice. In some spices, flavor is intensified through drying because of the elimination of most moisture. This leaves a greater concentration of the low volatile compounds that give stronger flavor but less aroma due to the loss of the volatile constituents. Dried spices can better withstand higher temperatures and processing conditions than fresh spices.

Some dried spices can be used to characterize an application's flavor and texture. Garlic and onion, which come powdered, granulated, ground, minced, chopped, and sliced, and in various-sized particles, characterize flavor and texture in garlic bread, onion bagels, or chips.

Whether a dried spice is used ground, granulated, cracked, or whole will depend on its use in specific applications. Many ground spices need to be "rehydrated" in order to develop their flavor, such as ground mustard that becomes pungent only when water is added. This addition of water triggers an enzyme reaction that releases the spice's aroma. Acidulants, oil, or vinegar are also added to preserve the pungency or intense flavors of the spice in the finished product.

In processed foods, dried spices can be more economical to use than fresh spices. For example, dried leafy spices do not require the cutting, chopping, or grinding preparation that the fresh forms do. Also, most dried spices retain a higher overall flavor concentration than fresh spices. For example, one pound of dried garlic has an equivalent flavor of five pounds of fresh garlic.

The sensory, physical, and chemical characteristics of dried spices are determined by environment, climate, soil conditions, time of harvesting, and postharvest handling. The same type of spice can have different sensory characteristics depending on where it was grown and how it was harvested, stored, and processed. For example, dried ginger from India has a subtle lemonlike flavor, dried ginger from southern China comes with slightly bitter notes, and ginger from Jamaica has more pungent flavor. Similarly, ground black pepper, which comes from a dried berry called peppercorn, varies in flavor intensity depending on its origins. Black peppercorns from Tellicherry are highly aromatic (India), while Lampong (Sumatra) pepper is less aromatic with more pungency. The Malaysian and Brazilian peppercorns, in contrast, have milder aromas with stronger bites.

For most spices, the time period between harvesting and storage and between when the spice is ground and added to a food are crucial for obtaining maximum potential.

The way a spice is treated or processed before being ground, and the conditions of storage before delivery to the food processor, create flavor and color differences. Spice flavor can be readily oxidized, and losses occur during milling and storage of spices.

Most spices such as cumin, coriander, and cardamom give more aroma and flavor when freshly ground than when bought as a preground spice. When spices are ground, the oils tend to volatilize, causing aroma losses. Anise, black pepper, and allspice lose their aroma quickly as soon as they are ground. To better retain color, flavor, and aroma, spices are sometimes milled using lower temperatures. While spices lose more aroma as they are ground more finely, the advantage is that finely ground spices blend better in finished products that require a smooth texture.

Dried spices can have some disadvantages. Some have poor flavor intensity, can cause discoloring in the finished product, and can thus create an undesirable appearance in the product. For example, dried ground cayenne can cause irregular variations in flavor and color, sometimes creating "hot" spots in food products. Anticaking agents are added to ensure better flowability of dried spices. In applications with high moisture content, such as salad dressings or soups, where particulates are desired for visual and textural effects, there is a great risk in using dried spices, unless they are sterilized.

SPICE EXTRACTIVES

Flavor is a combination of taste, aroma, and texture. The sensations of sweet, piney, sour, bitter, spicy, sulfury, earthy, and pungent are derived from an overall combination of aroma (due to volatile components) and taste (mainly due to nonvolatile components) in a spice. Crunchiness, smoothness, or chewiness, adds to a spice's overall flavor perception.

Spice extractives, which are highly concentrated forms of spices, contain the volatile and nonvolatile oils that give each spice its characteristic flavor. The volatile portions of spice extractives, also referred to as essential oils, typify the particular aroma of the spice. Most spices owe their distinctive "fresh" character to their essential oil content that generally ranges from 1% to 5% but even goes up to 15% in certain spices. The nonvolatiles include fixed oils, gums, resins, antioxidants, and hydrophilic compounds, and they contribute to the taste or "bite" of a spice.

Certain spices are prized for their bites and coloring, such as black pepper, chile peppers, ginger, saffron, and turmeric. These properties are due to the nonvolatile portions of spices.

Volatile oils contain several chemical components whose amount and proportion give rise to the spices' characteristic aromas. These can include one, two, or several components. The major chemical components of essential oils are terpene compounds—and depending on its molecular size, monoterpenes, diterpenes, triterpenes, and sesquiterpenes occur. Monoterpenes are the most volatile of these terpenes and constitute the majority of the terpenes in spices, and which give out strong aromas when spice tissues and cells are disintegrated through heating, crushing, slicing, or cutting. They are most concentrated in the mint and parsley family; sesquiterpenes in cinnamon and ginger family. Diterpenes and triterpenes are strong and bitter compounds.

The taste of a spice such as sweet, spicy, sour, or salty, is due to many different chemical components such as esters, phenols, acids, alcohols, chlorides, alkaloids, or sugars. Sweetness is due to esters and sugars; sourness to organic acids (citric, malic, acetic, or lactic); saltiness to cations, chlorides, and citrates; astringency to phenols and tannins; bitterness to alkaloids (caffeine and glycosides); and pungency to the acid-amides, carbonyls, thio ethers, and isothiocyanates.

The ratio of volatiles to nonvolatiles varies among spices causing flavor similarities and differences within a genus and even within a variety. Within the genus *Allium*, for example, there are differences in flavor among garlic, onions, chives, shallots, and leeks, which differ in this ratio. They vary depending upon the species of spice, its source, environmental growing and harvesting conditions, and storage and preparation methods. Even the distillation techniques can give rise to varying components—through loss of high boiling volatiles, with some components not being extracted or with some undergoing changes.

Nonvolatiles in a spice also vary with variety, origins, environmental growth conditions, stage of maturity, and postharvest conditions. For example, the different chile peppers belonging to the *Capsicum* group, such as habaneros, cayennes, jalapenos, or poblanos, all give distinct flavor perceptions, depending on the proportion of the different nonvolatiles, the capsaicinoids.

Spice extractives come as natural liquids (which include essential oils, oleoresins, and aquaresins) and dry encapsulated oils (spray-dried powders and dry solubles). Developed from fresh or coarsely ground spices, spice extractives are standardized for color, aroma, and, with some spices, for their antioxidant activity. They are more concentrated than dried or fresh spices and so are used at much lower levels. These extractives provide more consistency than dried spices in prepared foods.

ESSENTIAL (VOLATILE) OILS

Essential oils, such as oil of basil, oil of caraway, or oil of black pepper, are produced by grinding, chopping, or crushing the leaf, seed, stem, root, or bark; then cold expressing, dry distilling, or extracting through distillation (using water, steam, or steam and water) and recovering the distillate oil with a solvent. Sometimes the oil is distilled from a whole spice, such as the leaf or flower, or from a broken spice. Depending upon the method of extraction, the nature of the volatiles can differ with the same type of spice.

Essential oils are the major flavoring constituents of a spice. Each essential oil has many chemical components, sometimes even up to fifteen, but the characterizing aroma generally constitutes anywhere from 60% to 80% of the total oil (Table 2). The essential oils are composed of hydrocarbons (terpene derivatives) or terpenes (e.g., α-terpinene, α-pinene, camphene, limonene, phellandrene, myrcene, and sabinene), oxygenated derivatives of hydrocarbons (e.g., linalool, citronellol, geraniol, carveol, menthol, borneol, fenchone, tumerone, and nerol), benzene compounds (alcohols, acids, phenols, esters, and lactones) and nitrogen- or sulfur-containing compounds (indole, hydrogen sulfide, methyl propyl disulfide, and sinapine hydrogen sulfate).

TABLE 2
Examples of Characterizing Essential Oil Components in Some Popular Spices

Spice	Components in Essential Oils
Allspice seed	Eugenol; 1,8-cineol; humulene, α-phellandrene
Basil, sweet	Linalool; 1,8-cineol; methyl chavicol, eugenol
Cardamom	1,8-cineole; linalool; limonene; α-terpineol acetate
Dill leaf	Carvone, limonene, dihydrocarvone, α-phellandrene
Epazote	Ascaridol, limonene, para-cymene, myrcene, α-pinene
Fennel	Anethole, fenchone, limonene, α-phellandrene
Ginger	Zingiberene, curcumene, farnescene, linalool, borneol
Juniper	α-pinene, β-pinene, thujene, sabinene, borneol
Kari leaf	Sabinene, α-pinene, β-caryophyllene
Lemongrass	Citral, myrcene, geranyl acetate, linalool
Marjoram	Cis-sabinene, α-terpinene, terpinene 4-ol, linalool
Nutmeg	Sabinene, α-pinene, limonene, 1,8-cineol
Oregano	Terpinene 4-ol, α-terpinene, cis-sabinene
Pepper, black	Sabinene, α-pinene, β-pinene, limonene, 1,8-cineol
Rosemary	1,8-cineol, borneol, camphor, bornyl acetate
Star anise	Anethole, α-pinene, β-phellandrene, limonene
Turmeric	Turmerone, dihydrotumerone, sabinene, 1,8-cineol
Zeodary	Germacrone-4, furanodienone, curzerenone, camphor

Terpenes usually contribute to the aromatic freshness of a spice and can be termed floral, earthy, piney, sweet, or spicy. The oxygenated derivatives, which include alcohols, esters, acids, aldehydes, and ketones, are the major contributors to the aromatic sensations of a spice. The compounds with benzene structure provide sweet, creamy, and floral notes, while the sulfur- and nitrogen-containing compounds give the characteristic notes to onion, garlic, mustard, citrus, and floral oils.

Essential oils are soluble in alcohol or ether and are only slightly soluble in water. They provide more potent aromatic effects than the ground spices. Essential oils lose their aroma with age.

Essential oils are very concentrated, about seventy-five to one hundred times more concentrated than the fresh spice. They do not have the complete flavor profile of ground spices, but they are used where a strong aromatic effect is desired. Essential oils are used at a very low level of 0.01% to 0.05% in the finished product. They can be irritating to the skin, toxic to the nervous system if taken internally (by themselves), and can cause allergic reactions and even miscarriages.

Sometimes, alternative oils extracted from a different part of the same spice plant or from another variety are used to enhance or adulterate the more expensive essential oils, but suppliers need to meet the quality specifications that are required from manufacturers for these essential oils.

OLEORESINS (NONVOLATILES AND VOLATILES)

The nonvolatile and volatile flavor components of spices, also referred to as oleoresins, are produced by grinding or crushing the spices, extracting with a solvent, and then removing the solvent. Oleoresins have the full flavor, aroma, and pungency of fresh or dried spices because they contain the high boiling volatiles and nonvolatiles, including resins and gums that are native to spices.

The nonvolatile components create the heat and or pungency of black pepper, mustard, ginger, and chile peppers. These components can be acid-amides, such as capsaicin in red pepper or piperine in black pepper, isothiocyanates in mustard, carbonyls such as gingerol in ginger, and thioethers such as the diallyl sulfides in garlic or onion.

The different pungent and or heat principles give different sensations—spicy, hot, sharp, biting, or sulfury. The pungent sensation of onion or garlic is sulfury, while that of Jamaican ginger is spicy. Red pepper and white pepper do not contain much aroma because they have very little essential oils, whereas ginger, black pepper, and mustard contribute aromatic sensations with their bites because of a higher content of volatile oils. White pepper has a different bite sensation than black pepper because of their differing proportions of nonvolatiles, piperine, and chavicine.

Five types of capsacinoids have been isolated in chile peppers: capsaicin, hydrocapsaicin, homocapsaicin, dihydrocapsaicin, and dihydrohomocapsaicin, each with its own characterizing "bite" sensation in the mouth. In any particular type of chile pepper, the levels of capsaicinoids vary, causing varying heat levels. Each type of capsacinoid also creates a different perception of heat. Habanero has an initial sharp and violent bite that quickly disappears, leaving behind an aromatic sensation, whereas the cayennes give an initial burn that lingers.

TABLE 3
Hot and Pungent Nonvolatiles in Some Spices

Spices	Pungent Components of Spices
Red pepper	Capsaicin, hydrocapsaicin, dihydrocapsaicin, homocapsaicin, dihydrohomocapsaicin
Black or white pepper	Piperine, chavicine
Sansho pepper	Sanshool
Mustard	Allyl isothiocyanate
Horseradish	Allyl isothiocyanate
Ginger	Gingerol, shogoal
Garlic	Diallylsulfide
Onion	Diallylsulfide

Similarly, the release of heat sensation in mustard is different from wasabi. In wasabi, heat is immediate and in the front of the mouth, while with mustard and horseradish the release is delayed and comes at the back of the mouth, with a shooting sensation to the sinuses. Table 3 details some of the nonvolatiles that contribute pungency to a spice.

Oleoresins come as viscous oils and thick pastes and are more difficult to handle than essential oils. Usually, oleoresins are mixed with a diluent such as propylene glycol, glycerol, or other oils for better handling. An emulsifier is added to make it water soluble, or gum is added to make it into an emulsion for use in beverages, sauces, soups, pickles, and salad dressings.

Oleoresins are used at very low concentrations because they are highly concentrated. They have greater heat stability than essential oils. Oleoresins give more uniform flavor and color with less variability than their ground spice counterparts. They are typically used in high heat applications such as soups, salad dressings, processed meats, and in dry mixes and spice blends.

Aquaresins are water dispersable versions of oleoresins. They are convenient to use because of the ease with which they disperse into water-based foods such as soups, sauces, pickles, or gravies.

OTHER SPICE EXTRACTIVES

Liquid soluble spices are blends of essential oils or oleoresins that are made for aqueous systems. Fat-based soluble spice is made from essential oil or oleoresin blended with vegetable oil and used for mayonnaise, sauce, or soup. Dry soluble spices, usually used in dry blends, are prepared by dispersing standardized extractives on a carrier such as salt, dextrose, or maltodextrin.

Encapsulated oils are prepared from essential oils and/or standardized oleoresins with gum arabic or modified starch as the encapsulent. These have five to ten times the strength of dried ground spices. Spray-dried flavors are the traditional encapsulated products.

Spray-dried spice flavors or dry soluble spices are created to make liquid spices or extractives more convenient to handle and use in dry applications. They are a dispersion of up to 5% or more of spice oleoresin on a free-flowing carrier such as salt, dextrose, gum arabic, modified starch, or maltodextrin. These encapsulated spice extractives are used for high temperature applications, such as baked or retorted products. The spice flavors are slowly released into the product at the appropriate processing temperatures.

Essential oils or oleoresins are encapsulated to keep the full flavor impact of spices over an extended shelf life. This process grinds and encapsulates the spices in a closed system so no volatile oils escape. They are encapsulated by creating an emulsion with modified starch, dextrose, and maltodextrin or soluble gum (gum acacia) and spray-dried under controlled temperature and humidity conditions. The spice extractives are entrapped in this matrix that protects the flavor from oxidation and high heat and thereby provides an extended shelf life.

Encapsulated oleoresins tend to retain the fresh notes of spices better than the oleoresins. They have no particulates, are completely natural and, like essential oils or oleoresins, have a friendly ingredient label. They are water soluble and allow flavor to be liberated uniformly throughout the food. For application, a 1:1 or 1:2 ratio is used as a replacement for the noncapsulated extractives.

Encapsulation of an extractive renders it wettable and dispersible in water or oil and also decreases the dusting in production. The quantity of extractives that can be dispersed on the carrier varies with the type of carrier and the extractive. It is important to evenly disperse and blend the extractive onto the carrier.

There are new forms of encapsulation, such as coacervation, that show better heat stability and protect spice oils during high-heat cooking or extrusion.

Spice extractives are cost effective compared to fresh or dried spices because they are used at very low concentrations and provide similar or sometimes more acceptable flavor perceptions. One part oleoresin or aquaresin is equivalent to 20 to 40 parts of a ground spice. Also, the color, texture, and flavor of dried and fresh spices are altered through heat and freezing, while extractives have some heat and freezer stability. Extractives are available throughout the year and have standardized flavor and color, whereas the dried or fresh spices fluctuate in availability and quality. Extractives are also free of microbes and other extraneous matter, so they do not cause microbial contamination in the finished product.

Spice extractives are labeled as natural flavors, natural flavorings, or as spice extracts. Extractives are typically used by food developers because of their consistency in flavor and aroma, instant flavor release, uniformity of color, and stability in high-heat applications. By using extractives, the quality and consistency of products from development through production can be better controlled. Finished product quality, uniformity during mixing, and consistency can be maintained from batch to batch during production.

Proper usage levels in the finished product are very important to achieve the right flavor profile and to prevent bitterness. Usage level is generally 0.01% to 0.05% by weight in the final product but varies depending on the type of application. A more uniform dispersion of color and flavor is achieved with the liquid soluble extractives than with dried spices. This creates an acceptable appearance in finished products, unlike dried spices that sometimes mix unevenly in large quantities and leave pockets of flavor and color. When formulating a dry seasoning, oleoresins and essential oils need to be effectively dispersed on a soluble carrier, such as salt, dextrose, maltodextrin, or sugar, before being added with the other ingredients. Otherwise, they will not blend well in the finished product. Overoptimum use of the liquid extractives will create caking of the finished product during storage.

Today, the trend is to capture the varied flavor profiles from spices. For example, researchers are breeding habaneros to obtain their wonderful flavor profiles with less heat. Consumers are using fresh chile peppers, more for their flavor than heat, for example, the anchos and chipotles. Asians and Mexicans have traditionally used chile peppers for flavoring their dishes and have created unique flavors from chile peppers through different preparation techniques. The chile pepper's flavor is contained in its outer fleshy parts and is intensified when roasted or fried. Chile peppers vary in color, flavor, and texture profiles when they are fresh, dry, smoked, or grilled. As a result of the popularity of these ethnic cuisines, the flavor characteristics of chile peppers and many other spices are becoming a major portion of the profile sought.

Table 4 provides a summary of the advantages and disadvantages of various spice forms.

TABLE 4
A Summary of Advantages and Disadvantages of Different Spice Forms

Spice Form	Advantages	Disadvantages
Fresh whole	Fresh flavor	Variability in flavor and color
	Release of flavor slowly at high temperatures	Short shelf life
		High microbes
	Label friendly	Unstable to heat
		Seasonal Availability
Dried ground	Process friendly	Less aroma
	Longer shelf life	Hot spots—flavor and color
	Easy handling and weighing	Takes more storage space
	Stronger taste intensity than fresh	Variability in flavor and color
		Undesirable specks
		Seasonal Availability
		Spice dust contamination during production
Extractives	Standardized flavor and color	Difficult to handle and weigh
	Uniform appearance, color and flavor	Aroma and taste usually not typical of natural spice
	Low usage	Loss of volatiles at high temperatures
	Low microbes	
	Less storage space	
	Available throughout the year	

Spices should be stored in tightly closed containers and should not be exposed to light, high temperatures, or high humidity conditions. This maintains their freshness for longer periods of time. Heat, moisture, and direct sun accelerate the loss of their flavor and aroma components. Moisture and high temperatures will help mold growth that will cause spoilage. Generally, the moisture content of spice is 8% to 10%. High storage temperatures cause flavor loss, color changes, and caking or hardening of the ground spice. Spices need to be stored at 50°F to 60°F (10°C to 15°C) with a relative humidity (RH) of 55% to 65%. Spices should not be stored in the freezer as repeated removal from freezer creates condensation in containers which accelerates loss of their flavor and aroma. And always close containers tightly after each use. Ground spices generally keep for 2 to 3 years, whole spices for about 3 to 4 years, and seasonings for about 1 to 2 years if stored in ideal conditions. When exposed to steam, flavor and aroma are lost at a faster rate and caking occurs. Any moisture introduced into the bottle will result in caking.

FUNCTIONS OF SPICES

Edible spices serve many functions in food products. Their primary functions are to flavor food and to provide aroma, texture, and color. Spices also provide secondary effects, such as preservative, nutritional, and health functions (Table 5).

Spices are composed of fiber, carbohydrate, fat, sugar, protein, gum, ash, volatile (essential oils), and other nonvolatile components. All of these components impart each spice's particular flavor, color, nutritional, health, or preservative effects. The flavor components (volatile and nonvolatile) are protected within a matrix of carbohydrate, protein, fiber, and other cell components. When the spice is ground, cut, or crushed, this cell matrix breaks down and releases the volatile components.

TABLE 5
Primary and Secondary Functions of Selected Spices and Flavorings

Taste	Thai basil, black pepper, cardamom, jalapeno, asafetida, lemongrass, star anise, kokum, sorrel, chipotle, habanero
Aroma	Clove, ginger, kari leaf, mint, nutmeg, rosemary, cardamom, tarragon, cinnamon, sweet basil, mango, rose petal
Texture/Consistency	Mustard seed, onion, sassafras, sesame seed, shallot, peppercorn, ajowan seed, poppy seed, candlenut, almonds
Color	Annatto, cayenne, paprika, parsley, turmeric, saffron, basil, cilantro, mint, marigold
Antimicrobial	Cinnamon, clove, cumin, oregano, rosemary, sage, thyme, ginger, fenugreek, chile peppers
Antioxidant	Turmeric, rosemary, sage, clove, oregano, mace
Health	Chile pepper, cinnamon, fenugreek, ginger, turmeric, garlic, caraway, clove, sage, licorice

---

PRIMARY FUNCTION OF SPICES

Flavor, Taste, Aroma, and Texture

The overall flavor, taste, aroma, texture, or color that a spice contributes to food or beverage determines its effectiveness in a recipe or formula.

Every spice or flavoring contains predominating chemical components that create these sensual qualities. A spice's chemical compounds can contribute mild to strong flavors. The balance of these chemical compounds gives a spice its characteristic flavor profile.

Spices give characterizing tastes and aromas. They give six basic taste perceptions: sweet, salty, spicy, bitter, sour, and hot. The other descriptive terms include pungent, umami (brothy, MSG, or soy-sauce-like), cooling, and floral, earthy, woody, or green. The taste sensations are generally experienced at different locations of the tongue: sweet is detected at the tip of the tongue, salty at the frontal sides of the tongue, sour at the posterior sides of the tongue, bitterness at the back of the tongue, and heat, depending on the type, at different areas of the tongue.

Most spices have more than a single flavor profile. For example, fennel has not only sweet notes but also bitter and fruity notes, tamarind has fruity and sour notes, while cardamom has sweet and woody notes.

A spice's textural qualities are derived from its specific physical characteristics, the form in which it is used in a recipe (e.g., whole or ground), and the techniques used in its preparation. Most textural characteristics are obtained through the preparation and cooking techniques of spices, which are discussed later in this section.

Let's look at the typical sensory characteristic obtained from each spice or ingredient (Table 6). (Note, however, that spice sensory profiles are frequently described from a mainstream, Western perspective. Western foods are generally seasoned in a blander and milder fashion. Therefore, Western descriptions are not necessarily consistent with the taste experiences of other cultures. For example, cumin, coriander, or cloves are often described as spicy, but to Asians and many other cultures, who are accustomed to eating hot and spicy foods, they are not perceived as spicy. This book generally describes spice tastes from a mainstream U.S. perspective but takes other cultural taste descriptions into account.)

Coloring

Some spices, such as saffron, paprika, turmeric, parsley, and annatto provide color as well as flavor to foods and beverages. Spices can meet consumer's demands for "natural" colorings (Table 7).

The components responsible for the coloring in spices are oil soluble or water soluble. Some typical coloring components in spices are crocin in saffron, carotenoids in paprika, capsanthin in chile pepper, bixin in annatto, or curcumin in turmeric. The overall coloring given by a spice is sometimes a combined effect of two or more of its coloring components.

Saffron imparts a beautiful yellow color, which ranges from a deep red, yellowish orange to a reddish orange color, to paellas and pilafs. Its yellow color is primarily due to a terpene glycoside called crocin that is water soluble and not stable to acid

TABLE 6
Typical Sensory Characteristics of Spices

Sensory Characteristic	Spices and Other Flavorings
Sweet	Green cardamom, anise, star anise, fennel, allspice, cinnamon
Sour	Sumac, caper, tamarind, sorrel, kokum, pomegranate
Bitter	Fenugreek, mace, clove, thyme, bay leaf, oregano, celery, epazote, ajowan
Spicy	Clove, cumin, coriander, canela, ginger, bay leaf
Hot	Chile peppers, mustard, fagara, black pepper, white pepper, wasabi
Pungent	Mustard, horseradish, wasabi, ginger, epazote, garlic, onion, galangal
Fruity	Fennel, coriander root, savory, tamarind, star anise
Floral	Lemongrass, sweet basil, pandan leaf, ginger flower
Woody	Cassia, cardamon, juniper, clove, rosemary
Piney	Kari leaf, rosemary, thyme, bay leaf
Cooling	Peppermint, basil, anise, fennel
Earthy	Saffron, turmeric, black cumin, annatto
Herbaceous	Parsley, rosemary, tarragon, sage, oregano, dillweed
Sulfury	Onion, garlic, chives, asafetida
Nutty	Sesame seed, poppy seed, mustard seed, whole seeds (ajowan, cumin)

or light. Crocin is hydrolyzed to crocetin that gives a darker shade of color. Though it is a permitted natural color in the United States, its use is limited because of its cost. Safflower is frequently substituted for saffron in many countries, but it is not permitted in the United States.

Turmeric, from the ginger family, is often called "Indian saffron." Its root is dried and ground to give a yellow color with an orange tinge. It is used as a natural food coloring in salad dressings, pickles, mustards, soups, and condiments. Its coloring is due to curcumin, a diketone, that accounts for 3% to 7% of this spice. The curcumin content varies depending upon its source, with Allepey (India) turmeric having a higher amount of curcumin than other varieties.

Turmeric's color varies from a bright orange yellow to a reddish brown and is unstable to light and alkaline conditions. It can be used with high-heat products and in products with a pH of 2.5 to 6.5. Its color is yellow in an acid to a neutral pH, but reddish brown in an alkaline pH. Its color will break down when prepared in an iron utensil.

Paprika is produced from the mild to pungent dried red pepper. The United States generally uses the mild paprika that has a brilliant red color derived from many different carotenoids. Capsanthin accounts for about 35% of the total carotenoids, violaxanthin 10%, cryptoxanthin and capsorbin each 6%, and other carotenoids 2%. Capsanthin is oil soluble, stable against heat, and has a strong red color. Paprika powder loses its color through oxidation, catalyzed by light and high temperatures. Its oleoresin, which has a reddish orange shade, is more stable to light and heat and is used in snack products, spice blends, crackers, and salad dressings.

Annatto is ground from annatto seeds. It exhibits an orange yellow to a golden yellow shade and is used in cheddar cheese, bakery products, and sometimes in

TABLE 7
Coloring Components of Selected Spices

Spices	Coloring Component	Type of Color
Saffron	Crocin	Yellowish orange
	Crocetin	Dark red
	Beta-Carotene	Reddish orange
Paprika	Carotenoids:	
	Capsanthin	Dark red
	Violaxanthin	Orange
	Cryptoxanthin	Red
	Capsorbin	Purplish red
	Beta-Carotene	Reddish orange
	Lutein	Dark red
	Zeaxanthin	Yellow
Chile pepper	Beta-Carotene	Reddish orange
	Cryptoxanthin	Red
	Capsanthin	Dark red
	Capsorbin	Purplish red
Turmeric	Curcumin	Orange yellow
Parsley	Chlorophyll	Green
	Lutein	Dark red
	Neoxanthin	Orange yellow
	Violaxanthin	Orange
Annatto	Bixin	Golden yellow
	Norbixin	Orange yellow
Safflower	Carthamin	Orange red
	Saflor yellow	Yellow

combination with paprika and turmeric oleoresins. Its coloring is due to norbixin (water soluble) and bixin (oil soluble) which are stable at pH 5 to pH 14. Its color is resistant to heat but is less resistant to light.

The coloring components of essential oils are generally extracted with ethanol or organic solvents, and the solvents are removed to give the coloring matter. Extracts of paprika, saffron, annatto, and turmeric with emulsifiers (polysorbate 80) are used as natural colors to provide the bright yellow, red, and orange hues in processed foods and beverages. They are available as oil-soluble (oleoresins) and water-dispersible extracts. Trace metals, oxidized fats and oils, intense light, and exposure to oxygen will promote color losses in these extractives.

SECONDARY FUNCTIONS OF SPICES

Secondary effects of spices are becoming more important as the public's desire for natural or organic foods and natural ways of healing increases. Spices have been used traditionally to stimulate appetite, enhance digestion, relieve stress, and increase energy. Spices can also aid nutrition when they are used in lieu of salt, fat, or sugar to enhance taste in processed foods.

Spices may be used in food products as preservatives, which allows for a more "natural" or friendly label on processed foods.

Spices as Preservatives

Spices have long been known for their preservative qualities, as antimicrobials, and as antioxidants. They have been used by many ancient cultures—Egyptians, Romans, Indians, Greeks, Chinese, and Native Americans—to fumigate cities, embalm the royalty, preserve food, and prevent diseases and infections.

Spices as Antimicrobials

As early as 1500 BC, Egyptians used spices to preserve foods. In Europe, the Middle East, and Asia, before the days of refrigeration, spices were used to preserve meats, fish, bread, and vegetables. Spices were used alone or in combination with smoking, salting, and pickling to inhibit food spoilage. The Romans preserved fish sauce with dill, mint, and savory, and meats and sausages with cumin and coriander. The Greeks used garlic to prevent food spoilage, and in India, ginger, garlic, clove, and turmeric were used to preserve meats and fish. In ancient Egypt, cinnamon, cumin, and thyme were used in mummification. Spices are still used to preserve food in the villages of India, Africa, Indonesia, and Thailand.

Spices have also been used for bactericidal and health reasons. During the Middle Ages, spices such as cinnamon, garlic, and oregano were used to treat cholera and other infectious diseases. In the late nineteenth century, clove, mustard, and cinnamon were shown to have antimicrobial activity. In the twentieth century, new research on spices, including ginger, garlic, fenugreek, coriander, turmeric, and clove, as potential natural antimicrobials, continued. Today this research continues.

Aldehydes, sulfur, terpenes and their derivatives, phenols, and alcohols, exhibit strong antimicrobial activity. Spices have strong, moderate, or slight inhibitory activity against specific bacteria (Table 8). Cornell University studies have reported that garlic, oregano, onion, and allspice kill all bacteria; thyme, cinnamon, tarragon, and cumin kill up to 80% of bacteria; chilies up to 75% of bacteria; and black and white peppers, ginger, anise, and celery seed up to 25%. Kansas State University studies have reported that clove, cinnamon, oregano, and sage suppress growth of Escherichia coli O157:H7 in uncooked meats, which causes gastrointestinal disease. Other recent studies have shown that dodecenal in coriander leaf and seed kills Salmonella in meats.

A combination of spices can be more effective as preservatives than one spice. Microorganisms differ in their susceptibility to specific spices. Gram-positive bacteria are more sensitive to spices than gram-negative bacteria. *Bacillus(B) subtilis* and *Staphylococcus(S) aureus* are more susceptible than *Eschrichia (E) coli* bacteria. Certain spices can act as broad-spectrum antimicrobials, such as rosemary and sage, while others are very specific in their functions, such as allspice and coriander.

TABLE 8
Spices as Antimicrobials

Spice	Effective Component	Microorganism[1]
Mustard	Allyl isothiocyanate	*Escherichia(E) coli, Pseudomonas, Staphylococcus(S) aureus*
Garlic	Allicin	*Salmonella typhii, Shigella dysenteriae,* molds, yeasts
Chile pepper	Capsaicin	Molds, bacteria
Clove	Eugenol	*E. coli* 0157:H7, *S. aureus, aspergillus,* yeast, *acinetobacter*
Thyme	Thymol, isoborneol, carvacrol	*Vibrio parahemolyticus, S. aureus, Aspergillus*
Ginger	Gingerone, gingerol	*E. coli, Bacillus(B) subtilis*
Sage	Borneol	*S. aureus, B. cereus*
Rosemary	Borneol, thymol	*S. aureus, B. cereus*
Coriander	Dodecenal	Salmonella

[1] Kenji and Mitsuo, 1998.

The essential oils of some spices have an inhibitory effect on bacteria and fungi in meats, sausages, pickles, breads, and juices. Eugenol in clove; cinnamaldehyde in cinnamon; allyl isothiocyanates in mustard; thymol in thyme, carvacrol in oregano; linalool in coriander, and allicin in garlic, are some of the components used as antimicrobials.

The more pungent spice nonvolatile oils have also been shown to have strong antimicrobial properties, such as gingerol in ginger, piperine in black pepper, capsaicin in red peppers, and diallyl sulfide in garlic.

Spices must be used at high levels to be effective antimicrobials, but this will cause flavor issues in food products. The level of spices typically added to Western style foods is generally not enough to inhibit microorganisms completely but may inhibit spoilage to some extent. Spice extractives at the levels used are effective against certain bacteria. Thyme (thymol), anise (anetol), and cinnamon (eugenol) essential oils inhibit mold growth and aflatoxin production.

Spices as Antioxidants

Spices can be used in foods to help fight the toxins created by our modern world. Heat, radiation, UV light, tobacco smoke, and alcohol initiate the formation and growth of the free radicals in the human body. Free radicals damage the human cells and limit their ability to fight off cancer, aging, and memory loss. Many spices have components that act as antioxidants and that protect cells from free radicals (Table 9).

Some spices have more antioxidant properties than others depending on the food they are in. Combining spices with other spices or antioxidants such as tocopherols and ascorbic acid produces synergistic effects. The naturally occurring phenolic compounds (phenolic diterpenes, diphenolic diterpenes) in spices are

TABLE 9
Spices as Antioxidants

Spice	Chemical Component
Rosemary	Carnosol, carnosoic acid, rosmanol
Sage	Rosmanol, epirosmanol
Turmeric	Curcumin, 4-hydroxycinnmoyl methane
Clove	Eugenol
Oregano	Phenolic glucoside, caffeic acid, rosmarinic acid, protocatechuic acid
Mace and nutmeg	Myristphenone
Sesame seed	Sesaminol, δ-tocopherol, sesamol
Ginger	Shogoal, gingerol

effective against oxidative rancidity of fats and color deterioration of the carotenoid pigments.

Spices can prevent rancidity and extend shelf life by slowing the oxidation of fats and enzymes. Fats are broken down into peroxides (free radicals) with exposure to air or oxygen and finally into aldehydes and alcohols that give a rancid taste. Spices can halt the oxidative process by blocking or "scavenging" the free radicals.

Today, with consumer demand for "natural" products, spices can be used commercially as natural antioxidants in foods. Sage, rosemary, oregano, thyme, cilantro, and marjoram are found to have stronger antioxidant properties than other spices. Rosemary and sage are currently used as natural antioxidants in foods, while other spices such as cilantro are being explored.

Rosemary and sage antioxidants are available as oil-solubles, water-dispersibles, or dry-solubles, and can be used in seasoning blends, salad dressing mixes, lard, sausage, or instant potatoes. They are used as sprays, dips, or surface coatings in comminuted poultry, seafood, or meats before they are frozen to inhibit "warmed over" flavors that develop after cooking and reheating. For snack foods, they are added in the frying oil or atomized on the surface of snacks or put into the dough. They are also added to glazes and injection marinades for meats, and are extremely heat stable as they withstand extrusion, spray-drying, or baking temperatures.

Rosemary and sage are the most effective of all spices in retaining the red color of processed meats by inhibiting the flavor and color degradation of fats and oils in them. The flavonoids and diterpenes and triterpenes of rosemary and sage are responsible for their antioxidant properties. They exhibit antioxidant properties superior to BHA or BHT. Other effective spices include thyme, turmeric, oregano, ginger, clove, majoram, red pepper, mace, sesame, and nutmeg.

Rosmanol, caffeic acid, myristphenone, curcurmin, eugenol, thymol, and sesaminol in rosemary, clove, thyme, oregano, ginger, turmeric, nutmeg, sage, and sesame seed are found to be strong antioxidants with meat, lard, and soybean oil.

EMERGING FUNCTIONS OF SPICES

Spices as Medicines

The growing emphasis on healthy eating is drawing attention to spices as critical ingredients for not only creating tasty low-fat or low-salt foods but as a natural way for improving health and promoting wellness. Consumers prefer eating a "natural" food product to taking medicine or drugs. With greater research into their medical benefits, spices are becoming more attractive to consumers.

Spices can be used to create these health-promoting products. The active components in spices—phthalides, polyacetylenes, phenolic acids, flavanoids, coumarins, capsacinoids, triterpenoids, sterols, and monoterpenes—are powerful tools for promoting physical and emotional wellness.

Spices such as celery, parsley, ginger, turmeric, fenugreek, mint, licorice, garlic, onion, mustard, horseradish, and chile pepper can be used to stimulate production of enzymes that detoxify carcinogens, inhibit cholesterol synthesis, block estrogen, lower blood pressure, elevate immune activity, and inhibit tumor growth.

Since ancient era, spices have been used not only by Indians and Chinese, but also Latinos, Africans, Egyptians, and Greeks, to relieve ailments and prevent illnesses. In recent years, Western scientists have been isolating the active compounds in spices to study their therapeutic effects to promote wellness: rosemary, sage, and basil fight against tumors; chile peppers inhibit blood clotting, stimulate digestion and circulation, induce perspiration, and reduce pain; ginger prevents motion sickness and aids digestion and stomach ulcers; garlic lowers cholesterol and high triglycerides, prevents colds and flus, and prevents tumor growth; turmeric inhibits tumors, heals wounds, acts as an antidepressant, and fights against Alzheimer's; licorice treats gastric and duodenal ulcers and relieves chronic fatigue, coughs, and cold symptoms; peppermint combats indigestion, irritable bowel syndrome, and inflammation of gums.

Aromatherapy, using essential oils, can relax or stimulate the body, create certain moods, relieve cold symptoms and respiratory problems, and ease muscle pains. The vapors are inhaled to release neurochemicals in the brain, through receptors in the mouth and nose, that cause the desired effects. Cooking foods with spices is the oldest form of aromatherapy, since their aroma can stimulate gastric secretions that create appetites. Spices are also used as balms or massage oils and are applied on the skin, joints, and muscles to relieve stress and pain. Examples of the health benefits of spices are almost endless.

In the United States and elsewhere, the growing trend of using foods as nutraceuticals that boost energy and improve health will promote spices that were historically used to cure ailments and prevent diseases.

Let's look at traditional healing methods to provide an understanding of how and why spices are used by many global cultures as medicines. This will also help in understanding the basis for the contrasting flavors presented by spices and the many taste sensations experienced in an Asian meal.

Traditional Medicine

Spices are used in many traditional cultures to boost energy, relieve stress, improve the nervous system, aid digestion, relieve cold symptoms and headaches, and treat many diseases. A food creator can explore interesting food concepts that combine taste and "cure" by using authentic spices and methods of preparing and presenting them in meals.

Europeans have used oils from leafy spices or plants to provide physical, mental, and spiritual benefits, such as to provide a calming effect, to soothe muscle aches, or to cure many ailments such as colds and fevers. Similarly, Middle Easterners and Egyptians used spices for many therapeutic and cosmetic effects. The Middle Eastern system of medicine, called *Unani Herbal* Medicine, has many similarities to Indian and Chinese traditional medicine, with some roots to ancient Greek and Roman medicine.

Indian cooking is based on the therapeutic principles of ancient *Ayurvedic* medicine. This Ayurvedic philosophy of eating dictates the blending and preparation of spices, as well as how foods are balanced, to achieve well-being. *Ayurveda* combines two words, *ayu,* which means life, and *veda,* which means knowledge. This system of healing has been practiced in India for more than 5000 years. It is based on prevention and well-being and looks at the causes of disease and the ways they occur. It is based on the life force or *prana,* that flows easily into every cell of the body and which is accomplished by eating the right foods, by deep breathing techniques, and by following a healthy lifestyle. In an *Ayurvedic* meal, the ingredients are chosen not only for taste but also to assure physical and emotional harmony and well-being. *Ayurveda* emphasizes prevention of disease through the pursuit of mental, physical, and emotional harmony.

Similar to the ancient Greek and Roman theories of medicine, spices are classified into five elements: earth, water, fire, air, and ether. *Ayurveda* categorizes foods into six tastes or *rasas* (Table 10). Many foods and spices contain more than one

TABLE 10
Foods and Contributing Tastes (*Rasas*)

Tastes (Rasas)	Example Foods
Sweet	Anise, fennel, cumin, coriander, sugar, butter, honey, jaggery, rice, legumes, fruits, milk, cardamom, coconut, most grains, ghee
Sour	Kokum, pomegranate, tamarind, tomato, lemon, citrus, grapefruit, fermented foods (yogurt, pickles, miso, soy sauce), vinegar
Salty	Salt, seaweeds, vegetables (high moisture)
Bitter	Fenugreek, turmeric, clove, cinnamon, endives, lettuce, purslane, bay leaf, ajwain, chicory
Spicy/pungent	Chile peppers, ginger, garlic, horseradish, mustard, onion, black pepper
Astringent	Asafoetida, teas, licorice, legumes, fenugreek, cauliflower

TABLE 11
Doshas and Respective Personalities

Doshas	Qualities
Vata	Quick thinking, creative, flexible, nervous
Pitta	Determined, strong willed, fiery, passionate
Kapha	Good endurance and stamina, heavy, stable, relaxed, tolerant

taste. For example, fenugreek seed provides a bitter taste as well as astringency, and fennel seed has sweet and cooling tastes. Thus, there is more than one taste contributed by a spice (see Table 10). With regard to *rasas*, other spices can be explored to provide more variety. Understanding these *rasas* is essential to understanding and applying *Ayurvedic* medicine. They affect our digestion, disposition, and health.

To maintain perfect balance and be well nourished, all six tastes or *rasas* should be part of every meal. This explains the complex spice combinations and depth of flavor in Indian foods. To harmonize the body, these tastes must be balanced in the meal according to each person's constitution or *doshas*. Illnesses and diseases occur when there is imbalance with foods and a person's constitution. There are three basic constitution types or *doshas* that came from the five elements of energy—earth, fire, air, space, and water (Table 11). Generally, a person has influence from three *doshas* with one or two predominating *doshas*. The *dosha* or *doshas* that exhibit the stronger characteristics becomes or become the predominating *dosha* or *doshas* for that person.

These *doshas* regulate the functions in our mind-body system. They control the basic functions of all organs and systems in our body. Certain spices and flavorings are used in our foods to balance our doshas and create harmony (Table 12). These *doshas* also have their own inherent natural tastes. The six tastes or *rasas* have a

TABLE 12
Spices that Balance and Reduce the *Doshas*

Doshas	Natural Taste	Spices for Balance	Spices to Reduce
Vata	Bitter	Salty, sour, sweet	Pungent, bitter, astringent
	Astringent, pungent	Cardamom, fennel, anise, tamarind	Ginger, asafetida Chilies, fenugreek
Pitta	Sour, salty, pungent	Sweet, bitter, astringent Asafetida, ginger, cardamom, cinnamon, fenugreek	Pungent, sour, salty Chilies, salt, tamarind Garlic
Kapha	Sweet, salty, sour	Pungent, bitter, astringent Turmeric, fenugreek, mint, mustard, clove	Sweet, salty, sour Salt, anise, fennel, tamarind, cardamom

beneficial effect on the *doshas*, increasing one type and decreasing the others. *Ayurvedic* cooking calls on us to choose foods and spices with the tastes that balance our own *doshas* to maintain good health.

In *Ayurvedic* cooking, the art of using spices not only makes food taste better, but also increases the overall therapeutic value and negates any side effects in the body.

In Indian cooking, foods need to be well digested to be nourishing for the body. Foods can be hot, cold, moist, dry, heavy, or light. Hot food will speed digestion, while cold food will slow it. Every meal should be well balanced between hot and cold foods to promote digestion and avoid illnesses. The way Indian dishes are served in a traditional meal illustrates these principles. In a typical vegetarian meal, rice or bread (chappati, naan, or dosai) are placed on a thali or tray with an array of tiny silver bowls. They contain peppery sambar (lentil stew), spicy, sour rasam (spice broth), cool, minty raita (cucumber and tomato with yogurt), a couple of hot and spicy braised vegetables, and a crunchy, astringent mango pickle. The "cold" raita, the sour broth and the "hot" spicy vegetables balance each other, thus creating "harmony" in the body. The crunchy mango pickle balances the soft textured vegetables, while the hot lentil stew perks up the plain neutral bread or rice. Also, the order in which the different side dishes are eaten with rice creates a healthy balance in our body. Food developers can use these meal presentation concepts today to develop interesting and well-balanced meals.

In China, historically, there has been an integration of nutrition, medicine, and foods. Even today, combinations of spices are made into tonics that nourish and strengthen the body and cure illnesses. Similar to the Indian *Ayurvedic* medicine, in Chinese traditional medicine, there are five different tastes—sweet, salty, bitter, sour, and spicy hot. The proper balance of these different tastes with appropriate textures and colors creates good taste and health in Chinese cooking. The five tastes are also associated with major organs in the body. For example, sweet is associated with the spleen (yin organ) and stomach (yang organ); sour with liver (yin organ) and gall bladder (yang organ); and spicy with lungs (yin organ) and large intestine (yang organ).

Each taste affects the *Qi* or (*chi*), an invisible, vital force which, similar to the *prana* in Ayurveda or *ool* in Mayan practice of healing, circulates throughout the body along prescribed pathways. The proper balance of the five tastes is essential to creating harmony or good health. If *Qi* does not flow smoothly or "gets stuck," illnesses occur. *Qi* can be "unstuck" by stimulants such as food, acupuncture, or acupressure.

The different tastes as well as the movement of *Qi* are described by yin (cold) and yang (hot). Movement of *Qi* in our body is associated with yin and yang. Yin and yang are very basic to Chinese culture, just as hot and cold are to Indian culture. They are similar to water/fire, female/male, moon/sun, and so forth. They describe *Qi*'s location, movement, and functioning. For yin, the movement in the body is to contract and flow downward and inward; while for yang, the movement is to expand and flow upward and outward. One is not separate from the other, and in food combinations, yin follows yang and vice versa. Just as yin–yang creates movement, so do spices.

The yin foods (which are nourishing, cooler, softer, moister, and alkaline) and yang foods (which are spicier, hotter, drier, and acidic) are balanced in Chinese cooking. Yin foods include mild and sweet spices (mint, fennel, anise, and parsley), seafood, melon, fruits, asparagus, tofu, bland and moist vegetables (seaweed and watercress), steamed foods, and water. Some yang foods are chilies, ginger, garlic, sesame oil, fatty meats (mutton and pork), deep-fried foods, and strong alcoholic drinks. Yin spices are sedative and slow metabolism, while yang spices are active and increase metabolism. Balancing the yin and yang keeps the body in a state of equilibrium and good health. Rice is a neutral food, thus it serves as a centerpiece in a meal.

Foods are prepared and cooked with spices based on this theory of well-being. The balance of yin and yang with contrasting tastes and textures during cooking and at a meal table is the basis of a Chinese meal. Contrasting ingredients or spices are added to ingredients during cooking to manipulate their hotness or coldness. For example, chile peppers and sesame oil are added to seaweed or tofu to balance its effect on the body, and likewise, sugar and anise are added to ginger-flavored pork.

Cooking techniques are also classified under hot or cold. For example, grilling or deep frying, adds "heat" to foods, while steaming or slow simmering provides a "cooling" effect.

The state of the person's health is also a factor. A person with a fever will not eat warm or hot ingredients. She needs cooling ingredients to balance her system. Seasons also become important factors in balancing ingredients for health. During the cold season, spices that produce an upward and outward movement (yang spices) are more desirable for the flow of *chi,* while for the summer, yin spices which produce a downward and inward movement are more appropriate.

In addition to philosophies of balanced eating to promote well-being, traditional Asian cultures have been using spices for their healing properties. Unlike Western meals, spices are used daily in Asian meals and are taken in abundance to flavor dishes. Today, many of these spices are being examined and recognized by Western scientists for their positive therapeutic or pharmacological properties.

The healing and medicinal properties of spices in many traditional cultures are attributed to using the whole spice or a combination of spices that work synergistically. The medium in which it is to be consumed also gives a synergistic effect. Spices are not only used in cooking but are also added to milk, tea, vinegar, salt, hot water, or sugar to create healing effects. For example, chile peppers are added to milk as a drink to reduce swellings and tumors, cinnamon is added to sugar and taken to prevent tumors, while turmeric paste is added to milk and consumed to reduce coughs and colds.

Modern Medicine

With greater research into their medical benefits, spices are becoming more attractive to consumers. Using modern research methods, many Asian and Western studies have isolated active compounds in spices that have medical benefits (Table 13). The active components of spices (or phytochemicals) include sulfides, thiols, terpenes and their derivatives, phenols, glycosides, alcohols, aldehydes, and esters. Turmeric

TABLE 13
Reported Therapeutic Effects of Spices

Spice	Chemical Component	Medicinal Value
Turmeric	Curcumin	Anti-inflammatory; antitumor (inhibits tumor initiation and promotion); prevents Alzheimer's
	Curcumene	Antitumor
Ginger	Gingerol	Digestive aid for stomachaches, indigestion, stomach ulcers
	Shgoal	
	Gingeberane	Prevents bloating and vomiting
	Bis-abolenenan	
	Curcumene	
	Gingerol	Antitumor
	Shogoal	Enhances gastrointestinal mobility
		Inhibits cholesteral synthesis
Fenugreek Seed	Trigonelline	Arrests cell growth and prevents hypoglycemic effect
	Diosgenin	Synthesis of steroid drugs and sex hormones
Seed and Leaf	Soluble dietary fiber (galactomannan) saponins, diosgenin, protein	Improved glucose tolerance (reduces plasma glucose levels), lowers cholesterol and triglyceride (decreases bile secretion)
Garlic	Allicin	Breaks down blood clots, prevents heart attacks, prevents gastric cancer
	Glutamyl peptides	Lowers blood pressure and blood cholesterol
	Allicin, diallyl	Inhibits platelet aggregation
	Sulfide, s-allyl	Inhibits platelet aggregation
	Cysteine	
Licorice	Glycyrrhizin	Treats gastric and duodenal ulcers, prevents coughs and colds, treats chronic fatigue syndrome

* Data obtained from Mazza, G. and Oomah, B.D., Eds., *Herbs, Botanicals and Teas*, Technomic Publishing Co., Inc., Lancaster, PA, 2000.

is a wonder medicine and has been used and researched for its many healing properties. Curcumin and curcumene in turmeric are the active compounds. Turmeric has protection against free radical damage and cancer prevention; possesses anti-inflammatory properties by lowering histamine levels; protects the liver against toxic compounds; reduces platelets from clumping together thereby improving circulation and protecting against arteriosclerosis; prevents cancer; acts as an antipeptic ulcer and antidyspepsia agent; and heals wounds. Researchers at UCLA (University of California at Los Angeles) have shown curcumin to slow the formation of, and even destroy, accumulated plaque deposits that play a key role in development of Alzheimer's disease. Allicin in garlic lowers cholesterol, capsaicin in chile peppers prevents blood clotting, trigonelline in fenugreek seeds prevents rise in blood sugar, and gingerol in ginger aids digestion. Research is being conducted on many more spices.

While the healing effects of many phytochemicals in spices have been identified, questions still remain unanswered. Are phytochemicals by themselves the effective healing agents, or is the whole spice, as was traditionally used, the cure? Can the healing property of a spice be strengthened or lessened through a synergistic effect with other spices or through a medium such as water, sugar, tea, milk, protein, or starch? Does the fresh form of a spice give different healing effect than the dry form, as ginger is used in Ayurvedic healing?

Finally, it should be remembered that cultures that use spices as traditional cures consume them at much higher levels than Western cultures. For many cultures, "spiced up" foods are eaten at every meal on a daily basis. Moreover, "spice cures" are commonly taken through infusions in teas, milk, coffee, water, and other beverages, which is not a common practice in the West. To meet the demands of today's consumer, spices and other flavorings can be explored as integrative or complementary medicine.

SPICE PREPARATION

Spices need to be ground, sliced, roasted, toasted, fried, or boiled to release their characteristic flavors. Just as we roast garlic, saute onions, or mustard, we also need to apply appropriate cooking techniques to other spices to derive their optimal flavor sensations. Volatile oils are released from spices through grinding, cutting, and heating. Spices also release unique flavors when they are cooked. Several different flavor characteristics can be derived from the same spice by using different preparation methods. Around the world, many ethnic groups prepare spices to suit specific applications and to create totally different flavor profiles. Spices are dry roasted, fried in oil, deep fried, simmered, pickled, braised, barbecued, or boiled in water to reach their full flavor potential and to create a broad spectrum of flavors. The flavor of spices is generally enhanced, intensified, or simply becomes different through heat. Cooking techniques are also discussed in Chapter 6, "Emerging Flavor Contributors."

Certain spices are more stable to heat and can be cooked at higher temperatures, while others have to be added at the end of cooking or just before serving. Certain pungent spices do not give any hot or sharp sensations until the spice is cut or roasted, which triggers enzymatic and other reactions that release these characteristic flavors. For example, whole onion needs to be chopped, whole black pepper cracked, ginger cut, and ground mustard flour needs to be "wet" before their pungency can be experienced. Spices that don't give a noticeable flavor, such as sesame seeds, star anise, or bay leaf, develop intense flavors after dry roasting, braising, or simmering. Star anise, popular in Chinese cuisine, is added to simmering beef, steamed chicken, and braised fish. Similarly, cooking ground spices intensifies their flavors.

Spice volatiles are soluble in alcohol, so many cultures use wine or other alcohols to preserve the spice aroma. Because the volatiles are oil soluble, spices are often cooked in oils before they are added to the finished product for better flavor retention. For example, in Indian cooking, spices are sauteed, toasted, or roasted in oil before other ingredients are added. Dry roasting or toasting a spice removes any remaining moisture in a spice, which takes on a totally different flavor and texture profile, while adding a more intense flavor to the finished product, such as curries or chutneys. More intense aromas and notes are derived when whole spices are freshly ground before use.

European and North American cultures tend to add spices such as bay leaf, coriander, or cinnamon, in boiling or simmering water or by steaming, to release their flavors. But South Asians roast (dry or in oil), braise, or saute whole and ground spices, such as cumin, coriander, fennel, cardamom, mustard seeds, clove, cinnamon sticks, nigella, and ajwain, before adding them to curries and condiments. Even leafy spices, such as kari leaf, are sometimes fried in oil to give extra crunchiness and a roasted aroma. The pungent and aromatic sensations of many South Asian foods are due to these prepared spices. Indians traditionally roast mustard seeds in oil to intensify their flavor, braise fresh kari leaf to remove its bitter green taste, and saute asafetida to create a sweet flavor.

These preparation techniques tend to slightly crack open the whole spice, which releases its flavors into the cooking oil or, when added to foods, begins to slowly release its volatiles. Whole spices and ground spices are generally roasted for about one or two minutes, depending upon the spice, until they become slightly brown. Beyond their optimum cooking times, they become bitter and unacceptable.

Smoking, *tarkaring, tumising,* or *bagharing* jalapeno, mustard seeds, ground coriander, or cumin seeds, creates many desirable flavors. In India, spices are *tarkared* or *bargared* to obtain a broad spectrum of flavor profiles that enhance, intensify, or simply change flavors. For example, when preparing South Indian dals and *sambars*, cumin and fenugreek seeds are dry roasted, while mustard seeds are cooked in hot oil until they "pop." "Saute-ing" a whole spice in oil and adding this spice oil mixture to the food, called *tarkaring,* is a common cooking feature in the Indian kitchen. In *bargharing*, ground spice is cooked in oil or coated with oil, vinegar, wine, or any other liquid from recipe, to make a thick paste, which is then cooked till fragrant. Then other ingredients are added to create the unique, balanced flavors that characterize South Asian dishes.

Tumising is a slow stir-frying technique used with wet spice mixtures or spice pastes to create more aromatic and intense flavors. It is carried out at lower cooking temperatures with constant stirring. The cooking procedure is complete when the oil in which the spices are cooked begins to separate or seep out of the *tumised* mixture or sauce. For example, in Malaysia, Myanmar, Singapore, and Indonesia, chile peppers (dried or fresh) are blended with shallots, lemongrass, and other ingredients, which are then *tumised* in oil to create a fragrant "chili *(cili) boh.*" This condiment is then used to flavor everything from sandwiches, salads, and vegetable curries to *laksas* and soups.

Fire-roasted chile peppers, oven-roasted garlic, or grilled onions to provide enhanced flavor profiles have become the norm in the United States. Fire roasting is done by holding the chilies over an open flame until the outer skin blisters and chars. The charred skin is then removed leaving a flavorful soft product. This product is then added to sauces, soups, spreads, stews, or salad dressings to create an enhanced aroma and rich, smoky notes. Today, with the increasing influence of South Asian, Southeast Asian, and Mexican cuisines, the trend is on food preparation techniques with spices to obtain optimum flavors.

Some spices are more stable to heat while others are added at the end of cooking. Adding different spices at appropriate times or stages during cooking helps each spice to retain its individual flavor and balance the other spices. For example, when making a fish curry, mustard seeds are added to the heated oil first, followed by cumin, coriander, fennel, and turmeric; then garlic and ginger are added, after which onions are added. When the oil separates, tomatoes are added with fenugreek seeds (already dry roasted) and saffron is added toward the end, to avoid bitterness. This gives rise to a balanced layering of complex flavors in dishes of South Asia.

GLOBAL EQUIPMENT USED IN SPICE PREPARATION

In many cultures around the world, spices are prepared with certain basic tools. On a domestic level, mortars and pestles, woks, or *kwalis* are the basic spice tools needed for spice preparation. Mortars and pestles are used traditionally for crushing, pounding, and coarsely grinding whole dry spices, chile peppers, or garlic and for coarsely blending black peppercorns and other spices for sauces or condiments. (In the section in Chapter 6, "Emerging Flavor Contributors," the effect of spice preparation and cooking techniques on flavors is discussed in further detail.)

Mortars and pestles have different names in many cultures — *molcajete* and *tejolate* in Mexican, *krok* and *saak* in Thai, *batu lesung* in Malaysian, *idi kallu* in Tamil, *kootani* in Hindi, *wurk kacha* in Amharic, and *almirez* in Tagalog. They are made from volcanic or lava rock, stone, baked clay, wood, granite, marble, or ceramic. They come in varying sizes and are essential tools in Southeast Asian, Indian, and Mexican kitchens. In the United States, a spice mill or coffee grinder serves the same purpose.

Other tools used in spice preparation include the rectangular, flat stone called *metate* in Mexican, *batu giling* in Malaysian, *ammi* in Tamil, or *chakki* in Hindu. These are used to finely grind spices, such as chile peppers, onions, garlic, or sesame seeds. Water is added to the grinding process to create a paste form of ground spice. Therefore, some metates slope gently downward to allow the paste to be pushed forward, as it is worked, into a container placed under the front edge. The stone rolling pin or *metlapil* is used as a grinding tool. Mexicans use these tools to grind corn, chocolate, and roasted chile peppers. Asian Indians grind onions, ginger, garlic, chile peppers, soaked rice, and lentils. Today, a spice grinder, blender, or food processor is used on a domestic or commercial scale.

The flat griddle, made of clay or cast iron, called *comal* in Mexican or *tawa* in Hindi, is used to roast spices, tomatoes, garlic, corn, and chilies. In Mexico, the clay-type *comal* is rubbed with slaked lime to prevent ingredients from sticking to it and creating hot spots in the spice being roasted.

The wok-*kwali* (Malaysian), *tsao-guo* (Mandarin), or *kadai/karhai* (Hindi/Urdu) is made of iron, brass, aluminum, or cast iron and can be used to dry roast whole spices and saute whole spices or ground spices in oil.

SPICE APPLICATIONS

Spices provide savory, spicy, sweet, pungent, bitter, or sour notes to foods and beverages. A vast spectrum of tastes and aromas can be created by combining spices and other flavorings. Asian Indians use various spices to create uniquely balanced curry blends, while the Chinese use them to contrast sweet, sour, or pungent notes with vivid textures. Thais, Japanese, and Caribbeans create great visual appeal with spices. Whether we are creating authentic or fusion themes or merely adding ethnic zest to traditional American foods, spices form the principal basis of flavor, texture, and visual appeal in finished products.

MARINADES, RUBS, AND GLAZES

As consumers continue to seek tasty and healthy foods that are easy to prepare, unique seasonings introduced through marinades, rubs, glazes, or as sprinkle-on seasonings are becoming the hottest new trends. These marinades and dry seasonings can be used to create a variety of flavors and textures and to add convenience for consumers through easy-to-prepare or ready-to-eat products.

The perfect blends of dried spices or extractives that are balanced with acids, salt, sugar, starch, and oil can be used to develop dry rubs, emulsions, topical seasonings, glazes, and tumbling or injection marinades for foods.

Rubs and glazes provide flavor and texture to a product. Rubs in dry or paste forms containing marinade with flour, sugar, salt, and vinegar can be externally added to chicken, meats, or seafood. These contain spice particulates that give visual appeal. For example, blackened chicken or fish, which are popular Cajun dishes, have unique crunchy coatings from spice rubs. Unacceptable, charred flavors often develop when using spice marinades that contain high tomato solids or high D.E. (dextrose equivalent) maltodextrins. Thus, encapsulated spices are preferred for retaining flavors in processed foods.

Dry or liquid glazes are surface applications containing starch, gums, and spices, and they may or may not include spice particulates. Some glazes penetrate the product to provide flavor, while, at the same time, they stay on surfaces to create visual and textural appeal. In the processed meat industry, glazes are applied after the marinades; they are applied to meats in tumblers after the meat absorbs the marinade.

The earliest marinade was a mixture of salt and spices with vinegar or fruit juices that was added to flavor and preserve meats. Today, marinades typically contain coarsely or finely ground spices (with or without particulates), oil, vinegar or other acid sources, salt, sugar, and alkaline phosphates. A marinade can be either a tumbling or an injection marinade. In a tumbling marinade, meat is placed in a tumbler, and marinade is added. The meat pieces are tumbled under vacuum until the marinade is absorbed by the product. Injection marinade is an internal soluble spice extractive with no particulates or insoluble spices. Flavor is delivered by injecting the spice solution into a whole bird or meat, such as rotisserie chicken, resulting in uniform flavor and color. To avoid color streaks on products, colorless spice solutions need to be used, such as decolorized capsicums, black pepper, or turmeric. Sometimes, this injection is followed by a tumbling step.

Spice Blends, Seasonings, and Condiments

Aromatic vindaloos, sour colombos, peppery sambals, and spicy tagines owe their unique flavors to the spice blends contained in them. This is discussed in detail in Chapter 7, "Emerging Spice Blends and Seasonings." When creating or using ethnic spice blends, it is important to remember the great variety of spice blends that exist. Even for particular spice blends such as adobos, curry blends, *ras-el-hanouth*, or *recados*, flavor variations exist depending on regional and cultural preferences and on the availability of ingredients.

For example, no chili blend, pasta sauce, salsa, or curry blend is the same. Curry blends vary from region to region and even within a region (Table 14). Curry (or

---

TABLE 14
Global Curry Blends and Their Characterizing Flavorings

Global Region	Flavor	Characterizing Ingredients
India		
Basic curry blend	Mild, medium, or hot, aromatic	Turmeric, cumin, coriander, ground cayenne or black pepper
North India	Creamy, mild, nutty	Yogurt, almonds, bay leaf, mint, clove, cinnamon, cardamom, garlic, onions, paprika
South India	Hot, fiery, coconutty	Coconut, tamarind, fresh green chile pepper, fennel seed, dried red chile pepper, turmeric, mustard seeds, kari leaf, ginger
East India	Sweet, medium hot, aromatic	Mustard, tamarind, kalonji, fenu, Greek seed, coriander leaf
West India	Hot, sour	Vinegar, cayenne pepper, mint, Saffron, coriander leaf
Sri Lanka	Black, fiery	Coconut milk, toasted spices, cayenne pepper
Pakistan	Mild, creamy, nutty	Paprika, bay leaf
Southeast Asia	Aromatic, hot, pungent, coconutty	Coconut, star anise, lemongrass, coriander leaf, turmeric, galangal, cayenne, mint, ginger, tomato fish sauce, shrimp paste, kari leaf, peanuts
East Asia	Mild, sweet, starchy	Soy sauce, turmeric, corn starch, caramel, fish sauce, sugar
Caribbean	Slightly sour, hot, fruity	Habaneros, allspice, turmeric, vinegar, black pepper, fruits, onions
England	Sweet, fruity	Apples, raisins, cream, turmeric, sugar
Middle East	Spicy, nutty, intense	Black pepper, caraway, tomato, mint, olive, sumac, pistachio, sesame seed, nigella
South Africa	Mild, sweet, fruity	Turmeric, clove, mace, cinnamon, nutmeg, bay leaf

spice mixtures for sauces) originated in India and traveled to other parts of the world, and ingredients were added and deleted to suit local tastes. Curry blends from India, Thailand, Japan, the Caribbean, England, and Africa differ greatly because of what they contain. Even in India, curry blends vary in flavor depending on the geographic regions, who makes it, and what its application is, whether for fish, lentils, or lamb. Most curry blends have some of the basic spices that were originally present in the generic blend from India, such as turmeric, cumin, coriander, and dried red chile pepper. But, other spices and flavorings may be added, depending upon preferences, such as lemongrass, cilantro, habaneros, soy sauce, coconut, yogurt, allspice, or shrimp paste, that give them their distinct regional flavors.

The same principle holds true for *adobos*, *dukkahs*, *recados*, *pestos*, or *sofritos*. The basic ingredients of adobo are garlic, oregano, black pepper, and turmeric. Then, based on regional preferences, cumin, habanero, lime juice, or soy sauce are added to the basic adobo.

4 Spice Labeling, Standards, Regulations, and Quality Specifications

The consumption of spices in the United States has been increasing annually. Most spices come from the Far East—India, Southeast Asia, and China. Others come from the Middle East, Mexico, Central and South America, the Caribbean, and Europe. The U.S. domestic production supplies a major proportion of some spices, including dehydrated onions, garlic, chile peppers, paprika, and sweet basil.

Traditionally, the spice trade's channel of distribution has had importers, growers, brokers, agents, grinders, blenders, and processors, each having their special functions. But today, their roles have overlapped and become multifunctional. For example, many processors and blenders import spices from overseas without using an importer or broker, and grinders have become seasoning suppliers, doing product development in addition to supplying spices.

Quality and specifications play a significant role in the supply and buying of spices. Spices have to meet certain requirements to be acceptable to buyers. Safety of spices and bioterrorism also play a key role in buying spices. The pricing and trading of spices is unregulated and market prices are determined by supply and demand. Barriers in spice trade can affect a buyer's ability to procure spices from many regions, thus affecting product development. There are a number of reasons for this. Global differences exist, from defining "spice," to differences regarding quality specifications, labeling, regulations, and authentication.

SPICE DEFINITION AND LABELING

To establish appropriate standards of quality and authenticity for spices, we need to properly define spices. Standard definitions for spices will allow spice suppliers to write specifications on a global basis. The International Standards Organization (ISO) defines spices as "vegetable products used for flavoring, seasoning, and imparting aroma in foods." The Food and Drug Administration (FDA) defines spices as "any aromatic vegetable substance in whole, broken, or ground form, except for those substances traditionally regarded as foods, such as onions, garlic, and celery, whose significant function in food is seasoning rather than nutritional; that is true to name; and from which no portion of any volatile oil or other flavoring principle has been removed." The American Spice Trade Association (ASTA) defines spices as "any dried plant product used primarily for seasoning purposes" and includes tropical aromatics more commonly referred to in trade as "spices" (pepper, cinnamon, cloves, etc.), herbs (e.g., basil, oregano, mint), spice seeds (poppy, sesame, mustard), and dehydrated vegetables (onions, garlic, etc.). Webster's New World Dictionary defines spice as "a vegetable condiment, or relish, in form of powder or condiments."

These definitions are outdated, limited, and incorrect in today's more knowledgeable and sophisticated society. Chile peppers, star anise, nutmeg, and mace, which are fruits or parts of fruits used for flavoring, should not be defined as "vegetable products." Dehydrated onions, garlic, chives, and shallots are primarily used for flavor enhancement and should be included in the spice definition. While a spice may serve the purpose of condimenting a meal, it is not a condiment. Condiments include prepared sauces, dressings, dips, relishes, and spreads. In bygone days, leaves and seeds of temperate zone plants were called herbs while only tropical aromatics were called a spice. Over time, this definition changed. But herbs are still defined in many books as distinct from spices. Many edible herbs have been used since ancient times to flavor foods and beverages. Herbs should not be a separate category from spices, but spices should include edible herbs (leafy spices) that flavor or color foods. The definition of spice should include "all parts of a plant that provide flavor, color, and even texture," since all parts of a spice plant—leaf, seeds, root, fruits, bark, buds, rhizome, and stalks—are used.

The Indian Spice Board's (ISB) spice definition as "in various forms; fresh, ripe, dried, broken, powdered, etc. that contributes aroma, taste, flavour, colour and pungency to food. . ." and the International Pepper Community's (IPC) as "various parts of dried aromatic plants and relates to dried components or mixtures thereof, used in foods for flavoring, seasoning, and imparting aroma," appear more up to date.

For food labeling purposes, FDA permits spices to be declared by their common or usual name (e.g., black pepper) or generically as "spice" on food labels. See 21 Code of Federal Regulations (CFR) 101.22 (h)(1). Spices such as paprika, turmeric, saffron, and other spices that may be used for their coloring properties shall be declared for labeling purposes as "spice" or "coloring" or by their common or usual name. See 21 CFR 101.22 (a)(2).

The definition of spice excludes, for labeling proposes, substances that have been traditionally regarded as foods, such as onion, garlic, and celery, and these substances must be declared on the food label by their common or usual name (e.g., onion, garlic, celery). See 21 CFR 101.22(a)(2). Spice offered for sale at either bulk or retail must bear a label with its common or usual name (e.g., paprika, black pepper) or an otherwise accurate description of identity and is considered misbranded if it does not. See 21 CFR 101.3, 102.5. The U.S. Department of Agriculture (USDA) has similar requirements to FDA except that mustard and spices that impart color must be listed separately. Also, onion and garlic may be listed as natural flavors.

SPICE REGULATIONS

The U.S. government plays a prominent role in the import and supply of spices. They regulate spice sanitation, pesticides, and sterilizers used, labeling, tariffs, and now, bioterrorism. FDA and ASTA set guidelines and quality specifications for importing and trading of spices. ASTA specifications are used throughout the world today to provide cleaner and better quality raw spices from the producing countries. ASTA specifications place limits on extraneous matter (insects, insect excrement, stones, stems, sticks, molds, etc.) and set standards, sampling procedures, and testing procedures. Imported spices not meeting these specifications are reconditioned at the port of entry, while local spices are reconditioned (using fumigants) before they are processed for use in a product.

Another trade barrier with spices is regulations, which differ globally. For example, certain sterilization treatments for spices are allowed in some countries but barred in others. Use of ethylene oxide or irradiation, allowed in the United States, is prohibited in Japan. Turmeric oleoresin can be used as a spice but not permitted as a color in the European Union (EU), although curcumin is. In Europe itself, there is variation among member states in applying EU harmonized legislation.

There are other regulatory issues facing the U.S. spice industry today that would affect the global spice trade, including policies regarding FDA's bioterrorism regulations, allergy labeling, possible phasing out of methyl bromide fumigation, new treatment studies for ethylene oxide as a fumigant, irradiation safety concerns, flavor regulation labeling differences for active principles that are naturally derived (e.g., by fermentation) and chemically synthesized, and the use of genetically modified spices such as mustard seed and black pepper.

The safety of spices is addressed by the FDA through ASTA's conclusions that they are "generally recognized as safe" (GRAS). FDA regulates the marketing of spices regarding what may not be included in spices, how spices may be labeled, and what is adulteration and misbranding.

The Federal Food, Drug & Cosmetic Act (FFDCA) granted FDA specific authority related to safety of spices. FDA considers spices to be GRAS for use in foods (see 21 CFR 182.10). Many essential oils derived from spices are also considered GRAS (see 21CFR 182.20).

FDA also regulates some aspects of the marketing of spices regarding adulterated food (21 U.S.C §342) and misbranded food (21 U.S.C. §343). Under 21 U.S.C. §342(A), a food including a spice, is considered adulterated if (1) it contains any "added" poisonous or deleterious substance; (2) it contains filth; (3) it contains unapproved food or color additives; (4) any valuable constituent has been omitted or removed; (5) any substance has been substituted for it; (6) inferiority is concealed; (7) any substance has been added to increase bulk or weight to make it appear more valuable.

Under the general misbranding provisions of FFDCA (21 U.S.C. §343), a food, including a spice, is considered misbranded if (1) its labeling is false or misleading; (2) it is offered for sale under the name of another spice; (3) it is an imitation of a spice unless labeled as such; (4) its container is made or filled to be misleading; (5) it contains added color unless so declared.

SPICE AUTHENTICITY AND QUALITY CONCERNS

The terms quality and authenticity are rather confusing with regard to buying spices. Authenticity is defined as a spice that is free from adulterated materials including foreign bodies, extraneous matter, or fillers, or free from impurities within the spice itself. For reasons of regional preferences, availability, stability, or economics, many spices that are traded globally are not the classic spice species, but are often from lesser-known species, or from a blend of different species from different regions of the world.

Many commercially available dried spices are adulterated with cheaper spices or other parts of the same spice plant, or even bulked with fillers and dyes for cost and availability reasons. Since saffron is the most expensive spice, it is adulterated in many ways, including with safflower petals, marigold, coloring matter, gelatin, moisture, syrup, salt, and starch. Ground chilies and turmeric are frequently adulterated with cheap dyes, tapioca starch, cereal flour, and lead chromate all of which lower their coloring principles and create health concerns. Stems and other parts of the clove plant are added to clove buds; and bark of cassia and other inferior types are added to "true" cinnamon. Dried ground cinnamon often has carriers, such as nuts, sugar, rhizomes, and dyes, aromatized with cinnamaldehyde. Concerns about adulteration are not limited to ground spices but spice extractives as well. Also, as discussed above, spices are sometimes mixed from a variety of origins or bulked with foreign materials and other undesirable parts of plants, which make it difficult to authenticate a product unless an appropriate extractive or flavor is at hand.

When working with spices, food technologists and flavorists must remember the distinction between quality and authenticity. While ATSA and similar organizations set minimum quality standards for cleanliness and purity, the sensory profiles are left to the suppliers to define. Product developers need to set minimum criteria for a spice or flavor from suppliers, with more detailed information (source and varieties included). For example, whether fennel seed desired is the sweet or the bitter variety, or cinnamon is Ceylon cinnamon or Vietnamese cassia.

Thus, product developers must not only become familiar with ethnic cuisines but also the many varieties of spices, deal with quality specifications and potential adulteration, and conform to regulatory issues when developing products for local consumers with global tastes.

SPICE QUALITY SPECIFICATIONS

Consumers need the highest quality spices in products. An understanding of the measurements and procedures used for monitoring spice quality becomes a useful tool for the buyer and user. How do spice suppliers best achieve the quality and consistency they seek? When was the spice ground, what is its shelf life, and how long has it been stored before shipment? When a spice supplier provides ground spice, we need to learn its specifications, including its origins, how long it was stored before grinding, and if it was processed before grinding. When a spice extractive is provided, we need to learn its specifications, how it was extracted, and how it was stored. These are some of the many quality control factors we need to ascertain from the spice supplier.

When buying small amounts of spices, it is preferable to buy whole spices and then grind them just before use to obtain a fresher flavor and more intense aroma. Grinding or crushing spices releases their volatile flavors. When grinding small amounts of spice, a mortar and pestle or coffee grinder can be used; for larger amounts, a spice mill is used. The fresher the whole or ground spice, the more flavor it has.

When purchasing larger quantities of spices, spice specifications become important because the flavor and color of spices can vary with different batches depending on their origin, when they were harvested, how they were processed, and the conditions and length of time in storage. Keep in mind that spices come from many regions of the globe, with varying climatic conditions and quality standards. It is important to be aware of measures or controls taken in exporting countries. Cleanliness, insect and rodent infestation, and microbial quality are important criteria for the spice processor and the food developer. Spice specifications become important forms of communication between the food developer and the supplier for ensuring that the correct spice is delivered within economic limits. When spice specifications are developed and recorded, product consistency is maintained over time.

The two major international standards for quality specifications are those set by the United States and the European Union (EU). ASTA has long been involved in setting quality standards for spice imports. The European Spice Association (ESA) sets minimum specifications for spice quality in Europe. Many growing regions have their own quality specifications, including the Indian Spice Board (ISB) and International Pepper Community (IPC), which implements quality certifications in close association with ASTA and other international standard groups, as do importing countries, such as the British Standards Institute and the All Nippon Spice Association.

MAINTAINING SPICE QUALITY

When spices are exported into the United States, they must meet ASTA specifications. The general quality tests set by ASTA include cleanliness (foreign and extraneous matter), ash level (impurities), volatile oil (adulteration), moisture content (pricing, stability), water activity (microbial growth), pesticide levels, mycotoxin/aflatoxin levels, and particle size. Other tests include piperine levels for black and white peppers; ASTA color values; capsaicin level/Scoville units for chile peppers; and curcumin content for turmeric color. Using these methods, quality limits are set for moisture, pungency, or color values. Most times, pungency, color, and other sensory values are correlated with organoleptic evaluations with trained sensory panelists. Spices not meeting the U.S. quality standards set by ASTA and recognized by FDA and the U.S. Dept. of Agriculture have to be retreated and recleaned before distribution.

Several methods for treating spices for microbial contamination, and preventing microbial and insect growth and breeding during storage conditions have been used, including fumigation with methyl bromide, sterilizing with ethylene oxide, irradiation, and heat treatment. Ethylene oxide (EtO) has been banned in many European countries and Japan because of concerns that residues left after fumigation may be harmful to human health and cause cancer in workers who have prolonged exposure to it. Fumigants impart undesirable odors and colors and are not always effective with whole spices. The use of EtO does affect the sensory profiles of spices and destroys vitamin B1 and vitamin C. It reacts with chlorides in foods to form chlorohydrins that are toxic to humans. The Environmental Protection Agency (EPA) is reregistering ethylene oxide, and ASTA, which favors use of EtO, is conducting a new treatment method study to generate data in support of reregistration by EPA. The issue of phasing out methyl bromide by 2005 under the Montreal Protocol for Ozone Protection is under discussion again.

Irradiation, first approved by FDA for use on spices in 1983, exposes spices up to a million rads of ionizing radiation, the highest amounts allowed for any food. Approved by ASTA, this process more effectively kills microbes than EtO. Concerns have been expressed by EPA that irradiation changes the chemical composition of a spice, potentially creating toxic and carcinogenic by-products in the food. This method is banned in Japan. It also reduces the sensory and nutritional quality of the spices and gives a lower consumer acceptance.

Consumers' concerns with irradiation and chemical treatments have led to use of steam heat (e.g., in Japan) for sterilizing spices. High-pressure steam creates clumping of spices, dissipates aroma volatiles, and discolors spices to some extent. Controlled atmospheric storage with low temperature storage and controlled humidity conditions to prevent mold growth and aflatoxin production are possible ways of reducing contamination. Insects do not survive in an atmosphere with less than 2% oxygen, and so, nitrogen or carbon dioxide is used to replace oxygen in storage areas.

Microbiological requirements for "clean" spices include counts for total bacteria, yeast, mold, coliforms, and food pathogens such as *E. coli* and *Salmonella*. High microbial counts are caused by contamination during growing and postharvesting handling. Spore-forming bacteria, such as the *Bacillus* species or *Aerobacter aerogenes* found in the soil can be transferred to the spice during the drying process, especially with "under the ground" spices such as turmeric, ginger, galangal, and garlic. The type and amount of molds and bacteria on a spice depends on the type of spice and the conditions under which it is harvested and dried. *Staphylocccus* and *Streptococcus* bacteria species predominate, but pathogenic bacteria tend not to exist on spices. Spices that show strong antimicrobial properties tend to have low counts of microbes.

Molds, such as *Aspergillus,* that produce toxins are found on certain spices including red pepper, fenugreek, and ginger, so there are specification limits for these toxins. Molds tend to multiply during the drying process and during storage.

Good storage conditions, monitoring, and specifications are important in retaining quality attributes of spices. Proper packaging is essential for preventing oxidation and, thus, retaining color and flavor. To retain good aroma in spices, long-term storage is not recommended. Processing conditions such as grinding and sterilization can decrease the volatile oils in spices. During storage, microbial growth and insect infestation can occur to varying degrees, depending upon the extent of contamination during harvesting, transportation, and processing conditions. Filth levels include foreign materials such as insect fragments (moths, mites, beetles), small stones, metal fragments, and glass pieces. Insects and mold growth can change the color and, to some extent, the flavor of the spice.

Dried spices generally do not spoil but lose their strength in aroma, flavor, and color over a period of time. Whole spices last longer than ground or crushed forms. Commercially available whole spices have from 2 to 4 years, ground spice from 6 months to 2 years, and leafy spices from 3 months to 2 years, depending upon type of spice, process, and storage conditions. Extracts last up to 4 years except vanilla which has longer shelf life. Spice blends and seasonings last about 1 or 2 years depending on contents.

Light will fade the color and character of spices, and heat volatilizes and dissipates the essential oils in ground spices, and moisture or high humidity tends to cake ground spices. High levels of moisture in ground or whole spices are potential for mold and microbial growth. Exposure to light, humidity variations, air, and certain metals can discolor many spices such as paprika, turmeric, or the green leafy spices. Dry, ground chile peppers turn from a natural green or red color to an olive or dirty reddish brown color when exposed to light. Flavor and aroma losses as well as insect and rodent infestation occur when spices are not stored in airtight containers.

Spices or spice extractives should be stored in tightly closed containers in cool, dark, dry conditions below 68°F and 60% humidity. Some spices need cooler refrigeration temperatures, such as 32°F to 45°F, to prevent mold infestation (capsicum peppers), color deterioration (paprika), and to avoid rancidity (in high fixed oil seeds, such as sesame seeds). Colder temperatures also help preserve volatile oil flavor and aroma, freshness, and sanitary quality. Refrigeration slows microbial growth in ground or whole spices.

The control of insects and microbes is important in receiving a quality spice. Spices need to be free of microbes to reduce the initial bacteria or mold content in processed foods. Spice extractives and sterilized spices tend to meet these objectives. Ground spices for minimally processed foods such as salad dressings, condiments, or "sprinkle on" seasonings should be well cleaned and sterilized.

In summary, proper storage and use of spices will maintain spice quality:

- Store in a cool dry place, away from heat (oven, stove), light (near window or in transparent packaging), or moisture (steam from cooking near spice container or use of a wet spoon into container). All this will hasten the loss of spice aroma and flavor and cause caking.
- Store spice in airtight containers to maintain freshness. After each use, close container tightly. Exposure to air accelerates flavor loss.
- Store spices at cool temperatures as they help retain flavor of spices.
- Do not store spices in freezer as repeated removal for use results in condensation in the containers, resulting in loss of flavor and aroma.

To help assure quality of incoming spices, spice suppliers often put in place Hazard Analysis and Critical Control Point (HACCP) systems in addition to meeting ASTA standards.

5 A to Z Spices

AJOWAN

Sometimes mislabeled as lovage seeds, ajowan, also referred to as "royal" cumin, Ethiopian cumin, Egyptian black caraway, or caraway. In India, similar names are given to ajowan, nigella and celery. Ajowan is an essential ingredient in a Bengali seasoning called *panchphoron*. *Omam* water, an infusion of ajowan seeds, has been used since ancient times in India for stomach pains, colic, diarrhea, and other disorders.

Scientific Name(s): Trachyspermum (T) ammi, T. copticum or *Carum copticum.* Family: Apiaceae (parsley family). Ajowan originated from the Eastern Mediterranean region and came to India with the Greeks, who were called *Yavanas* by South Indians. Ajowan originated from the Sanskrit word *yavanaka* or *ajomoda.*

Common Names: ajwain, carom, Ethiopian cumin, wild parsley, bishop's weed. It is also called netch (white) azmad (Amharic), ajwan, kamun al-mulaki, taleb el koub (Arabic), joni-gutti (Assamese), jowan, yamani (Bengali), yan-jhon-wuih-heung (Cantonese), nanava (Farsi), ajowan (Dutch, French, German, Italian), ayamo, yavan (Gujerati), ajwain, carom omum (Hindi), ajamoda, oma (Kannada), ayowan (Korean), ajowan (Japanese), ayamodakam (Malayalam), yin-dou-zeng-hui-xiang (Mandarin), javano (Nepali), oregano-semente, ajowan (Portugese), ajavain (Punjabi), assamodum (Singhalese), ajowan (Spanish), omam (Tamil), omamu (Telegu), chilan (Thai), and misir anason (Turkish).

Origin and Varieties: Brought by the Greeks to India from the eastern Mediterranean, ajowan is today cultivated in South India, Europe, Egypt, Pakistan, Iran, and Afghanistan.

Form: it is a small, caraway-like seed that is used whole or ground.

Properties: ajowan is a close relative of caraway, dill, and cumin. It has large curved and ridged oval, celery-like seeds that are light brown to purplish red in color. Ajowan seed when bruised, has a flavor similar to thyme but stronger. Its volatile oil, thymol, has a piney, phenol-like and slight lemony notes. When crushed or ground, it has a more intense flavor. It can be bitter and slightly spicy. Its leaves, stems, and roots are aromatic.

Chemical Components: ajowan contains 2.5% to 5% essential oil, mainly phenols-thymol (35% to 60%) and carvacrol (11%) along with non phenols, beta-pinene, para-cymene, alpha-pinene limonene with gamma- and beta-terpinenes. South Indian ajowan has mostly thymol, about 98%.

Ajowan contains iron, niacin, and calcium.

How Prepared and Consumed: ajowan is commonly used by North Indians, Pakistanis, North Africans, and Iranians. It is used whole or ground and has a natural affinity with starchy foods, such as root vegetables, legumes, breads, snacks, and green beans. Ajowan makes starch and meats easier to digest and is added to legumes to prevent flatulence. It goes well with cumin, ghee, garlic, ginger, and turmeric.

In North India (Gujerat and Punjab), ajowan seeds are popular with vegetarian cooking. It is fried in ghee with other spices, and this aromatic mixture is added to cooked legumes and vegetables. Ajowan seeds are added to flatbreads called par-athas, to snacks (pakora), and pastries, and served with nuts. In Bengal, it is used as part of a seasoning called *panchphoran*, which flavors fish and vegetable curries. To enhance its flavor, ajowan is roasted or fried in oil until it becomes light brown. It then provides a more intense aroma to fish curries, lentil stews, and potatoes.

In Ethiopia, ajowan is an integral part of a spice blend called *berbere*, which is used for meat stews and vegetables.

Spice Blends: berbere, chat masala, panchphoran, and pakora filling blend.

Therapeutic Uses and Folklore: ajowan is highly valued in India as a gastrointes-tinal medicine and an antiseptic. It is combined with salt and hot water and taken after meals to relieve pain in bowel or colic pain, and to improve indigestion. Ajowan was also a traditional remedy for cholera and fainting spells. Westerners generally use it against coughs and throat issues. Ajowan is an ingredient in mouthwashes and toothpastes because of its antiseptic properties.

ALLSPICE

An essential ingredient for Jamaican jerk paste or seasoning, allspice is native to the Caribbean and the Americas. The English gave the name allspice because it has a flavor that combines the flavors of several spices, cloves, cinnamon, nutmeg, and black pepper. The Spanish explorers named it *dulce pimienta* or sweet pepper, because of the berry's resemblance to the black peppercorn. Also called aromatic pepper and Jamaican pepper, allspice is not related to the peppercorn. It was first imported into Europe in 1601 as a substitute for cardamom.

Scientific Name(s): Pimenta dioica, formerly *P. officinalis.* Family: Myrtaceae (myrtle family).

Common Names: Jamaican pepper, pimento, clove pepper, English spice. It is also called baharat, bahar halu (Arabic), do heung guo, duo xiang go (Cantonese, Mandarin), piment (Dutch), pimento (English), toute-epice, poivre de Jamaique, poivre aromatique (French), piment neugewurz, nelkenpfeffer (German), pilpel angli (Hebrew), bahari (Greek) orusupaisu (Japanese), pepe di Giamaica, pimento (Ital-ian), kappalmulagu (Malayalam), pimento de Jamaica (Portuguese), kryddpeppar (Swedish), pimienta gorda, pimento dulce, pimienta de Jamaica (Spanish), kat-tukaruva (Tamil), and yeni bahar (Turkish).

Origin and Varieties: there are many types of allspice, each with varying tastes. It is indigenous to the Caribbean Islands, specifically Jamaica, South America (Brazil, Leeward Isle), Central America (Guatemala, Honduras, Belize), and Mexico. Allspice is also grown in India and Réunion. The United States buys mainly from Central America and Mexico. This spice is often adulterated with ground clove stems or a closely related species, P. racemosa.

Form: the berry/seed of the pimiento tree is picked green/unripe and then dried until it turns dark reddish brown in color. It is globular and has a rough textured surface. It is slightly larger than the black peppercorn. The Mexican type is the largest and darkest in color. Jamaican allspice berries are smaller. Allspice is used whole or ground. The aromatic leaves and bark can also be used to provide an allspice-type flavor to foods, especially to smoked meats and beverages.

Properties: allspice has a warm, pungent taste and the aroma of cloves with sweeter, floral background notes. Its flavor has a hint of cinnamon, mace, and nutmeg with peppery overtones. Jamaican allspice which is superior to all other varieties, is the most aromatic. The Mexican variety is less sweet and mellower than the Jamaican or Central American types. Allspice berries lose their aroma upon ripening, so they are collected when unripe and dried in the sun until they turn dark reddish brown in color. The leaf has a different flavor, a woodier aroma with less intense but coarser notes. The bark has more coarse and woodier notes than the leaf.

Chemical Components: the allspice berry contains 1.5% to 5% essential oil, which is colorless to reddish yellow. The Jamaican type has up to 5% essential oil, Guatemalan—3%, Mexican—1.4% to 3%, and Honduran—1.3% to 4%. Jamaican allspice has a minimum of 65% phenols, mainly eugenol (68% to 78%), methyl eugenol (2.9% to 13%), 1,8-cineol, α-phellandrene, humulene, terpinolene, and caryophyllene. The fixed oil content is about 6%. The Mexican variety has a high myrcene content (5%). Allspice has over 8% quercitannic acid that gives it its astringency.

The Jamaican variety produces the most leaf oil—its fresh leaf contains 0.35% to 1.25% essential oil (dried leaf has higher oil, from 0.7% to 2.9%). The essential oil contains 80% to 90% phenols, mainly eugenol, 65% to 90% higher than in the berry oil. Others contributing non phenols are α-pinene, caryophyllene, limonene, 1,8-cineole, and good amounts of tannin. The bark contains small amounts of eugenol and higher levels of tannin.

Its oleoresin is brownish green to dark green in color. About $2^{1}/_{2}$ lb. of essential oil will replace 100 lb. of freshly ground spice, while 5 lb. of oleoresin will replace 100 lb. of freshly ground spice.

Dried allspice has calcium, potassium, sodium, manganese, and beta-carotene.

How Prepared and Consumed: the Aztecs and Mayans flavored their chocolate drink with allspice seeds. Caribs and other indigenous Americans used it for preserving fish and meat. This practice was learned by the Spanish who also used allspice to preserve meats. During the seventeenth century, pirates in the Caribbean smoked and barbecued meat with allspice, which they called *boucan*.

Allspice is popular with Western (British, Scandinavian, German, and American), North African, and Caribbean cuisines and not with Asian cooking. Nowadays, allspice seeds are typically used whole as part of a spice blend for pickling and marinating fish and meats. The British add it to their stews, sauces, and pickled vegetables while the Scandinavians enjoy it in meat patties and sausages. In the United States, allspice is ground for use in seasonings and sauces, and its extracted oils are used in sausages. Allspice is also used in ketchup, jams, pumpkin pies, gravies, roasts, and ham. It goes well with smoked pork, beef, and fish and with habaneros, cumin, onions, tamarind, cinnamon, and cloves. Allspice leaf is used in baked goods, chewing gum, candy, ice cream, fruit soups, teas, and liqueurs.

Allspice is an important spice in Caribbean cooking and is added to curries, stews, barbecues, and sweet potatoes. Jamaica's popular drink, *pimento dram*, has whole berries while its signature seasoning, jerk, has ground allspice as its main ingredient. It is rubbed over pork, chicken, or fish that are then cooked over a fire. It gives a smoky and spicy flavor to the barbecued product. Allspice leaves are sometimes stuffed into the meat which is then barbecued over allspice wood or bark to give it the typical flavor of "jerk."

In Oaxaca, Mexico, allspice is used in certain mole sauces. In Kerala India, it becomes part of some curry blends. Scandinavians use it to preserve herring. The English use ground allspice in cakes, puddings, mincemeats, pickled vegetables, sausages, and cured meats. In North Africa, allspice is used in Ethiopian berbere and Moroccan *ras-el-hanout* spice blends. Middle Easterners flavor stews, *kibbeh* (ground lamb with cracked wheat) and pilafs with ground allspice.

Spice Blends: jerk seasoning, berbere, ras-el-hanout, quatre-epices, fish pickling blend, ketchup blend, Jamaican curry blend and Kerala fish curry.

Therapeutic Uses and Folklore: the Aztecs and Mayans used allspice to embalm bodies because of its preservative qualities. It was also considered an aphrodisiac.

Allspice has been used to promote digestion and remove gases from the upper intestinal tract. It is used as a mild anaesthetic for aching gums and teeth and as a mild pain reliever for muscles and joints.

Allspice has bactericidal, fungicidal, and antioxidant properties.

ANISE/ANISEED

Anise/aniseed, first used by early Egyptians as early as 1500 BC, is a popular spice used throughout the world. Called *anysum* by early Arabs, *anison* by Greeks, and later anise by the English, it was used by Europeans as an aphrodisiac and as a charm to prevent nightmares. Ancient Assyrians used anise as a medicine, Greeks found it to be a digestive aid, and the Romans used anise to soothe sore throats. They ate anise spice cakes to soothe digestion.

Associated with the taste of licorice, the Portuguese call anise, *erva doce*, the "sweet herb," the Indonesians call it *jintan manis*, the "sweet seed," and the Arabs, *kamun halu* or sweet cumin. Thought to be a foreign variety of fennel, Asian Indians often confuse anise with fennel because of its similar flavor and name, *saunf*. To distinguish they called it *patli saunf* or thin fennel. In China, anise is commonly used with fennel and star anise to add a savory, sweet flavor to barbecues.

Scientific Name(s): Pimpinella anisum. Family: Umbelliferae or Apiaceae (parsley family).

Common Names: sweet cumin, aniseed, and common anise. It is also called yansoon, kamun halu, habbet hilwa (Arabic), sulpha (Bengali), dai wuih heong, huei xiang (Cantonese, Mandarin), anijs (Dutch), anisun (Farsi), anis vert (French), anis (German, Swedish, Danish, Russian), anison (Greek), anis (Hebrew), patli saunf (Hindi), anice (Italian), anisu (Japanese), sutha koppa (Malayalam), jintan manis (Malaysian, Indonesian), erva doce (Portuguese), anis (Spanish), anis (Tagalog), anisu (Tamil), sompu (Telegu), anason (Turkish), and cay vi (Vietnamese).

Origin and Varieties: it is indigenous to Greece, Egypt, Crete, Turkey, and Lebanon. Anise is also grown in Mexico, Chile, Argentina, Syria, Spain, Italy, India, Pakistan, China, Russia, Japan, and Germany.

Form: anise is a dried ripe fruit or seed. It is small, oval, greenish gray to yellow brown, with a ridged or ribbed surface. Anise is sold whole, cracked, or ground. When it is ground into powder, anise quickly loses its flavor.

Properties: anise seed has a sweet licorice-like taste and is warm, fruity, and camphoraceous. Anise's flavor is similar to fennel and star anise, but it is more camphor–like and delicate. Its leaves are also aromatic.

Chemical Components: depending on its source, anise seed has 1.5% to 6% essential oil, mainly trans-anethole (80% to 90%), methyl chavicol (10% to 15%), iso-anethole (2%), ketone, and anis aldehyde (less than 1%). It has 8% to 20% fixed oil. The leaf has a much lower level of essential oil.

About $2^1/_2$ lb. essential oil (yellowish green to orange brown) is equivalent to 100 lb. freshly ground spice, and $8^1/_4$ lb. oleoresin (has 15% to 18% volatile oil) will replace 100 lb. freshly ground spice.

Anise contains iron, potassium, phosphorus, and calcium.

How Prepared and Consumed: the early Romans used anise to flavor a special cake called *mustaceum* that was served as a dessert to aid digestion. They also mixed anise with vinegar and honey and used the mixture as a tonic to soothe sore throats. This spice tends to be used in sweet foods in Europe, while in Asia, anise is combined with pungent, spicy ingredients for curries and savory applications. Anise goes well with fruits, sugar, fennel, wine, and cinnamon. It is a popular spice in Chinese cooking. Anise leaves and stalk can garnish fruit salads and are sometimes added to fish soups and cream sauces of Europe. They are roasted or sauteed in oil with other spices to enhance stewed vegetables, roasted meats, curries, and tomato sauces.

The Portuguese, Germans, Scandinavians, French, and Italians use anise to flavor cakes, sweet rolls, cookies, sweets, applesauces, rye bread, churek, pancakes, cheeses, relishes, marinated meat and fish, beef stew, salad dressings, sausages, and luncheon meats.

Europeans flavor many liqueurs and spirits with anise, such as anisette, *raki* (Turkey), *ouzo* (Greece), *arrack* (Arab regions), *kibib* (Egypt), *sambuca* (Italy), pernod and *pastis* (France), *ojen* (Spain), *anesone* (Italy), and even juice drinks and teas.

Middle Easterners use anise in sweet and savory dishes, and it is the fundamental ingredient in their local spirits, *ouzo* and *raki*. Syrians use it in a beverage called *miglee* and in their popular fig jams.

Spice Blends: curry blends, hoisin, tomato sauce blends, sausage blends, and betel leaf mixture.

Therapeutic Uses and Folklore: traditionally, Europeans used anise to treat epilepsy and to ward off evil. The Aztecs drank tea made from its flowers and leaves to relieve coughing and to dispel gas. Anise aids digestion, improves appetite, alleviates cramps and nausea, and soothes colic in infants. Anise is commonly used in lozenges and cough syrups because it is a mild expectorant. It also soothes insect bites and is chewed to induce sleep. In India, anise seeds are served after meals to aid digestion and sweeten breath.

Anise shows antimicrobial and antioxidant properties.

ANNATTO

Annatto, called *urucul* by the Tupi-Guarani Indians of the Amazon region, achiote in the Nahuatl language of the Aztecs in Mexico, annatto by the Caribs, and *achuete* by Filipinos, is better known today as achiote by Mexicans and Caribbeans. They are an important coloring and seasoning for Latin American, Native American, Spanish, and Filipino cooking.

The Caribs, Mayans, and other Native Americans dyed their bodies with annatto oil to protect against the sun, thus giving rise to the term "redskin" by the early European settlers in the Americas. Aztecs used annatto seeds to intensify the color of their chocolate drink. This practice of coloring traveled to Europe, where annatto is now used to give a deep yellow color to butter and cheese. Often called "saffron" by Puerto Ricans, this spice was introduced to India by the Portuguese and to the Philippines by the Spanish.

Scientific Name(s): Bixa orellana. Family: Bixaceae.

Origin and Varieties: Annatto is indigenous to South America, the Caribbean, Mexico, and Central America. It is today cultivated in Brazil, Peru, Guatemala, Philippines, India, and western Africa.

Common Names: orellana, achiote, bija, and bijol. Also called latka (Bengali), yin ju suih, yan zhi shu (Cantonese, Mandarin), anatto, rocou (Dutch), lipstick tree (English), roucou (French, Caribbean), annatto (German), anato (Hebrew), kesumba (Indonesian), beninoki (Japanese), sa ti (Laotian), jarak belanda (Malaysian), urucum (Portuguese-Brazil), orellana, achiote, bijol (Spanish), annatto biksa (Russian), achuete (Tagalog), kongaram (Tamil), kam tai (Thai), arnatto (Turkish), and hot dieu do (Vietnamese).

Form: annatto are small dark-red seeds in a prickly, heart-shaped fruit. It is sold as a paste, as oil (extracted from seeds), or ground (from the whole seed).

Properties: annatto is deep golden yellow to orange red in color. It has a delicate, slightly sweet and mild peppery flavor with flowery and earthy undertones.

Chemical Components: bixin, an oil-soluble apocarotenoid, is the main coloring pigment, with norbixin, other carotenoids, and apocarotenoids making up 7% of the dry seed. Its flowery scent is due to tricyclic sesquiterpene hydrocarbons called ishwarane.

How Prepared and Consumed: annatto is used as a cooking oil and a gentle flavoring and coloring agent in Central and South America, the Caribbean, Mexico, and the Philippines. It is popular in Puerto Rican and other Caribbean cooking and added to rice, polenta, beans, chicken and fish stews, soups, pork, okra, yuca, and tomatoes. In Jamaica, the popular salt cod and ackee dish, fried chicken, and pork are colored with annatto.

Annatto combines well with cumin, garlic, oregano, and coriander. Typically, it is used to color or flavor food in the form of *aceite* (oil) or *manteca de achiote* (annatto lard) in Latin American and Filipino rice dishes, stews, and meats. Filipinos, Vietnamese, and Chinese color batters, Peking style duck, coconut based curries, and marinate pork and fish with annatto. The seeds are fried in oil or lard that becomes a golden orange in color. Then, the seeds are discarded, and this colored oil is used to fry vegetables, rice, chicken, or meats. It is also blended with other ingredients to create the Puerto Rican sofrito. In the Yucatan and in Guatemala, the whole annatto seed is ground into a paste and is used with other spices as a rub and to provide a deeper flavor to barbecued pork, poultry, and fish dishes.

In Europe, annatto extract colors butter, margarine, ice cream, confectionary, sausages, and many cheeses including red Cheddar, Muenster, Edam, Chesire, Livarot, and Leicester. In the United States, annatto is used in relishes, snacks, beverages, gravies, seasonings, baked goods, and margarine.

Spice Blends: sazon, recados, achiotes, sofritos, and adobos.

Therapeutic Uses and Folklore: the ancient Mayans and Caribs used annatto to paint their faces and bodies in religious ceremonies and in preparation for wars. To them, it represented courage and strength. It has been used to control fevers and dysentery.

ASAFOETIDA/ASAFETIDA

The name asafoetida is derived from the Persian word *aza,* meaning resin, and the Latin word *foetida,* meaning fetid or bad smelling. Asafoetida was known to early Persians as "the food of the Gods" and to the Romans who used it to flavor sauces and wines, as Persian *sylphium.* Europeans equated its smell to truffles and the French flavored mutton during the early Middle Ages, after which, its use declined. In ancient India and Iran, asafoetida was used as a condiment and as a medicine. Today, asafoetida is commonly used in the vegetarian cooking of South India and Bengal.

Scientific Name(s): Ferula asa-foetida L (hing type); *F. foetida* (hingra type). Family: Umbelliferae (carrot family) or Apiaceae (parsley family).

Common Names: asafetide, stinking gum, and devils dung. It is also called haltit, abu kabeer (Arabic), ah ngaih, ah wei (Cantonese, Mandarin), asa foetida, duivelsdrik (Dutch), retshina fena, anghuzeh (Farsi), assa foetida, ferula persique (French), asant/stinkasant (German), aza (Greek), hiltit (Hebrew), hing (Hindi, Bengali), assafetida (Italian), agi asahueteida (Japanese), ma ha hing (Laotian), kaayam (Malayalam), asafetida (Russian), asafetida (Spanish), mvuje (Swahili), dyvelstrack (Swedish), perungayam (Tamil), inguva (Telegu), seytanterin (Turkish), and anjadana (Urdu).

Origin and Varieties: asafoetida is indigenous to Iran, India, Pakistan, and Afghanistan. It is also found in Russia and China. There are many varieties, but the two most commonly sold varieties are called in Hindi language, *hing* (water soluble) and *hingra* (oil soluble). The former is more popular because of its aroma. Each type shows more sweetness or bitterness depending on its country of origin.

Form: it is a congealed, dark brown to black resinlike gum obtained from the juice of the rhizome of the ferula or giant fennel plant. After drying, it becomes a darker brown mass. It is sold as different grades of resin, dried granules, chunks, or powders.

Properties: the resin is a pale brown color that darkens after drying. It is acrid with a strong garliclike odor and bitter, unpleasant back notes. If it is used sparingly and fried in oil, pleasant shallot and garliclike notes develop. The powder is less intense and can be added to dishes without prior frying. The commercial paste and powdered forms are resin mixed with diluents such as rice flour, cereal flour, or gypsum and are, thus, less intense in taste. The *hing* type is white or pale in color, while the *hingra* type is darker, almost black, in color.

Chemical Components: its repugnant strong smell is due to sulfur compounds (that disappear during cooking) and ferulic esters. Dried asafetida consists of resin (comprising 25% to 60% of total) which consists of essential oil (10% to 17%), gum (25% to 48%), esters (40% to 60%) and ash (1% to 10%). The essential oil has an abundance of sulfur compounds, mainly 50% of 2-butyl-1-propenyl disulfide, 1-1-methylthiopropyl disulfide, and 2-butyl-3-methylthioallyl disulfide with some terpenes (α-pinene, phellandrene), and farnesiferoles.

How Prepared and Consumed: Iranians and Asian Indians use it abundantly in vegetarian dishes. Asian Indians fry the resin in oil for a few minutes to disperse it well before it is mixed with other ingredients. The unacceptable smell disappears when it is cooked. At very low or "pinch" levels, asafoetida enhances many dishes such as fish curries, brined or pickled fish, spiced legumes, vegetables, *chewda* (Indian snack), relish, and even Worcestershire sauce.

South Indians use it in sambar podi, a spice blend added to legume dishes to enhance their flavor and to prevent flatulence. Jains, a religious group in India, do not eat root vegetables or root spices such as garlic, onion, ginger, or turmeric, for fear of killing living organisms. Therefore, they rely on asafoetida as an alternative flavoring. The Brahmins, who will not eat garlic or onions because they consider them aphrodisiacs, also use asafoetida as a substitute flavoring.

It pairs well with nuts, grains, legumes, mushrooms, vinegar, and grilled, barbecued, or roasted meats. Iranians rub asafoetida on warmed plates before placing meat on them. Afghans and Persians also eat the stem and the leaves as vegetables, the odor disappearing once they are boiled.

Spice Blends: sambar podi, dal podi, chat masala, and chewda blend.

Therapeutic Uses and Folklore: Romans used asafoetida to aid digestion and as an aphrodisiac. In India and Iran, it is used to treat hysteria and taken as an antispasmodic, anticoagulant, and sedative. Asafoetida is also used to reduce flatulence and to treat nervous disorders. In India, singers take asafoetida before singing because it supposedly mellows their voices and produces a sensation of warmth.

Asafoetida has antibiotic and antimicrobial properties.

BASIL

The name basil is derived from the Greek word *basileus,* meaning "king" because of its wonderful "royal" fragrance. The French call it "herb royale." In ancient times, basil was considered to have magical properties. The Greeks gave basil the name *basilisk* because it was reputed to provide protection from a half-lizard, half-dragon monster of the same name.

Holy basil is used in India mainly for religious purposes for Hindu temple ceremonies. This basil, called *tulsi,* is woven into a garland to grace the Hindu God, Lord Vishnu. *Tulsi* is named after Lord Vishnu's wife, Goddess Tulasi, who took the form of this leafy spice when she came down to earth.

Today, there are many emerging basil varieties in the U.S. from Mexico, Africa, and Asia. Sweet basil is the most popular type in North American, Italian, and other European and Mediterranean cuisines. Thai basil (also called anise basil), lemon basil, and holy basil are popular in Southeast Asian and South Asian cooking.

Scientific Name(s): Sweet basil: *Ocimum basilicum,* holy basil: *Ocimum sanctum,* lemon basil: *Ocimum basilicum citriodorum,* cinnamon basil: *Ocimum basilicum cinnamon,* curly basil: *Ocimum basilicum crispum,* and Japanese basil or shiso/jiso: *Perilla frutescens.* Family: Lamiaceae (mint family).

Common Names: sweet basil, herb royale, and great basil. Also called besobila (Amharic), habaq, reehan (Arabic), lo lak/lo le, yu xiang ca (Cantonese, Mandarin), basilicum (Dutch), reihan (Farsi), basilie/basilic commun (French), basilikum (German), vasilikos (Greek), reihan (Hebrew), barbar (Hindi), basilico (Italian), bajiru (Japanese), paqe i tou (Laotian), daun selasih/kemangi (Malay, Indonesian), manjericao (Portuguese), bazilik (Russian), suwndutala (Singalese), albahaca, alfabega (Spanish), mrihani (Swahili), basilika/basilkort (Swedish), balanoi (Tagalog), tiruniripacha (Tamil), rudrajada (Telegu), reyhan (Turkish), and e tia (Vietnamese).

Anise basil or true Thai basil—bai horapha (Thai), daun selaseh, (Malaysian, Indonesian), and rau que (Vietnamese).

Lemon basil—daun kemangi (Malaysian, Indonesian), and bai maenglak (Thai).

Holy basil or sacred basil—babui tulasi (Bengali), laun (Burmese), tulsi (Hindi), sulasi, ruku-ruku (Malay), Sivatulasi (Malayalam), sapha (Laotian), madurutala (Singhalese), sulasi (Tagalog), tulasi (Tamil), bai krapao (Thai), oddhi (Telegu), jangli tulsi (Urdu), and e do (Vietnamese).

Japanese basil or shiso/jiso-Also called ban tulsi (Bengali), sou yihp, xiang su (Cantonese, Mandarin), shiso blad (Dutch), Chinese basil, sesame leaf, beefsteak plant (English), sesame sauvage (French), perilla (German) perila (Hebrew), bhanjira (Hindi), jiso (Japanese), tulkae (Korean), daun shiso (Malay, Indonesian), nga chien chin (Laotian), perilla (Russian), bladmyanta (Swedish), nag mon (Thai), perilla and la tia to (Vietnamese).

Origin and Varieties: basils are indigenous to Europe, India, and Southeast Asia. They are cultivated in Iran, Africa, Seychelles, Southeast Asia, Greece, Italy, France, Egypt, Hungary, Morocco, southern Europe, Japan, and the United States (California). Many types of basils exist. They vary in size, color, and flavor intensity based on their origins and climatic and soil conditions, all of which affect their chemical components. Sweet basil is most commonly used in the United States, but there are many other emerging varieties, including holy, lemon, Thai, dark opal, shiso, Cuban, West African, cinnamon, East Indian, purple ruffle, minty Egyptian, and many more. Even with sweet basil, flavor variations occur, depending on its country of origin.

Form: basil comes fresh, dried, or as a paste in oil. Fresh basil is used whole, chopped, or pureed. Dried basil is used as ground and as particulates of varying sizes. The dried form is less aromatic than the fresh form.

Sweet basil has bright green leaves. Thai basil is similar in size to sweet basil, but with purplish stems and veins. Holy basil is smaller and narrower, with a dark green to almost reddish purple tinge. Lemon basil is paler in color than Thai basil. Japanese basil (shiso) is light to dark green.

Properties: the Mediterranean or European-type sweet basil has a sweet and floral, anise-like aroma with cooling clovelike undertones. Its taste is delicate and fresh with slight minty notes. Its delicate aroma decreases with cooking. Thai basil has a sweet aniselike aroma and licorice-like notes with a spiciness that sweet basil does not have. It has strong, phenolic notes, with a lingering aftertaste. Holy basil has a strong, pungent clove and allspice-like, slightly musky taste with a camphoraceous aroma. Lemon basil has a slightly spicy, lemony taste with a distinct fruity aroma. Cinnamon basil has overtones of cinnamon, while curly basil has sharper and harsher notes than sweet basil.

Japanese basil or shiso is aromatic with a flavor that is a cross between basil and mint. Eaten by the Chinese in the past, it fell out of favor, and when it was introduced to Japan, it became popular in fresh or pickled forms. It is sold in Japan as a green-type (ao-jiso) and a red-type (aka-jiso).

Chemical Components: the differences in aroma among basils are due to their differing chemical components, especially methyl chavicol (or estragole), linalool, citral, methyl cinnamate, eugenol or 1,8-cineole. Monoterpenes (ocimene, geraniol, camphor), sesquiterpenes (bisabolene, caryophyllene), and methyl eugenol influence their overall flavor. The dominant aroma component in sweet basil is linalool, in holy basil is eugenol, and in Thai (anise type) basil is methyl chavicol. Sweet basil has about 0.5% to 1.1% essential oil, mainly linalool (40%) and methyl chavicol (25%), with the remainder consisting of eugenol, 1,8-cineole, and geraniol. In anise or Thai basil, 85% of the essential oil is methyl chavicol (which oxidizes when exposed to light and air), less than 1% is linalool, and the rest consists of camphor, borneol, eugenol, and 1,8-cineole. Perilla/shiso has 0.2% essential oil, the main component, about 75% being perillaldehyde, with limonene, linalool, β-caryophyllene, and α-pinene. Mexican basil has methyl cinnamate, 1,8 cineol, estragole, and bisabolene, cinnamon basil mainly methyl cinnamate, lemon basil citral, and African basil, camphor.

Sweet basil oleoresin is dark green and viscous, and 0.75 lb. are equivalent to 100 lb. of freshly ground basil.

Fresh sweet basil contains folic acid, potassium, magnesium, and calcium, while dried basil contains vitamin A, potassium, vitamin C, sodium, phosphorus, calcium, iron, folic acid, and manganese.

How Prepared and Consumed: basil is widely used in Italian, Southeast Asian, and Mediterranean foods. It is mainly used as a garnish—whole, chopped, or minced. Basil turns black in an acid medium and loses its aroma easily when heat is applied. Because sweet basil's delicate aroma is easily destroyed during cooking, it is frequently added whole or chopped to cold or warm dishes just before serving. Basil combines well with tomatoes, garlic, nuts, olive oil, olives, ginger, capers, pungent sharp cheeses, coriander leaf, garlic, galangal, lemongrass, and white and black pepper.

The popular *pesto alla Genovese*, from Liguria, Italy, made with sweet basil, olive oil, garlic, Parmesan cheese, pine nuts, and olive oil is tossed with pasta. Another pesto from southern Italy, called *pesto rosso*, is made with sun-dried tomatoes, fresh sweet basil, chilies, pine nuts, cheese, and olive oil. Italians also enjoy sweet basil in *insalata caprese* (tomato slices topped with mozzarella and sweet basil leaves, drizzled with olive oil), as pizza toppings, in bouquet garni, with capers in tomato sauces, in stews, sausages, pasta salads, and beans. The French use sweet basil to create their version of a pesto called pistou, which is served in soups, omelets, sauces, or stews. Sweet basil goes well with rosemary, parsley, tomatoes, mozzarella, cilantro, and sage. Lemon basil goes well with fish and lobster dishes.

Basils are also commonly used in Thai and Vietnamese foods, but these basils have more pungent notes. Each type of basil is chosen to flavor different dishes. Thai basil adds pungency to stir-fries (*gai prad krapao* or chicken with basil and chilies), salads, and curries; anise basil to *tom yam* or hot and sour soup (it is steeped for couple of minutes to get maximum flavor); holy basil to stir-fries where its flavor is developed during cooking; while lemon basil goes well with fish dishes. They generally par well with coriander root, galangal, turmeric, garlic, coconut milk, and chicken and pork dishes. Asian Indians use holy basil mostly in teas or as a garnish on meats and vegetables. It is not generally used in cooked foods.

The Vietnamese, who top every dish with fresh fragrant leafy spices, use all kinds of fresh basils in soups, condiments, stews, and freshly made salad rolls. Perilla leaves are used as fragrant garnishes for spring rolls and noodle dishes called *phos*. In Japan, green shiso is used as a garnish, in sushi and sashimi rolls, in salads or fried with tempura batter. The red type, which has larger leaves, is used to make a salty pickle from dried plums or apricot, called *umeboshi*.

Spice Blends: pestos, pistous, green Thai curry, Vietnamese phos, umeboshi, Malaysian kurma, bouquet garni, and pizza sauce blends.

Therapeutic Uses and Folklore: while the Greeks and French called basil the herb fit for a king, in many other parts of Europe, it symbolized hatred and hostility. For Romans, it was a symbol of love and fertility. The Italians wore a sprig of basil as a sign of interest in marriage. And, it was used as an antidote to venom or bites.

Holy basil is revered by Hindus in India and is grown near every temple and is planted around many homes to ensure happiness. Basil leaf is often given with gifts as a sign of hospitality. In some Hindu weddings, parents of a bride give her away by presenting a basil leaf, sometimes made of solid gold, to the groom.

In Asia, basil is used for most stomach disorders, cramps, diarrhea, headaches, whooping cough, and head colds. It is also used as a diuretic. In China, it was used as an antidote to fish poisons. The Italians believe that basil keeps insects away.

BAY/LAUREL LEAF

Greek legend says that the gods turned a beautiful nymph named Daphne into an evergreen laurel tree when she was fleeing from Apollo's (Greek god of prophecy, medicine, and poetry) love. The Greek name for bay leaf is *daphnee*. In ancient Greece, the winners of Olympic games were decorated with laurel wreaths, and these leaves became an immortal symbol of victory and courage. When Greek physicians completed their studies, they were crowned with laurel branches called the *baca lauris,* and which later gave rise to the term baccalaureate, which means completion of a degree. The Romans, who used these wreaths in honor of Apollo, made laurel leaf a popular spice in their cooking. The word bay is derived from Latin *baca,* meaning berry.

Scientific Name(s): Laurus nobilis. Family: Lauraceae (laurel family).

Common Names: true laurel, sweet bay/sweet laurel (English), warak al ghar, rand (Arabic), yuht gwai, yeuh kuei (Cantonese, Mandarin), laurebeer (Danish), laurier (Dutch), barg ebu (Farsi), laurel (Tagalog), laurier (French), lorbeer blatt (German), daphnee (Greek), aley dafna (Hebrew), tejpatta (Hindi), taglia de alloro, lauro (Italian), gekkeiju (Japanese), yueh kuei (Mandarin), loureio/louro (Portuguese), lavr (Russian), hoja de laural (Spanish), lager (Swedish), bai kra wan (Thai), defne yapregi (Turkish), and la nguet que (Vietnamese).

Origin and Varieties: native to Asia and the Mediterranean, bay leaf is grown in Mexico, Guatemala, Turkey, France, Greece, Russia, and North America. There are many "other" leaves that closely resemble the laurel leaf in flavor and are thus called bay leaves in other regions of the world. West Indian bay leaf (from the allspice family), Indian bay leaf, California bay leaf, Indonesian bay leaf, and boldo leaf are not true bay leaves, even though their flavors may be similar or sometimes stronger than bay leaf. The United States buys bay leaves mainly from Turkey and Greece.

Form: bay leaf is a thick, leathery, aromatic leaf with a bright green, glossy upper surface and a pale green color beneath. If the leaf has a brownish green color, it is overaged and is bitter with no aroma. It is elliptical in shape, tapering to a point at the base and tip. Bay leaf is used whole or crushed, fresh, or dried. The dried leaf is also ground into powder.

Properties: bay leaf has a strong, spicy, bitter, and pungent flavor with cooling undertones. Its flavor is described as piney, nutmeg, and clovelike with slight camphorlike notes. It has a slightly bitter aftertaste. The crushed leaf releases a strong aroma and has a grassy, sweet flavor. When the whole dried leaf is cooked in foods, its aroma is slowly released and becomes more intense. When it is used in ground form, bay leaf releases its aroma quickly and loses it quickly the longer the food is cooked. When the dried leaf is fried in heated oil, it develops a more intense flavor.

Chemical Components: bay leaf has 0.8% to 3% essential oil, mainly 1,8-cineole (35%), with methyl eugenol (4%), α-pinene (12%), α-pinene (6%), linalool (11%), α-terpineol (6%), limonene (4%), α-terpinyl acetate (10%), sabinene (5%), and eugenol (2%).

The dried fruit has higher essential oil (0.6% to 10%), mostly 1,8-cineole, α-terpineol, and β-pinene, cinnamic acid, and methyl ester. It is used for industrial purposes and used to adulterate the leaf oil.

The leaf contains calcium, iron, potassium, beta-carotene, and vitamin C.

How Prepared and Consumed: bay leaf is indispensable to Mediterranean cooking, especially in North African, Turkish, Greek, and Armenian dishes. Europeans commonly add it to soups, stews, soup stocks, pickles, and in meat and fish marinades. The French use it in their popular bouquet garni, bouillabaisse, *bourride*, and bean soups. Turks use it to flavor grilled fish, fish casseroles, and kebabs. Moroccans add it to their chicken tagines, stews, and pickled fish. Bay leaves can be cooked a long time before they lose their aroma. Steaming tends to bring out more of their flavor into the dish. Too high a level of ground bay leaf can create a bitter flavor. The whole leaf is usually removed from the cooked food before serving.

The fruits of bay leaf are sometimes used to season wild game or potatoes in European regional cooking.

Spice Blends: bouquet garni, North Indian curry blends, rice and beans, bean dishes, bouillabaisse, fish marinade, and ras-el-hanouth.

Therapeutic Uses and Folklore: ancient Greeks considered bay leaf holy. For Italians, it signifies good luck and protection. Bay leaf helps relieve pain in joints, chest, womb, and stomach. It also eases cramps and earaches. Bay leaf aids digestion by stimulating gastric secretion. Recent data show its hypoglycemic function to control diabetes and to have antiulcer activity. It has also been shown to have strong antimicrobial activity against pathogens in the gastrointestinal tract.

"OTHER" BAY LEAVES

There are many other so-called bay leaves that are not related to true bay leaves, such as the Indian bay leaf (cinnamon relative), West Indian bay leaf (leaf of allspice), the Indonesian bay leaf, boldo leaf, California, or many other varieties. Boldo leaf, which tastes similar to but stronger than bay leaf, is a distant relative of the bay leaf. The California bay leaf is milder and less pungent than the "true" bay leaf from Turkey or Greece.

Indian Bay Leaf

Indian bay leaf is not related to the true laurel leaf, but is a cinnamon relative, and it is popularly used in North Indian meat curries and vegetables. Called *tamala pattra* in Sanskrit, meaning dark leaf, or *tejpat* in Hindi, meaning pungent leaf, Indian bay leaf was well known to Romans who used it in brewing and in perfumes.

Scientific Name(s): Cinnamomum tamala/tejpat. Family: Lauraceae.

Origin and Varieties: Indian bay leaf is the leaf of a tree that is closely related to cinnamon and is cultivated in the slopes of the Himalayas facing North India.

Common Names: also called tejpata (Bengali), thitchabo (Burmese), Indisk laurbaerblad (Danish), tejpat (Hindi), talishapattiri (Tamil), indisches Lorbeerblatt (German), kanelilaakeri (Finnish) and laurier des Indes (French), tamal patra (Gujerati, Marathi), tezzi patto (Japanese), tejpatra (Punjabi), malabars kaya (Russian), and talisha (Telegu).

Form : slender whole fresh or ground, dried-crushed or ground.

Properties: Indian bay leaves are tough and leathery, very aromatic and have a flavor somewhat similar to cinnamon with cloves.

Chemical Components: has mainly monoterpenes, with 50% linalool, others being α-pinene, ρ-cymene, β-pinene, limonene, and cinnamic aldehyde.

How Prepared and Consumed: Indian bay leaf is popular in North Indian meat and lentil curries. Indians fry it in oil with other spices and curry powder before adding meat, lamb, or chicken. Indian bay leaf is commonly used in Mogul-type biryanis and kormas. It is an important ingredient in the garam màsalas of North India. In the United States, since this is not easily available, these same dishes can be prepared using the Mediterranean bay or laurel leaf.

Spice Blends: garam masala, korma, biryani, and dal curry.

West Indian Bay Leaf

A distant relative of allspice, it is sometimes referred to as the leaf of allspice in the Caribbean. It is used for smoking and cooking meats as well. Its essential oil is used in the industrial production of sausages.

Scientific Name(s): Pimienta racemosa. Family: Myrtaceae.

Origin and Varieties: it is cultivated in the Caribbean islands

Form: the leaves are large and long and available fresh or dried whole, or ground.

Properties: It has a harsh, penetrating, clovelike flavor.

Chemical Components: it has an essential oil of 1% to 2%, mainly phenols (50% to 60%). The essential oils are eugenol (38% to 75%), myrcene (14% to 32%), chavicol (11% to 21%), linalool (92% to 93%), limonene (1.5%), 1,8-cineole (0.2% to 2.0%) and 3-octanone (1%).

How Prepared and Consumed: West Indian bay leaf flavors rice and beans, stewed beans, meats, curries, relishes, and beverages in the Caribbean.

Indonesian Bay Leaf

Called *daun salam* in Indonesian, meaning "peace leaf," it is a popular flavoring in Indonesian and Malaysian cooking. Indonesian bay leaf is closely related to the cassia family.

Scientific Name(s): Eugenia polyantha. Family: Myrtaceae.

Origin and Varieties: Indonesia and other regions of Southeast Asia.

Common Names: daun salam (Indonesian, Malaysian), Indonesisk laurbaerblad (Dutch), and Indonesische lorbeerblatt (German), daeng klua, mak (Thai), and san thuyen, tram (Vietnamese).

Form: The leaves are small and thin and turn brown on drying. It comes as fresh or dried whole or ground.

Properties: the leaves are small and turn brown in color when dried. It has an aromatic and slightly sour taste.

Chemical Components: it has mainly flavanoids, tannins, and alkaloids. It contains 0.2% essential oil which has eugenol, methyl chavicol, and citral.

How Prepared and Consumed: Indonesian bay leaf is commonly used with meat dishes in the Indonesian islands of Bali, Java, and Sumatra. It is initially cooked or stir-fried to release its flavor into the dish. It pars well with chilies, galangal, lemongrass, and turmeric.

Boldo Leaf

Scientific Name(s): Peumus boldus Molina. Family: Monimiaceae. This is a close relative of the bay leaf family, Lauraceae.

Origin and Varieties: Chile, Bolivia, Peru, and other regions of South America and North Africa.

Common Names: boldina (English), boldoblatter (German), boldo (French, Greek, Hebrew, Portuguese, Spanish, Italian), and borudo (Japanese).

Form: the leaves are broad and medium-sized and come as fresh or dried, whole or ground.

Properties: boldo leaf has a strong warm flavor that is slightly bitter. It has a camphoraceous and cinnamon-like aroma.

Chemical Components: its essential oil is about 2%, with terpene derivatives such as ascaridol (40%), ρ-cymene, 1,8-cineole, eugenol, cinnamic aldehyde, and linalool. Boldo has alkaloids such as boldine, sparteine, and isocorydine.

How Prepared and Consumed: boldo leaf enhances meaty flavors and mushroom notes. In South America it is added to meats and fish. It also goes well with pickled vegetables, soups, and stews.

CAPER

Caper is derived from the Latin word *capra,* which means "goat," a name that reflects its strong smell. Thought to originate from the Near East or Central Asia, it has been used by Arabs for medicinal purposes. Other than Europe, caper is not well known in Asia or Latin America, though it is used in some Spanish style dishes in Mexico.

Scientific Name(s): Capparis spinosa. Family: Caparidaceae (closely related to the cabbage family).

Origin and Varieties: caper grows wild in the Mediterranean and is cultivated in Spain, France, Italy, Greece, Algeria, Cyprus, and Iran. There are some wild varieties that are used in northern regions of south Asia.

Common Names: caper is also called kubar (Arabic), kabra (Bengali), chee san gam, shi shan (Cantonese, Mandarin), (Dutch), capres (French), kaper (German, Norwegian), kappari (Greek), tzalaf kotsani (Hebrew), kiari, kabra (Hindi), cappero (Italian), melada (Malaysian), keipa (Japanese), alcaparra (Portuguese), kabarra (Punjabi), kapersy (Russian), alcaparra (Spanish), mruko (Swahili), kapris (Swedish), alcapparas (Tagalog), kokilakshmu (Telegu), gebre kapari (Turkish), kabar (Urdu), and cap (Vietnamese).

Form: caper is the green, dried bud of an unopened flower. It is graded based on its size—the smaller, the higher the grade. Usually, it is cured with brine, vinegar, or oil.

Properties: caper has a sharp fermented bitter taste, and its characteristic taste is developed when placed in vinegar or brine. Pickled capers have an acrid, tart, and pungent taste with a lemony tang.

Chemical Components: caper contains mainly water (85%), bitter glycosides (such as rutin and glucocapparin), pentosans, rutic acid, pectin, and saponin. Similar to mustard or wasabi, upon enzymatic action, methyl glucosinate releases methyl isothiocyanate which gives capers its pungency. Rutin is the whitish spots (crystallizes during pickling process) on pickled capers. It has high sodium content.

How Prepared and Consumed: capers have been pickled by Southern Europeans for over 2000 years. Today, they are consumed abundantly in the Mediterranean regions of Sicily, Apulia (in Italy), France, Spain, and Greece. Sicilians add capers to tomato sauces and wines with onions, garlic, green olives, and fresh leafy spices (such as basil, oregano, and chervil), game, pizzas, chicken, *caponata* (a salad that includes eggplant and tomatoes), tartar sauce, and fish. Apulians in Italy use caper with meatballs, string beans, and other boiled vegetables. The Spanish crush it, combine it with almonds, garlic, and parsley which is then served over fried fish.

Tapenade, a salty pungent spread with capers, black olives, garlic, anchovies, black pepper, mustard, and other ingredients, is a popular appetizer in Provence, France.

Capers pair well with fish, olives, chicken, basil, mustard, black pepper, garlic, oregano, and tarragon. Because heat easily destroys its aroma, caper is added to cold dishes of fish, meat, and vegetables. In the United States and northern Europe, it is served as a garnish for cold fish, roasts, and salads, as a spread, and added to pickles and relishes. Capers are also used to add tartness to the curried dishes of northern India.

Spice Blends: tapenade, pickling blend, caponata blend, and pizza sauce blend.

Therapeutic Uses and Folklore: capers have been used to aid digestion, prevent diarrhea, and increase appetite. In India, they were used as a traditional treatment against scurvy.

CARAWAY

Called in Sanskrit as *karavi*, caraway is an ancient spice, used at least 5000 years ago during Mesopotamian times. Its name is derived from the Arabic word *karawya* through the Latin word *carum*. Many global cultures call it a relative of cumin. In Hindi, it is called *vilayati jeer* or "foreign cumin," and in French, *cumin de pres* or "meadow cumin," because of its similar flavor profile to cumin. In ancient times, Romans seasoned sausages with caraway seeds. Hungarian herdsmen used them to flavor goulash. Midwives in Europe have used caraway to stimulate production of breast milk in nursing mothers. Today, caraway seeds characterize South German, Austrian, and Arabian cooking.

Scientific Name(s): Carum carvi. Family: Umbelliferae or Apiaceae (parsley family).

Origin and Varieties: caraway is native to central Europe and Asia minor and today cultivated in North Africa (especially Egypt), the Mediterranean, Germany, and Holland. It is also grown in North India, Britain, Canada, Russia, Poland, and North America.

Common Names: wild cumin, Roman cumin, German cumin and Persian cumin. It is also called karawya (Arabic), chaman (Armenian), ziya (Burmese), goht leuh ji, gel u zi (Cantonese, Mandarin), kommen (Danish), karwij, komijn (Dutch), miweh zireh (Farsi), carvi, cumin des pres (French), kummel (German), karokawi (Greek), karvya (Hebrew), shia jeera (Hindi), kyira wei (Japanese), karve (Norwegian), alcaravia (Portuguese), shima jirakam (Malayalam), tmin (Russian), alcaravea (Spanish), kummin (Swedish), shimai shembu (Tamil), hom pom (Thai), and frenk kimyonu (Turkish).

Form: caraway is a dried, ripe fruit that is used whole or ground. Its leaf is feathery and used as a garnish, while its root has a crispy texture, like parsnips.

Properties: caraway seed is curved with tapered ends and is dark brown with light-brown ridges. It has a sweet, warm, slightly dill and aniselike flavor. Its aftertaste is somewhat sharp, bitter, and soapy. If cooked too long, a bitter taste develops. The leaves have a milder taste than the seeds.

Chemical Components: caraway seeds have about 4% to 7.5% essential oil, which is pale yellow in color. Its aroma is mainly due to *d*-carvone (50% to 85%)

and limonene (20% to 30%). Other essential oil components are carveol, dihydro-carveol, α- and β-pinene, and sabinene. Caraway's fixed oil content is 15%. The oleoresin is greenish yellow, and 5 lb. are equivalent to about 100 lb. of freshly ground spice.

Caraway contains calcium, potassium, magnesium, phosphorus, and beta-carotene.

How Prepared and Consumed: caraway pairs well with garlic, vinegar, juniper, breads, pork, vegetables, and fruits. It tends to dominate the flavor in foods. It adds a savory sweetness to cakes, crackers, cheeses, applesauce, onion bread, potato soups, and cabbage. In Europe, it is a popular spice used in meats, vegetables, condiments, ice cream, marinade,s and pickling solutions. Germans use caraway abundantly in their cooking to flavor breads, meats, vegetables, cheeses, sauces, sauerbraten, and sauerkraut. It is an indispensable ingredient in rye bread and roast pork. In Dutch foods, it is a popular flavoring for cheese. The British serve tea and seedcake with cinnamon and caraway seeds. Central Europeans coat caraway with sugar or use caraway candies to sweeten their breath after meals. Caraway is the essential flavoring in a European liqueur called kummel, consumed and produced in Germany and Scandinavia. Italians boil chestnuts with caraway before roasting them. Alsatians serve a bowl of caraway seeds with Muenster cheese.

Caraway use is popular with North Africans. Tunisians combine caraway with dried chile peppers, garlic, and other spices to make their famous fiery condiment called harissa. Yemenites use caraway with coriander and other spices in a blend called zhug. The Nigerians savor a crispy caraway twist called *chin chin*. In North India, caraway is used to season *dahi wada*, a dumpling that also has lentil flour, chile peppers, and almonds, and which is eaten as a snack.

Caraway leaves are used in soups and salads, while the root is stewed or boiled and eaten in salads.

Spice Blends: pickle blends, harissa, zhug, sauerkraut blend, and dahi wada blend.

Therapeutic Uses and Folklore: ancient Egyptians placed caraway on their dead to keep bad spirits away. The ancient Arabs, Greeks, and Romans used caraway not only as a flavoring, but also as a medicine. Caraway had magical powers of protecting prized possessions, and it was also used in love potions to keep a lover from wandering away.

In Europe, caraway was chewed after a starchy meal to help digestion. Simmered with milk and honey, caraway was a remedy for colic. It has also been used as a sedative, to relieve uterine and intestinal cramps, and to help calm a queasy stomach after taking medicines that cause nausea.

CARDAMOM/CARDAMON

Known as the "queen of spices," green cardamom is the world's third most expensive spice, after saffron and vanilla. The name cardamom comes from the Greek word *kardamom* (from Persian origins), and was used by Indians, Greeks, Romans, Persians, Egyptians, and Chinese over 2000 years ago in foods, beverages, medicines, and perfumes. Originating in South India, the Vikings from Northern Europe bought cardamom from traders who obtained it through Constantinople, and introduced it to Scandinavia. To this day, it is an essential ingredient in Danish pastries. Used by Cleopatra to fill her chambers with the sweet smell of cardamom smoke before Mark Anthony's visit to Egypt, Egyptians chewed cardamom to cleanse their teeth.

It is a popular ingredient in Indian, Iranian, and Arabian cooking. An essential ingredient in Arab and Turkish coffees, the Bedouin Arabs place cardamom pods in the spouts of coffee pots so that when coffee is poured, it takes the flavor of the cardamom. Its Indian names were derived from the Dravidian language.

There are two types of cardamom, true, green or lesser cardamom, and false or greater cardamom. There are many grades of true or green cardamom, depending upon their origins. The cheaper or false cardamom is from Nepal, Vietnam, West Africa, and Bengal. It is bigger and dark brown or blackish brown with a very different flavor profile from true cardamom.

Scientific Name(s): True or "lesser" cardamom: *Elettaria cardamomum* (Malabar, Mysore*);* false or "greater" cardamom: *Amomum (A)* or *Afromomum subulatum* (North India, West Africa, Cambodia); *A. korarima/kewrerima* (Ethiopian type); *A. compactum* (Java cardamom); *A. globosum* (Chinese round cardamom); and *A. melegueta* (grains of paradise). Family: Zingiberaceae (ginger family).

Origin and Varieties: true (or lesser) green cardamom is native to India (Kerala, Karnataka, Tamil Nadu) and Sri Lanka. It is also grown in Thailand, Sumatra, and Guatemala. The United States buys the green cardamom from Guatemala, India, and Sri Lanka. Other green varieties are cultivated and consumed in Indonesia and Thailand. False (or greater) cardamom is cultivated in Bengal (India), Madagascar, Nepal, Laos, Vietnam, Cambodia, and West Africa. Grains of paradise (discussed later), a relative of cardamom, is relatively unused today except in parts of Africa. Also called guinea pepper or Guinea grain, it is common in West Africa (Nigeria, Benin) and North Africa. The Yorubas of Nigeria call it *atare* and use it in many of their food products.

Common Names: true or lesser cardamom, also called korerima hel (Amharic), hale, hab el-hal (Arabic), bak dan kou, shiou dou kou (Cantonese, Mandarin), kardemomme (Danish), kardemom (Dutch), hel (Farsi), cardomome (French), kardamom (German), kardamomum (Greek), hel (Hebrew), chotta elaichi (Hindi), kau blong (Hmong), krako sbat (Khmer), kapulaga (Indonesian), cardamomo (Italian, Spanish, Portuguese), karudamon (Japanese), hmak heng (Laotian), yelakai, elathari (Malayalam), buah pelaga (Malaysian), velachi (Marathi), kardamon (Russian), enasal (Singhalese), kardemumma (Swedish), elakai (Tamil, Telegu) luk krawan (Thai), sugmel (Tibetian), kakule tohomu (Turkish), elichi (Urdu), and truc sa (Vietnamese).

False or greater cardamom, also called Indian cardamon, Nepal cardamom, hal aswad (Arabic), boro elach (Bengali), sort kardemomme (Danish), zwarte kardamom (Dutch), chou gwo, cao que (Cantonese, Mandarin), cadamome noir (French), Nepal cardamom (German), elchi (Gujerati), kali elaichi, bada elaichi (Hindi), cardamomo nero (Italian), perelam (Malayalam), elaichi (Nepal), kardamon chyornyz (Russian), cardamomo negro (Spanish), and periya yalam (Tamil).

Form: cardamom is the dried, firm unripened fruit packed with many seeds. It is used as pods (whole or crushed), as whole seeds (decorticated cardamom), or as ground seeds. The pods are handpicked and are generally sticky to the touch. South Indian and Sri Lankan pods are slightly smaller but more aromatic.

Properties: true cardamom pods are small and dark green, light green, or white, depending upon how the pods have been processed. They are dark green when the pods are oven-dried, light green when the pods are air-dried, and white or buff colored when they are bleached. The seeds in all of these pods are reddish brown to black and are highly aromatic. The green pods have a delicate clean, sweet, and spicy floral flavor with a lemony scent. They have overtones of mint, eucalyptus, and black pepper. The white pods have very little flavor. The ground seeds have an intense flavor when just removed from the pods, but they lose their flavor quickly. Toasting the whole pod intensifies cardamom's sweet flavor.

False cardamom pods are larger-sized, and when dried are dark reddish brown, brownish black, or grayish black in color (depending on the variety) and coarsely ribbed. They give a smokey, camphorous aroma and a harsher, more pungent flavor. The seeds are darker and have a menthol-like taste.

Grains of paradise (discussed later) have a pungent and peppery flavor that is in between the flavor range of black pepper and dried chile pepper.

Chemical Components: true cardamom has 2% to 10% volatile oil depending on its regional origins and storage conditions. The major aroma contributing components are 1,8-cineole (25% to 35%), and α-terpinyl acetate (28% to 34%). Other minor ones being linalyl acetate (1% to 8%), sabinene (2%), limonene (2% to 12%), myrcene (2%), and linalool (1% to 4%). Again, components vary depending on its origins. The essential oil is spicy, sweet, citrusy, and musty, while the oleoresin is dark green, pungent, cool, and burning.

False cardamom has 2% essential oil which has more than 70% of 1,8-cineole. Its camphor taste is due to bornyl acetate. About 4 lb. of the oleoresin are equivalent to 100 lb. of freshly ground cardamom.

Cardamom contains calcium, magnesium, potassium, and manganese. Brown cardamom has high crude fiber.

How Prepared and Consumed: true cardamom enhances sweet and savory dishes. It pairs well with fruits, almonds, saffron, butter, parsley, cilantro, clove, and cinnamon. It is a very important flavoring in Turkish and Saudi Arabian foods and beverages. Arabs add cardamom to meat and rice dishes. In Saudi Arabia, dark-roasted coffee mixed with ground cardamom, called *gahira/gahwa*, is served to guests as a symbol of hospitality. It is added to fiery Arabic spice blends such as *baharat* and *jhoug* and to rice-meat dishes with nuts, dried fruits, and saffron, called *kabsah*, similar to the biryanis of India. North Africans add it to local seasonings such as *ras-el-hanout* and *berbere*. Ethiopians roast coffee with cardamom pods and cloves.

Green and large brown cardamoms are commonly used in Asian Indian cooking. In India, the green pods are used whole, or its seeds are ground and added during cooking or sprinkled on savory dishes to give a sweet aroma. Green cardamom is an essential ingredient in Indian sweets (*halwa*), puddings, yogurt (*shrikand*), custards (*kesari*), and ice creams. Green cardamom gives the distinct flavor to India's famous creamy kormas and Mogul-style biryanis. It is also an essential ingredient in Southeast Asian curries, pilafs, and desserts and other foods that have an Indian influence. Whole green cardamom is an important ingredient in chai masala, a spiced, hot tea beverage that is consumed all over India and that is now emerging as a popular drink in the United States. The Kashmiris enjoy a fragrant green tea flavored with cardamom pods.

The false cardamom pods are coarsely crushed for use in curries, vindaloos, pilafs, biryanis, snacks, and teas of North India. The whole or coarsely crushed cardamoms are braised before they are added to North Indian and Pakistani curries and pilafs to intensify their flavors. Sri Lankans and South Indians dry roast or toast pods in oil to enhance their fiery chicken and meat curries. They are usually removed before serving. Their flavor is too strong, so generally they are not added to sweet dishes, unlike green cardamom.

Cardamom is not commonly used in Europe except in Scandinavia, where ground cardamom is used instead of cinnamon in Danish pastries, biscuits, cookies, cakes, breads, *kaffekage*, *julekake*, *glogg* (hot spiced wine punch), apple pie, sausages, and meatballs.

Grains of paradise (discussed in detail on page 119) are used in the spice mixture *ras-el-hanout* and other savory dishes of North and West Africa.

Spice Blends: garam masala, curry powder, korma, berbere, baharat, zhoug, ras-el-hanout, kabsah blend, kesari blend, biryani blend, and coffee blends.

Therapeutic Uses and Folklore: Indians chew cardamom pods to sweeten and clean their breaths after meals and also after dinner to help settle their stomachs and aid in digestion. Cardamom also prevents nausea and vomiting. It soothes colicky babies, induces sweating, and cools the body during summer months. Arabs traditionally used cardamom as an aphrodisiac. In Scandinavia, it is used to mask the smell of alcohol, fish, and garlic.

CELERY

Derived from the Latin word *sedano*, celery has been used by the Greeks historically as a medicine and as a sign of victory. The Romans were the first to value it as a seasoning, and later, it became a delicacy for Italians and French. It was only in the nineteenth century that North Americans began to use celery seed, mainly in pickling solutions. Today Europeans commonly use the leaves for soups and sauces and as a garnish, and the stalks and roots as vegetables or salads. Bengalis use the ground seeds of a related species while the Chinese and Southeast Asians use a local celery leaf to flavor many of their dishes.

Scientific Name(s): Apium graveolens. Family: Apiaceae (parsley family).

Origin and Varieties: celery seed is native to eastern and southern Europe and the Mediterranean. It is cultivated in India, France, Britain, Japan, China, Hungary, and the United States. The seeds are cultivated from the original wild celery variety. Another cultivated celery variety is eaten for its stalk, leaves, and seeds. Celeriac or celery root, yet another celery variety, is savored for its root.

Common Names: garden celery and wild celery. Many global names call the seed, leaf, and stalk by similar names. Celery seed is also called karaf (Arabic), lakhod garos (Armenian), chiluri (Bengali), kun choi, qin chai (Cantonese, Mandarin), selleri (Danish), selderij (Dutch), karafs (Farsi), celeri (French), sellerie

(German), bodia jamoda (Gujerati), seleri (Hebrew), ajmud (Hindi), serori (Japanese), sedano (Italian), aipo (Portuguese), si sang (Laotian), ajmoda (Marathi), seldjerey (Russian), apio (Spanish), selderi, daun sop (Malaysian, Indonesian), selleri (Swedish and Norwegian), kinchay (Tagalog), kheun chai (Thai), kereviz (Turkish), ajmod (Urdu), and can tay (Vietnamese).

Celeriac is also called turnip root or knob celery.

Form: celery seeds are tiny globular seeds that are sold whole, slightly crushed, or ground. The leaves, which are light green, are used whole (fresh or dried), flaked, or ground. The stalks or stems and root are sold fresh.

Properties: The dried seed is dark brown with light ridges. It has a harsh, penetrating, spicy aroma and a warm bitter taste that leaves a burning sensation. The seeds have a stronger and more intense flavor than the leaf, stem, or root. The French type of celery seed is herbal and sweet with a citrus bouquet, while the Indian type is more herbal with a slight lemonlike aroma.

Celeriac root, leaf, and stalk have strong celery, herbaceous, and parsley-like tastes.

Chemical Components: the seeds have 2% to 3% essential oil, which is yellow to greenish brown. It consists primarily of terpenes, mainly limonene (68%), with sesquiterpenes such as 8% β-selinene, 8% *n*-butylidene phthalide, and myrcene. The characteristic aroma is due to the phthalides. The fixed oil is 16%.

The seed has calcium, vitamin A, vitamin C, iron, magnesium, sodium, potassium, and phosphorus.

Celery stem has 37% limonene, 19% *cis*-β-ocimene, 12% *n*-butyl phthalide, including myrcene, γ-terpinene, and *trans*-ocimene.

Celery root has 18% γ-terpinene and *trans*-β-ocimene, 16.7% 3-methyl-4-ethylhexane, and 13.2% ρ-cymene, including limonene, *cis*-β-ocimene and α-terpineol.

Celery leaf has about 0.1% volatile oil, mainly 33.6% myrcene, 26.3% limonene, 14.2% *cis*-β-ocimene, 6.2% *n*-butyl phthalide, and 3.7% β-selinene.

The seed oleoresin is green, with 4.75 lb. equivalent to 100 lb. of freshly ground celery seed.

How Prepared and Consumed: celery is a popular spice in European and North American foods and beverages. The seeds are used in fresh tomato juices, chicken soups, pickles, salad dressings, cole slaw, breads, and meats. Scandinavians and Eastern Europeans add celery seeds and leaves to sauces, soups, stews, and salads. The ground form or extractives are used in salami, bologna, frankfurters, knockwurst, sausages, corned beef, and Bloody Mary drinks.

Cooking tends to reduce its bitterness and enhance its sweetness. North Indians and Bengalis add the seeds to curries, pickles, and chutneys. Celery pairs well with chicken, turmeric, sage, cumin, soy sauce, ginger, and vinegar.

Celery stalks and roots are not spices but are discussed because they add flavor to many foods and beverages. Celery stalks are braised to give distinct flavorings and crunchy textures. Leaves of celery are chopped and used as a garnish for soups and sauces, while stalks of celery are cut and used to flavor soups, stuffings, and casseroles. Celery root is eaten raw in salads or is cooked and served as a vegetable. In North America, the leaves are added to Creole gumbos and soups. In East Asia and Southeast Asia, it flavors soups, stir-fries, and sauces, and is used as a garnish for Chinese-style rice dishes.

Celeriac root is eaten like the turnip—raw, blanched, pureed, stir-fried, or boiled. It provides a clean celery-like flavor and a crunchy texture.

Spice Blends: bouquet garni, gumbo blends, curry blends, stuffing blends, pickling blends, and tomato juice blends.

Therapeutic Uses and Folklore: Romans used celery seeds in herbal tonics as an aphrodisiac, while the Greeks used it in love potions. Celery was traditionally used as a sedative for nervousness or to promote sleep. It reduces swelling and was used to treat gout and arthritis. In India, it was taken as a remedy for rheumatism.

CHERVIL

Called French parsley by Europeans, chervil is a symbol of new life for them, because of its resemblance to myrrh given to baby Jesus. Romans and Greeks used these leaves in tonics (with watercress and dandelions) more than 2000 years ago to rejuvenate the body. Because of their pleasant sweet aroma, it was introduced to France and the rest of Europe, where it is used as fresh salads to accompany meals. Derived from the Latin word *cherifolum*, chervil is now widely used by central and western Europeans. One of the most common leafy spices used in French cooking, it has become an essential part of the local spice blends, bouquet garni, and fines herbes.

Scientific Name(s): Anthriscus cerefolium. Family: Apiaceae (parsley family) or Umbelliferae (carrot family).

Origin and Varieties: chervil is native to Russia, Europe, and Northwestern Asia. It is now cultivated around the Mediterranean regions of Greece, France, Spain, Italy, in Britain, and in the United States. Another sweet chervil, also called Spanish chervil, cicely, or garden myrrh has a strong anise and licorice-like aroma and is used in Scandinavian cooking.

Common Names: garden chervil, gourmet parsley, French parsley. It is also called maqdunis afranji (Arabic), chan lo bok, san lo boh (Cantonese, Mandarin), korvel (Danish, Swedish), kervel (Dutch), cerfeuil (French), kerbel (German), anthriskos (Greek), tamcha (Hebrew), cerfoglio (Italian), chabiru (Japanese), kzorvel (Norwegian), cerefolho (Portuguese), kervel (Russian), perifollo (Spanish), and frenk moydanoz (Turkish).

Form: chervil has small and delicate, feathery light-green leaves. It is used fresh or dried, whole, chopped, crushed, or as a paste in oil.

Properties: it has a sweet, aromatic, anise-like flavor with slight hints of pepper and parsley. The dried leaf is less aromatic than the fresh leaf. It loses its aroma with heat and drying. White vinegar preserves well its flavor.

Chemical Components: the leaf has 0.03% essential oil, mainly methyl chavicol or estragole, while the seed has 0.9% essential oil. The volatile oil smells like myrrh.

Dried chervil contains potassium, calcium, iron, sodium, magnesium, and phosphorus.

How Prepared and Consumed: chervil is usually added at the end of cooking or sprinkled over dishes to retain its delicate flavor. A staple in French cooking, chervil can be an alternative to parsley. It combines with tarragon, parsley, and chives in fines herbes to flavor eggs, dressings, and light sauces for fish and chicken dishes. The French and Scandinavians serve it as a salad with meals. It is also used in many other European cuisines, where it is chopped fine and used as garnish over eggs, soups, meats, cheese, potato salads, and fish. It gives the distinct taste to bearnaise sauce. It is also pureed with nuts, olive oil, and sun-dried tomatoes for a pestolike sauce.

It pairs well with cheese, peppermint, dill, stuffings, creamy soups, béarnaise sauce, mustards, potatoes, and pastries.

Spice Blends: fines herbes, bouquet garni, béarnaise sauce blend, and pesto sauce blend.

Therapeutic Uses and Folklore: the ancient Greeks used chervil as a spring tonic to rejuvenate the body after the winter months. In ancient times, the Romans used it as a cure for hiccups. It was also used to perk up memory, stimulate digestion, lower blood pressure and help menstrual cramps. But till today, it is still recognized for its role in lowering blood pressure.

CHILE/CHILI/CHILLI PEPPERS/CHILIES

The "mother" chile/chili/chilli pepper is thought to have originated in the Andean region of Bolivia, Peru, and Ecuador almost 10,000 years ago. Since 7000 BC, chile peppers have been part of the diets of Mayans and Aztecs in Central Mexico and the Yucatan. They were widely cultivated when the Spanish first landed in the Americas.

Chilies or chile peppers add flavor as well as heat to foods. They tend to enhance and provide a background note for other spices and flavorings. Mexicans have mastered the knowledge of use of different types of chilies to achieve that characteristic flavor, aroma, mouth feel, color, and bite in their foods. The sophisticated use of different chilies that began with the ancient Mayans and Aztecs continues around the world today. They were taken to India by the Portuguese and to Southeast Asia by the Arabs, Indians, and Portuguese. Chilies dominate the flavors of many cultures—South Indians, Sri Lankans, Southeast Asians, Latin Americans, and the Caribbean islanders.

Scientific Name(s): Within genus *Capsicum (C.)*, there are five species: *C. annum, C. frutescens, C. chinense, C. pubescens,* and *C. baccatum.* Family: Solanaceae.

Globally, there are more than 3000 known varieties of chile peppers that differ in shape, color, size, flavor, and degree of pungency. There are at least 150 known types in Mexico and about 79 in Thailand. Most chile peppers belong to the *C. annum* species, which are the principal chilies of Asia and other regions outside of Latin America and the Caribbean. New varieties of chile peppers are continually evolving.

Origin and Varieties: chile peppers are indigenous to South and Central America, Mexico, and the Caribbean but are now cultivated in India, the United States, Mexico, China, Africa, Japan, and Southeast Asia. Pods of different species have characteristic

TABLE 15
Chile Pepper Species and Varieties

Species	Varieties
C. annum	Ancho, Banana wax, cabe merah/hijau, cascabel, cayennes, chilacate, chili de arbol, chili guero, chili pasado, chilhuacle negro, chipotle, Cubanelle, exotics, Fresno, guajillo, Guntar red, Hungarian wax, jalapeno, Kashmiri, lal mirch, lombok, mirasol, mulato, New Mexican, pasilla, poblano, pungent bells/chili dulce, Sante Fe, serranos, tabia Bali, tipico, xcatik, shi feng jiao, Cherry-sweet/hot, chi yang jiao, kochu, pepperoncini, prik chee fa
C. annum (varietyaviculare)	Bird pepper, cili padi, cili rawit, ata wewe, chiltepins, piquin, tabia kerinji, devil peppers, ululte, parado, costeno, uvilla garande, max-ic, pili-pili, prik khee noo, Dundicut, tien tsin
C. chinense	Chinchu uchu, habanero, Scotch bonnet, Bahama Mama, Uxmal, Congo pepper, Bonney pepper, datil, ata rodo, ixni-pec, rocotillo, rosas-uchu, piment bouc, cachucha, Trinidad seasoning pepper, mak phet kinou, adjuma, aji panca, fatalli, safi, ose otoro, Jamaican hot
C. pubescens	Rocot uchu, rocoto, locoto, chili manzano, chili peron, canario, chili caballo
C. baccatum	Aji/kellu uchu, puca uchu, cusqueno, aji amarillo, Escabeche, cuerno de oro
C. frutescens	Tabasco, malagueta pepper, tezpur/Naga jolokia

shapes, sizes, colors, and pungency, and there are many different varieties within each species. Some of the more popular chile peppers are listed by species in Table 15.

These five species are crossed with each other to produce many new varieties. Even within each variety, such as jalapeno or cayenne, there are several cultivars with different pungencies, flavors, textures, and colors. Variations occur because of differences in breeding, environmental growing conditions, stage of maturity when picked, and postharvest conditions such as storage and processing. Some of the different cultivars are listed in Table 16.

Chile pepper (Capsicums) terminology:

The Nahuatl word for chile pepper is chilli. Chile, chili, or chilli pepper are the commonly used terms for hot peppers in the United States, Canada, Central America, and Mexico. Chilli is the word for the hot chile pepper in Asia, England, and other English-speaking regions, while the term capsicum is used for the nonhot sweet bell peppers. The Food and Agricultural Organization (FAO) refers to the hot varieties as chillies. South Americans refer to chile pepper as aji which came from the Arawaks. Hungarians call any Capsicum paprika, but to the world paprika is the ground red powder that provides mostly color with some flavor.

Common Names: (in General): berbere (Amharic), fulful haar (Arabic), gdzoo bgbegh (Armenian), jolokia (Assamese), morich (Bengali), nga yut thee (Burmese), lah jeeui (Cantonese), chili (Danish), spaanse peper (Dutch), chile (English), pilpil (Farsi), sili/siling (Filipino), piment (French), roter pfeffer (German), piperi kagien (Greek), pilpel harif (Hebrew), mirch (Hindi), cayenne bors (Hungarian), cabe/tabia/lombok (Indonesian), peperone, peperoncino (Italian), togarashi

TABLE 16
Chile Pepper Varieties and Cultivars

Variety	Description and Cultivars
Cayennes	Cayennes were first referred to as milchilli in 1542. They appear to have originated in French Guiana, South America, and were named possibly after the Cayenne River or capital of that region. Cayenne was derived from Tupi Indian word 'kian.' Introduced by the Portuguese all over the world, they are widely grown throughout Asia. In the United States, they are grown in Louisiana and New Mexico. Cultivars include Hot Portugal, Ring of Fire, Hades Hot, and Exotics. The Exotics are cayenne types and include santaka, Hontaka, Thai chilies, and Chinese kwangsi. The dried forms of cayennes are called cili kering, lal mirchi, or Ginnie peppers.
New Mexican	These peppers were originally called Anaheim. They are mild to hot. Cultivars include Big Jim, Sandia, Fresno, Nu-Mex No. 6-4, Nu-Mex sunrise, Nu-Mex sunset, Nu-Mex Eclipse, Sandia, Fresno, TAM Mild chili, long green/red chili, prik chee fa, la chiao, chimayo, chili verde, chili Colorado, and California long green or red. In Mexico, the dried green New Mexican chilis are called pasados, and red ones are called chili Colorado.
Jalapeno	Jalapeno peppers originated in Jalapa, Mexico. The name means "looking at the sun." Some cultivars are San Andres, TAM mild jalapeno-1, 76104 Jumbo Jal, Morita, Tipico, Peludo, and Espinalteco. In Mexico, the short fat version is called chili gordo. In Oaxaca, the smoked or dried jalapeno (or chipotle) is called chili huachinango, and the pickled jalapeno is called morita.
Serrano	Serrano peppers originated in Puebla, Mexico. Serrano means "from the mountains." When fresh green, they are called chili verde, and when dried, they are called balin/chico, tipico, and largo. Commercial cultivars are Tampiqueno 74, VeraCruz S69, Altamira, Panuco, and Hidalgo.
Ajis	This is the general term for chile peppers in Spanish-speaking South America. They were cultivated as early as 2500 BC in Peru. They were an integral part of the Inca diet. Ajis come in diverse shapes, sizes, and colors, including aji amarillo (or kellu uchu as the Quechua Indians call it) and cuerno de oro in Costa Rica. They are commonly used in South America (Peru, Bolivia, Colombia) and Central America.
Rocotos	Rocoto has its origins in Bolivia, was domesticated in 6000 BC and was commonly used by the Incas. These fruit-shaped chilies are called rocotos by Peruvians, caballo by Guatemalans, locoto in other parts of South America, and chili manzano, canario or chili peron in Mexico.
Poblano	Poblano means "people chili," and its name was derived from Puebla, Mexico, where it was first cultivated. The dried versions are ancho and mulato. Mulato is popular in Mexico and is known as chili negro, Verdano, Esmeralda, or morita. In Baja, Mexico, fresh and dried forms are called ancho or pasilla.
Chiltepin	Chiltepin peppers are also known as piquins, pequins, and tepins. They were part of the historic migration of peppers from the Andes. Because birds were responsible for their migration, chiltepins were called bird peppers. They are also called chili mosquito, chili de pajaro, or chili bravo (macho) in Mexico, huarahuao (in Haiti), ululte in Yucatan, parado in Oaxaca, and bird pepper, chili padi, and chili rawit in Southeast Asia.
Habanero	Its name comes from the word "Havana-like" and is thought to have originated in Cuba. They are now used abundantly in the Caribbean, Mexico, Brazil, and Belize. They are also called Scotch bonnet (smaller size), Congo, or Bahama Mama, Bonney pepper, pimient, Jamaican hots in the Caribbean, and Uxmal and INIA in Yucatan.

(Japanese), molaku (Kannada), ma-tek (Khmer), mak phet (Laotian), cili/ladah/ cabai (Malaysian), la jiao (Mandarin), mirchya (Marathi), pimiento, pimentao picante/piripiri (Portuguese), mircha (Punjabi), chili (Russian), miri (Singhalese), chile/aji/pimenta picante (Spanish), pilipili (Swahili), chilipeppar (Swedish), siling (Tagalong), mulagu (Tamil/Malayalam), prik (Thai), sipen (Tibetan), biber (Turkish), marach (Urdu), ot (Vietnamese), and ata (Yoruba, West Africa).

Form: the fruit is used ripe or unripe, fresh (whole, sliced, pureed), smoked, or dried (whole, crushed, or ground). The fresh and dried forms can be made into a paste or pureed (as *Cili Boh* in Malaysia/Indonesia).

Whole fresh or dried chile peppers can be roasted to develop more intense flavors. They are called by different names when the chile pepper is smoked or dried. For example, when green or red jalapeno is smoked or dried, it becomes a dark red chipotle, and when green poblano is dried, it becomes a mahogany ancho or reddish black mulato. Examples of fresh and dried chile peppers are listed in Table 17.

Dried chile peppers are available commercially as ground or crushed red peppers, ground chile pepper, and chili powder. Ground or crushed red pepper usually consists of blended red peppers from any variety of hot, dried, red chile peppers. Heat rather than flavor or color is expected from these products. Ground chile pepper in the United States is ground New Mexican, cayenne, or ancho chile peppers. In the U.S., it is used in chili powder, a blend of ground, dried cayenne/New Mexican chile peppers, cumin, oregano, paprika, and garlic powder. This blend forms the basis of the flavor for chili con carne or chili from Texas. Chili powder has varying degrees of pungency and different shades of color ranging from bright red to deep mahogany. In Asia and many other countries, chili or chilli powder means only ground chile pepper, usually cayenne.

Properties: chile peppers have many shapes, sizes, colors, flavors, and heat. They can be hot, sweet, fruity, earthy, smoky, and floral. They are variable in their heat levels. Fruits harvested from the same species but from different regions can differ greatly in heat. Chilies can enhance and round up other ingredients in a dish. General physical descriptions of several of the more popular chile peppers are listed in Table 18.

Chemical Components: Chile peppers contain 0.2% to 2% capsaicinoids (vanillylamides of monocarboxyl acids), which are responsible for the pungency or bite in capsicums. Heat varies widely among the different chile peppers depending on their chain length. Capsaicinoids are mostly found in the white "ribs"/placenta that runs down the inside of the pepper, in the seeds, and in the skin. Most of the overall heat is due to two capsacinoids, capsaicin and dihydrocapsaicin.

Capsaicin (C), an alkaloid, accounts for about 50% to 70% of the total capsacinoids. It gives the bite but has no odor. The other bite contributing components are 20% to 25% dihydrocapasaicin (DHC), which together with capasaicin provides the most fiery notes from midpalate to throat; 7% nordihydrocapsaicin (NDHC), which is fruity and sweet and has the least burning sensation; about 1% homocapsaicin (HC) and 1% homodihydrocapsaicin (HDHC), both of which give a numbing and prolonged burn.

The degree of heat or pungency of chile peppers varies based on the varieties, origins, growing conditions, and drying conditions. Heat is measured by using

TABLE 17
Chile Peppers: Fresh and Dried Versions

Fresh Chilies/Chile Frescos	Dried or Smoked Chilies/Chile Secos
Jalapeno	Chipotle, morita, tipico, huachinango, mora, chili ahumudo
Serrano	Balin, tipico, largo
Poblano	Ancho, mulatto, chino
Anaheim	Chili pasado
Cayenne	Sanaam, tien tsin, cili kering
New Mexican/Anaheim	Chili pasado, chili Colorado, lal mirchi
Chilaca	Pasilla (chili Negro)
Cili padi	Cili padi kering
Chiltepin, piquillo	Cascabel, chiltepin, piquin, coban
Aji amarillo	Cusqueno

TABLE 18
Physical Description of Chile Pepper Varieties

Variety	Description
Cayenne	Red, orange, green 3–10 long, 1 wide, narrow and elongated, sometimes curved, ending in a sharp point; hot to very hot.
New Mexican	Green to red, smooth, elongated, 6–12 long; mild to hot.
Jalapeno	Medium size (2–3 long), bright green, red, or purple with tapered to blunt ends, and conical and cylindrical in shape; slightly hot to very hot. (Red, ripe jalapeno is sweeter than the immature green.)
Chipotle	Wrinkled dark brown skin with a smoky, sweet chocolaty flavor.
Serrano	Dark green to red orange, yellow, or brown, 1–3 long or tubular in shape; very hot.
Poblano	Green, dark green, 3–6 conical or heart shaped; slightly hot.
Ancho	Flat, wrinkled, and heart shaped, red to reddish brown; mild to slightly hot.
Mulato	Reddish brown to blackish brown, mild to slightly hot with chocolaty notes.
Chiltepins	Tiny and ovoid; very hot, but the heat dissipates quickly.
Ajis	Diverse in size and shape, typically 3–5 long, elongated, orange, red, yellow, or brown; medium to hot.
Habanero	1–2.5 long, 1–2 wide, bell shaped, red, green, yellow, white, or orange, fruity taste and great aroma; extremely hot.
Tabasco	1–1.5 long, 3/8 wide and yellow when unripe, red orange when ripe; hot.

organoleptic tests or high-performance liquid chromatography (HPLC). The capsa-icinoids are measured in parts per million and are then converted to Scoville Units (Table 19).

Heat values can vary with the same type of chile pepper, such as jalapenos or cayennes, based on its origins, breeding, and climatic and growing conditions. Today there is a growing trend toward using chilies for flavoring foods as well.

TABLE 19
Scoville Units of Chile Pepper Varieties

Chile Pepper Variety	Scoville Units
Mild bell pepper, sweet banana, pimiento, sweet paprika, Hungarian sweet pepper, Anaheim, pepperoncino	0–100
Mexi-bell, cherry, romesco, nora, piquillo	100–200
New Mexican, prik num, guajillo, Hungarian hot pepper	200–500
Poblano, ancho, pasilla, cayenne, cascabel, yellow hot wax	1000–1500
Mirasol, jalapeno, cachuacha, aji dulce, mirasol	2500–5000
Jalapeno, chipotle, sandia, pasado, guero	2500–10,000
Serrano, jalapeno, Kashmiri, sriracha, aji Amarillo, crushed red pepper	10,000–25,000
Cayennes, prik khee fah, chili de arbol, Tabasco, cabe merah, rocotillo, ajis, Pakistan dundicut	30,000–40,000
Chiltepin, aji, rocoto, santaka, piquin, malagueta, Chinese kwangsi	40,000–50,000
Prik khee nu, chili padi, cabe rawit, bird peppers	50,000–100,000
Habaneros, Scotch bonnets, piment, Bahamian, Congo pepper, red savino habanero, Naga jolokhia	100,000–750,000
Pure capsaicin	16 million

Chile peppers are high in vitamins A and C. Green chilies have double the vitamin C of a regular-sized orange, while red chilies have more vitamin A than carrots. They are low in calories and sodium and contain potassium, phosphorus, magnesium, and folic acid.

How Prepared and Consumed: the ancient Mayans and Aztecs used chilies as food and medicine and in religious rituals and ceremonies. Nowadays, they provide not only heat but also flavor, color, and visual appeal to foods around the world. Caribbeans, Mexicans, South Americans, Mediterraneans, Asian Indians, and Southeast Asians use chile peppers to add zest or flavor to their cuisines. These ethnic cuisines use chile peppers as vegetables when fresh and as a spice when dried or smoked.

Chilies are also made into hot sauces or signature spice blends that are characteristic of a region. The early hot sauces were made by the Caribs and Arawak Indians, who mixed chile peppers with cassava juice to create coui.

Today, hot chile peppers are pickled or pureed and are used as table condiments or sauces all over the Caribbean. The fiery Bajan sauce contains scotch bonnets with other ingredients such as fruits, mustard, garlic, thyme, green onions, and clove. Similarly, Louisiana hot sauce, *sambals* of Indonesia, salsas and moles of Mexico, *goit chu jang* of Korea, *balachuang* of Myanmar, *harissa* of Tunisia, romesco of Spain, Matouk sauce from the Antilles, *berbere* of Ethiopia, *shichimi togarashi* of Japan, *rouille* of southern France, and the *chaat masalas* of India contain chile peppers as an essential flavoring.

Many dried chilies such as tien tsin, chipotle, Guntur red, or chili de arbol are ground and added to spice mixtures and made into hot chili oils. Japanese use these spice mixtures, such as *schichimi togarashi*, to add heat by sprinkling over foods.

In Asian cooking, dried whole red chilies are fried in hot oil until they turn dark brown. This hot oil is added to Szechwan and Hunan stir-fries, and, in India, a mixture of chile pepper and oil is added to spice up many curries and dals.

Chile peppers pair well with garlic, fermented beans, ginger, coconut, shallots, fermented fish or shrimp, galangal, turmeric, sesame oil, and fruit sauces. In Southeast Asia, Korea, and Szechwan region, they are added to fermented soybeans and seafood to make fiery hot pastes for many dishes. Dried or fresh chilies are a must in Thai salads and curry pastes, Korean *kimchis*, Indonesian *rendangs*, and Malaysian *sambals*. In India, dried red chile peppers and fresh green chile peppers are popularly used. Sri Lankans and South Indians use whole cut chilies abundantly in snacks, chutneys, and curries. The black curries of Sri Lanka contain bird peppers and cayennes with toasted spices.

In Eastern Europe, chile peppers are not commonly used, except in Hungary. In the Mediterranean regions, the North Africans usually use the hotter chilies, while the Spanish use the mildly hot ones.

In Latin America, chilies add heat to bland potatoes, yuccas, salsas, ceviches, moles, and condiments. Mexicans, Ecuadorians, Brazilians, and Peruvians use many types of fresh and dried chilies (e.g., jalapenos, habaneros, ajis, rocotos, malaguetas) to create moles, ceviches, and salsas. *Ocapa* (potato, peanut, cheese with ajis), stuffed rocotos, *pebre* (with olive oil, cilantro, and red chilies), *aji molida* (with fresh chilies, herbs, and onions), *salbutes* (puffed tortillas) topped with a hot sauce of habaneros, lime juice, and onions, and chimmichurri with ajis, cilantro, garlic, and lime juice are common applications.

Some examples of how chile peppers are used in ethnic cuisines are listed in Table 20.

Spice Blends: chili powder, jerk, sambal trassi blend, mole negro blend, pebre, berbere, periperi sauce, romesco sauce, jhoug, kaeng ped blend, nuoc cham blend, nam prik blend, ma pla blend, rendang blend, kochujang, and kim chee blend.

Therapeutic Uses and Folklore: the Incas from South America regarded chile peppers as holy plants and used them in offerings to appease the gods. They were valued as currency and were used as decorative ornaments. They were also used to ward off spells. Chile peppers have long been recognized by many cultures around the world for their medicinal qualities. Pre-Columbian Indians mixed pepper with other ingredients for sore throats, coughs, arthritis, acid indigestion, earaches, to promote better eyesight, and to expedite childbirth. The Caribs rubbed chilies (*axi*) into the wounds of would-be young warriors to strengthen their courage. The Andean cultures used it to treat severe headaches and strokes.

When chile peppers are eaten, capsaicin stimulates the release of endorphins, which give a pleasurable feeling. Chile peppers are believed to increase circulation, relieve rheumatic pain, treat mouth sores and infected wounds, reduce blood clots, and aid digestion by stimulating saliva and gastric juice flow.

CHIVES

Chives are a very popular garnishing spice in French and Chinese cooking. The name "chives" is derived from the Latin word *cepa,* meaning onion, which later

TABLE 20
Chile Peppers and Applications

Variety	Description of Applications
Rocotos	Popular in Andes and Mexico—salsas, stuffed with cheese or meat, rellenos, empenadas
Chiltepin	Used in soups, ceviches, salsas, bean dishes, and pickles of South and Central America, Mexico, and Texas
Serrano	Popular in Mexico, Tex-Mex, and Southwest cooking, good for pickling—escabeche, fresh salsas, table sauces, relishes
Tabasco	Added to Tex-Mex, Cajun, Creole cooking, and Mexican sauces.
Cayenne	Popular in Indian curries, chatnis, vindaloo, in Chinese, Korean, and Japanese stir-fries and condiments, in Southeast Asian and African curries and sambals, in Cajun and Creole cooking, used as sculptured toppings for Thai cuisines
Jalapeno and chipotle	Used in Mexican and Central American salsas, snacks, stuffings, toppings, and pickles; in the United States, popular in Tex-Mex and Southwest cooking—in enchiladas, nacho toppings, pickles, sauces, and chili con carne
Habaneros	Popular in hot sauces of the Caribbean, Yucatan salsas, and pickles—they are the principal ingredient in Pickapeppa (Jamaica), Yucateca (Mexican), Pica rico (Belize), curry colombo, Jamaica Hell Fire, Bajan (Barbados) sauces and jerk seasoning
Ajis	Found in many South American sauces—aji molida, salsas, ceviche, pebre, ocapa, and chimmichurri
New Mexican	Used in Southwestern salsas, stuffings, and in sauces; used in Chinese and Indian cooking; also used in the Southwest as decorative ristras
Piquillos	Sweet, slightly hot peppers found in Spanish cuisine—they are fire roasted or smoked and used to stuff fish, meat, or vegetables
Exotics	Are Thai peppers, hontaka, santaka, red chilies—they are dried and used for Szechwan stir-fries, sambals, as chili bohs, for curries or ground into chili powder in Asia

became *cive* in French. Called little onion in European languages, chives are related to onion, leek, and garlic and used as spice or vegetable, depending on the region.

Scientific Name(s): regular chives: *Allium (A) schoenoprasum;* garlic chives or Chinese chives: *A. tuberosum:* Nepalese chive or Himalaya onion*: A. walchii.* Family: Alliaceae (onion family).

Origin and Varieties: chives are native to Central Asia and now cultivated in Scandinavia, Germany, Britain, China, and Southeast Asia. There are many varieties from the West and the East.

Common Names: onion chives. They are also called warak basal (Arabic), manr sokh (Armenian), div chesun (Bulgarian), purlog (Danish), bieslook (Dutch), tareh (Farsi), civette/ciboulette (French), schnittlauch (German), praso (Greek), irit (Hebrew), erba cipollina (Italian), asatsuki (Japanese), cebolinha (Portuguese), lukrezanets (Russian), cebollino (Spanish), graslok (Swedish, Norwegian), and frenk sogani (Turkish).

Chinese Chives: garlic chives, spring onions, kow choi (Cantonese), kuchai (Malaysian, Indonesian), jui tsai (Mandarin), jimbu (Nepalese), ku chai (Thai), and ka choy (Khmer).

Form: chives are hollow, long, narrow stems attached to a bulb on one end and tube-shaped green leaves at the other. Chives are used fresh or dried. The light purplish flowers of chives have a light, delicate chivelike flavor. The entire length of the tubular leaf is used in foods. Chinese chives have flat and wider stems than regular chives.

Properties: the stems of chives are bright green. Regular or onion chives have a mild, oniony flavor because of their sulfur content. They have a more delicate flavor than scallions. Chinese chives or garlic chives are slightly garlicky and stronger in flavor than onion chives. The bulbs of chives have a delicate onionlike flavor. They tend to lose most of their flavor through drying and when exposed to heat.

Chemical Components: the essential oil of chives consists mainly of dipropyl disulfide, methyl pentyl disulfide, and penthanethiol.

Chives contain vitamin A, vitamin C, potassium, and folic acid.

How Prepared and Consumed: chives are a popular ingredient in European cooking because of their delicate flavor. They are an essential flavoring of the spice blend, fines herbs, that is commonly used in French cooking. The bright green leaves of chives are good as a garnish in cold salads, stews, and soups or when sprinkled over cooked sauces, soups, and finished meals. The finely chopped leaves and bulb are used to garnish vichyssoise, cheese and cream sauces, gravies, dips, and baked potatoes.

Chinese chives are chopped or sliced and are used as a common garnish for all dishes, stir-fried vegetables, fried noodles, fried rices, grilled meats, fish dishes, and dumpling fillings. Chives pair well with lemon, tarragon, parsley, sesame oil, vinegar, garlic, soy sauce, ginger, potatoes, and chile peppers. Their bulbs are also pickled. In the United States, chives are used in cottage cheese, egg dishes, cocktail sauces, dried soups, and sour cream. The Nepalese and Tibetans add it to dals and *momos* (dumplings). It is fried in butter fat to get a more intense flavor.

Spice Blends: remoulade blend, fines herbes, dumpling sauce blend, and vichyssoise blend.

Therapeutic Uses and Folklore: chives have been used to help lower blood pressure and aid digestion. Chives also stimulate the appetite and possess some antiseptic properties.

CINNAMON

Known as "sweet wood" in Malaysian, Arab traders brought cinnamon from the Far East to Egypt and the West, where it was greatly treasured. Cinnamon is one of the oldest known spices. Through trading by the Khasi tribe in North East India, cassia was used by the Chinese as a medicine as early as 2500 BC. It is mentioned in the Bible and was used by Egyptians in embalming mixtures to preserve their pharaohs.

Cinnamon's name is derived from the Greek word *kinamon*. The Greeks and Romans used it as body perfume and medicine and burned it as incense. It was one of the ingredients in ancient European love potions. The Dutch, Portuguese, and

British explorers' goal in searching for a quick route to the Far East in the fifteenth Century was to control the trade in spices, including cinnamon.

There are several varieties of cinnamon, but two types are commonly used: Cassia (also called Chinese cinnamon) and Ceylon cinnamon (also called canela). Cassia is referred to as "Chinese wood" by the Farsis, Indians, and Arabs. The former is used widely in East Asia, Southeast Asia, and the United States (sometimes mixed with Ceylon cinnamon), while the latter is most frequently used in Mexico, England, Europe, and South Asia. Indonesian cinnamon, called *kayu manis padang* (sweet wood), which comes from Padang, Sumatra, is sometimes sold as Ceylon cinnamon in the United States because its aroma closely resembles it.

Scientific Name(s): Chinese cinnamon/cassia: *Cinnamomum cassia;* Ceylon cinnamon/canela: *Cinnamomum zeylanicum;* Indonesian cinnamon: *C. burmannii,* Vietnamese cinnamon: *C. loureirii.* Family: Lauraceae.

Origin and Varieties: cassia is native to China, Indochina, Burma, and North Vietnam and is also cultivated in Indonesia and Malaysia. There are mainly four types: Indonesian cinnamon [korintje (or Padang) and cassia vera], Vietnamese cinnamon (cassia lignea), and Chinese cinnamon (Tunghing, Sikiang, Canton). Korintje, classified as cassia, has a flavor that is closer to Ceylon cinnamon (and sometimes traded as Ceylon cinnamon). It comes in many grades; the lower grades are commonly sold in the United States.

Ceylon cinnamon/canela is native to Sri Lanka and India, and is also cultivated in Mexico, South America, Seychelles, and Madagascar. There are three main types: Ceylon, Seychelles, and Madagascar cinnamon. Ceylon cinnamon is used abundantly in South Asian, English, Mexican, Central American, and South American cooking.

Common Names: cassia/Chinese cinnamon: false/bastard cinnamon. It is also called kerfe (Arabic), gun kwai (Cantonese), kinesisk kanel (Danish), kassie (Dutch), darchi (Farsi), cannelle de Chine/saigon (French), kassie (German), kasia (Greek), Chinesisener zimt/Vietnamesisches/Saigon zimt (German), kasia (Hebrew), cassia (Italian), kashia keihi (Japanese), kyae-pi (Korean), sa chwang (Laotian), kayu manis (Malaysian/In-donesian), you qui (Mandarin), kassia (Norwegian), canela de China (Portuguese), korichnoje derevo (Russian), canela de la Chine (Spanish), op choey (Thai), cin tarcini (Turkish), darchini (Urdu), and que don (Vietnamese).

Indonesian cinnamon/kayu manis padang: Java cassia, cassia vera, Timor cassia, Korintje, and Padang cinnamon. It is also known as Padang zimt (German), canelle de Padang (French), and jaavankaneli (Finnish).

Ceylon cinnamon/canela: true cinnamon. It is also called kerefa (Amharic), kerfa (Arabic), ginamon (Armenian), dalchini (Assamese), dalchini (Bengali), yuk qwai (Cantonese), kurfa kanel (Danish), kaneel (Dutch), darchin (Farsi), cannelle (French), zimt (German), kanela (Greek), tuj (Gujerati), kinamon (Hebrew), darchini (Hindi), fahej (Hungarian), cannella (Italian), seiron nikkei (Japanese), lavanga patta (Kannada), chek tum pak leong (Khmer), kayu manis (Malaysian, Indonesian), lavanga patti (Malayalam), rou kui (Mandarin), dalachini (Marathi), kanel (Norweigian), koritsa (Russian), dalasini (Swahili), kurundu (Singhalese), canela (Spanish, Portuguese), kanel (Swedish), kanela (Tagalong), patta/ lavangum (Tamil), op chuey (Thai), lavangamu (Telegu), seylan darcini (Turkish), and que srilanca (Vietnamese).

Indonesian cinnamon:yeh qwai (Cantonese), Indosisk kanel (Danish), Indoosische kaneel (Dutch), shiwanikei (Japanese), shan yue gui (Mandarin), canela de java (Spanish), falsa cunforeira (Portuguese), suramarit (Thai), and que ranh (Vietnamese).

Vietnamese cinnamon: Saigon cinnamon, que qui, canella de Saigon (French), Saigon-zimt (german), khe (Laotian), Saigon fahej (Hungarian), canela de Saigao (Portuguese), Vietnams kaya koritsa (Russian), and canela de Saigon (Spanish).

Form: every part of the cinnamon tree—the bark, leaves, buds, roots, and flowers, has some use. The bark oil is used extensively in commercial foods. The dried scented bark, sold as whole sticks called "quills," is ground. Ceylon cinnamon has slender and thin quills with smooth surfaces, while Chinese cinnamon has thick quills with coarse surfaces. The bark of Ceylon cinnamon has a dark reddish brown interior color and a light brown or tan outer color, whereas cassia has a much darker outer surface. Cinnamon bark is sold whole, in chunks, or ground.

The highest quality cinnamon comes from the trunk, and lower quality cinnamon comes from the side branches. Finely ground cinnamon is smooth, not gritty, and this is indicative of good quality cinnamon. Cinnamon is graded according to its length, breadth, and thickness. The quills are the best grade, with lesser grades being quillings, featherings, and chips. Cassia bark is more mucilaginous than canela bark.

The unripe fruits of cassia, called buds, look like cloves and are sweetish and less aromatic than the bark. The yellow flowers of cassia are very fragrant.

Properties: cassia has a coarser and thicker bark than Ceylon cinnamon. Cassia has a sweet spicy aroma, a harsh bitter taste, and is more pungent than Ceylon cinnamon. Ceylon cinnamon has a warm, sweet, slightly clove- and citruslike delicate flavor. Ceylon cinnamon has a more delicate flavor than cassia. Chinese and Vietnamese cinnamon are more intense in flavor than the other cassias. Sweeter and more delicate varieties of cassia are emerging today. Korintje cinnamon has a delicate flavor and is the least bitter. Ceylon cinnamon has a lighter colored and much thinner bark than the cassia.

Indonesian or Korintje cinnamon's aroma is similar to Ceylon cinnamon: slightly more intense but not harsh or bitter like Chinese or Vietnamese cinnamon. Its quills are thicker than those of Ceylon cinnamon so are less fragile, with an outer reddish brown color and a darker interior. It has lower slime than Chinese or Vietnamese cinnamon.

Chemical Components: depending on its origins, cassia has 0.9% to 7% essential oil, and canela has 0.5% to 2%. Cassia oil contains mainly cinnamic aldehyde (65% to 95%), cinnamyl acetate, cinnamic acid, benzaldehyde, small amounts of coumarin, and trace amounts of eugenol. Canela oil mainly consists of cinnamic aldehyde (65% to 75%) with eugenol (5% to 10%), linalool (2.3%), cinnamyl acetate (5%), safrol, 1,8-cineole, and benzyl benzoate. The oil is yellowish brown to brown in color and darkens as it ages.

Canela leaf contains about 1% essential oil, mainly eugenol (70% to 95%) with linalool, cinnamic aldehyde, and caryophyllene. Indonesian bark oil has 1% to 4% oil, mainly cinnamic aldehyde with no eugenol. Indonesian leaf oil has mainly cinnamaldehyde.

The oleoresin from the bark is dark reddish brown in color, and 2 lb. are equivalent to 100 lb. of freshly ground spice.

Cinnamon has calcium, potassium, manganese, iron, magnesium, and vitamin C.

How Prepared and Consumed: European and Mediterranean regions primarily use the ground form of cinnamon. Asian and Latin American countries use whole and ground forms. Cinnamon pairs well with tomatoes, onions, yogurt, nuts, bananas, oranges, peaches, apples, grapes, chocolate, corn, and cauliflower. Overcooking makes it bitter.

Cassia is the preferred cinnamon in the United States because of its spicy-sweet flavor. It is used in its ground form in fruit-based confections, cakes, pies, cookies, buns, and breads. The cinnamon stick flavors mulled cider. In East and Southeast Asia, cassia is used in savory applications—five-spice blend, sauces, curries, soups, and desserts. In Hunan, China, the ground form by itself or in a five-spice blend with Hoisin, soy sauce, and rice wine is popular for braised meat and poultry dishes, soups, and sauces. The Vietnamese add it with star anise for their long-simmering beef soups, called *pho bos*, which are served hot with noodles and herb garnishes.

Sri Lankan or Ceylon cinnamon is popular in European, Latin American, North African, Middle Eastern, and English cooking. In England, Ceylon cinnamon is mainly used in sweet applications, such as fruitcakes, stewed fruits, and pastries. In India and Sri Lanka, it is an essential spice in the fiery curries, pickles, teas, garam masalas, and fragrant biryanis. For curries, cinnamon stick is fried in hot oil, whereby it unfolds to release its aroma. It is then added to the curry or rice being cooked.

Ceylon cinnamon is also popular in Arab and North African cooking: it is added with paprika in *baharat*, with rose petals in *ras-el-hanout*, and with black pepper in *berbere*. It is also indispensable in Latin American confections and chocolate beverages, and in the moles of Mexico.

Cinnamon buds are dried and used for flavoring confections and pickles in China. The leaves of Ceylon cinnamon are sometimes used as a substitute for Indian bay leaf to flavor curries or as a garnish in Indian and Southeast Asian dishes. The cassia flowers perfume the sweets, soups, pastries, wines, and teas in China.

Spice Blends: garam masala, mole negro, Panang curry blend, baharat, ras-el-hanout, berbere, Indian curry blends, pho bo blend, and Chinese five-spice.

Therapeutic Uses and Folklore: as early as 2700 BC, Chinese herbalists used cinnamon to treat diarrhea and to help circulation and depression. Indonesians use a pair of cinnamon buds (growing side by side) in a drink for the bride and groom at weddings as a symbol of harmony and togetherness. Cinnamon is also used to relieve head colds and nausea. Its leaves are used as a treatment for diarrhea.

Since ancient times, cinnamon was commonly used as a flavoring for meat because of its antibacterial properties which helped to prevent spoilage.

CLOVE

Clove is native to the Moluccas (or Spice Islands), Indonesia, where traditionally, when a child was born, a clove tree would be planted. If the tree grew straight and strong, then so would the child. Children were also given clove necklaces to protect them from illnesses.

Cloves are an ancient spice but used extensively in meat curries. The Chinese used cloves as early as 200 BC, when court officials chewed cloves to sweeten their breaths in the Emperor's presence. Arab traders introduced clove to the Romans and Greeks, who used it in their love potions. Later, Europeans fought each other to monopolize the clove trade in the Spice Islands because of its value. The name clove derived from the Latin *clavus*, and the Spanish *clavo*, French *clou* or German *nelke* terms, all mean "nail," because of its resemblance to the shape of a nail.

Scientific Name(s): Syzyium aromaticum (formerly *Eugenia caryophyllata* or *E. aromaticus*). Family: Myrtaceae.

Origin and Varieties: cloves are native to the Moluccas islands of Indonesia and are now produced in Brazil, Madagascar, Pemba, Zanzibar, Sumatra, Malaysia, Sri Lanka, Grenada, and Jamaica.

Common Names: cloves are also called krinful (Amharic), kurnful (Arabic), chorpogpach (Armenian), laung (Assamese), lavanga (Bengali), ley nyim bwint (Burmese), ding heung (Cantonese), kruidnagel (Dutch), mikhak (Farsi), clou de girofle (French), nelke (German), garifalo (Greek), tziporen (Hebrew), lavang (Hindi, Gujerati, Marathi), garofano (Italian), choji (Japanese), khan plu (Khmer), kan phou (Laotian), karayanpu (Malayalam), bunga cinkeh (Malaysian, Indonesian), ting hsiang (Mandarin), nellike (Norwegian, Danish), cravinho (Portuguese), gvozdika (Russian), clavo (Spanish), karabu nati (Singhalese), kryddnejlikor (Swedish), clovas de comer (Tagalog), karambu (Tamil), kan ploo (Thai), karanfil (Turkish), and hanh con (Vietnamese).

Form: cloves are dried unopened flower buds that are sold whole or ground. The rose-colored buds are picked just before opening and are then dried in the sun. The color of the clove changes to a dark reddish brown stem with a light reddish brown bud or crown. The yellowish wrinkled cloves are immature and yield a low eugenol content. Defective cloves such as "mother cloves," headless cloves, or fermented cloves and stems are impurities. The ground form is from the clove crown.

Properties: cloves have a spicy, woody, burning, sweet, and musty aroma. Their taste is sharp, bitter, and pungent with a numbing feeling.

Chemical Components: the essential oils from the clove buds are superior to those found in clove stems or leaves. The clove buds have from 5% to 20% essential oil, which is yellowish green in color. It darkens upon aging. It contains mainly eugenol (81%) that gives its characteristic pleasant and sharp burning flavor. The other components are eugenyl acetate (7%), β-caryophyllene (9%), α and β-humulene (1.67%), traces of benzaldehyde and chavicol. Clove stems have 6% essential oil (about 83% eugenol), and leaves have about 2% essential oil (with 80% eugenol). Clove also has 13% tannins (quercitannic acid) and 10% fixed oils. Its oleoresin is brownish green to greenish brown in color; 6 lb. of oleoresin are equivalent to 100 lb. of freshly ground spice.

Clove contains vitamin C, calcium, vitamin A, sodium, manganese, potassium, magnesium, and calcium.

How Prepared and Consumed: clove goes well with sweet, fruity, caramelized, chocolaty, and meaty notes. In the United States, ground cloves or oil are used in baked ham, red cabbage, carrots, pickled beets, sausages, pot roasts, luncheon meats, cured meats, salad dressings, desserts, fruitcakes, and puddings. It contributes to the flavor of Worcestershire sauce and ketchup. It blends well with cinnamon, fish or meat extracts, tamarind, soy sauce, coffee, chocolate, chile peppers, and fruits.

Sri Lankans and North Indians use cloves whole or ground in garam masala, biryanis, meat dishes, and pickles. Clove is also an ingredient in the betel nut mixture (used as a breath freshener) that is chewed after meals.

Europeans like clove in sweet dishes. In England, it is added to apple tarts, pickles, and mincemeat; in France, it is used in quatre epices, soup stocks, and stewed fruits; in Germany, clove is mixed with sauerbraten and breads.

Clove is an important ingredient in Chinese five-spice blend that is used for barbecued meats or pork. In North African cooking, cloves are used in many spice blends, such as *baharat*, *berbere*, *galat dagga*, and *ras-el-hanout*, that are used to flavor meat and rice dishes. The Ethiopians add clove to coffee before roasting. In Indonesia, cloves provide a sweet aroma to local cigarettes called kretek. They are also added to their *rendangs*, *gulais*, and spicy condiments. In Mexico, clove enhances the local moles and other sauces.

Spice Blends: Chinese five-spice, garam masala, curry blends, quatre epices, baharat, berbere, ras-el-hanout, rendang blend, gulai ayam blend, Worcestershire sauce blend, and ketchup blend.

Therapeutic Uses and Folklore: ancient physicians recommended clove as an aid to digestion and to promote circulation. Portuguese women in the East Indies distilled a liquor from cloves for heartaches. In India, it was combined with nutmeg and black pepper to create an aphrodisiac for men. Clove is a pain reliever, so it is used as a mild anesthetic for toothaches. Clove causes the blood vessels near the skin to dilate and bring blood to surface, thus giving a warm feeling to the skin. Clove helps to get rid of intestinal gas and fights nausea. Its leaves are used to soothe skin burns.

Clove has antibacterial and antifungal properties because of its eugenol content.

CORIANDER (SEED, LEAF CILANTRO)

Since ancient times, coriander has been enjoyed by many cultures for its culinary and medicinal values. Coriander is mentioned in Sanskrit literature as far back as 5000 BC and in Greek Eber Papyrus as early as 1550 BC. About 400 BC, Hippocrates, the Greek physician, recommended coriander for its medicinal value. Coriander was found in Egyptian tombs dating from 1090 BC. As early as fourth century BC, the Chinese ate it to attain immortality. The Arabs used it as an aphrodisiac, while the Romans used it as a seasoning.

Coriander in the culinary world includes the fruit (seed), leaf, stem, and root, each having its own distinct flavor. Its name is derived from the Greek word *koris* meaning bedbug, due to their perception of a "buggy" odor from the unripe seeds. Coriander is a versatile ingredient, with its seed, leaf, and root commonly flavoring Latin American, Asian, and Middle Eastern foods. In the Americas, the leaves and stems are called cilantro, while the seeds are called coriander. Latins, Caribbeans, and Asians use a more pungent variety called leafy coriander, culantro, long coriander, or shado beni.

Scientific Name(s): Coriandrum sativum. Long coriander: *Eryngium foetidum.*
Family: Umbelliferae (carrot) or Apiaceae (parsley family).

Origin and Varieties: native to Asia, now grown in India, the Middle East,
Romania, Morocco, Mexico, Argentina, the United States (Kentucky), the Caribbean,
and Southeast Asia.

Common Names: coriander seed is also called dimbilal (Amharic), kuzbara hab
(Arabic), kinj (Armenian), dhoney (Bengali), nan nan zee (Burmese), heung soi
(Cantonese), geshniz (Farsi), coriandre (French), coriander (Danish), koriander (German,
Swedish, Dutch, Norwegian, Hungary), koliandro (Greek), dhane (Gujerati),
zir'ey kusbara (Hebrew), dhania (Hindi, Urdu), coriandolo (Italian), koyendoro
(Japanese), van sui (Khmer), pak hom pam (Laotian), ketumbar (Malay, Indonesian),
hui sui (Mandarin), coentro (Portuguese), koriandr (Russian), kothamallie (Singhalese), coriondro (Spanish), giligilani (Swahili), kothamalli (Tamil, Malayalam), phak
chee met (Thai), kishnish (Turkish), dhania (Urdu), and ngo (Vietnamese).

Coriander leaf—coriander leaf, cilantro, cilantrillo, Mexican parsley, and Chinese parsley. It is also called kuzbara warak (Arabic), nan nan bin (Burmese), heung
choi (Cantonese), persil arabe (French), Indische Petersilie (German), kusbara
(Hebrew), hara dhania (Hindi), koendoro (Japanese), chee van sui (Khmer), daun
ketumbar (Malaysian, Indonesian), yuen sui (Mandarin), kinza (Russian), kottham-
mallie (Sinhalese), cilantro (Spanish), yuan suey (Tagalog), kothammalli elay (Tamil,
Malayalam), phak chee (Thai), and ngo gai (Vietnamese).

Long coriander leaf—thorny coriander, Mexican coriander, Japanese saw leaf
herb, fitweed (Jamaican), Puerto Rican coriander, culantro, recao leaf, shado beni
(in the English-speaking Caribbean), chandon beni (French), langer koriander (German), ketumbar Java (Malaysian, Indonesian), kulantro (Tagalog), pak chi farang
(Thai), and ngo gai (Vietnamese).

Form: it is a globular-shaped seed (or fruit) with the mature form having a
brownish yellow color. It is sold whole or ground. The leaf is used fresh (whole or
chopped) and dried (whole or crushed). The stem and roots are used fresh or dried
and are used whole, chopped, shredded, or minced.

Properties: the many parts of the coriander plant exhibit different sensory characteristics. Coriander seed is spherical to slightly oblong and ribbed. Green unripe
seeds have a haylike strong odor while the ripe seeds have a sweet, spicy, and nutty
flavor with a hint of bitter orange. Coriander seed has tart, cedar, and floral-like
undertones. Its color varies from a brownish yellow to a deeper brown. The ground
Indian variety has sweeter notes and is lighter in color than the Moroccan or
Romanian versions. Moroccan coriander is bigger than Indian coriander, while the
Romanian type is smaller in size.

Coriander seed contains calcium, phosphorus, sodium, and magnesium.

There are many types of coriander leaf (or cilantro), but two types are most
commonly known and used. The flat-leaf, parsleylike variety has a refreshing, soapy,
piney, aniselike, and slightly lemony flavor, with mint and pepper overtones. A
second variety, known as thorny coriander, Japanese saw leaf, and Vietnamese *ngo*,
is also commonly called "*culantro*" or "*recao*" by Latin Americans. It is deep green,
6 to 10 inches long with serrated edges. It has a stronger and more pungent flavor
than the parsleylike cilantro.

Coriander root has stronger and more pungent notes than the leaf or seed.

Chemical Components: depending on regional varieties, coriander seed has 0.2% to 2% essential oil, mainly *d*-linalool (60% to 70%), α-pinenes (6.5% to 10%), β-terpinene (9% to 10%), camphor (5%), limonene (1.7%), ρ-cymene (3.7%), geranyl acetate (2.6%), and geraniol (1.7%). The aldehydes, *trans*-2-tridecenal, and decanal are found in the unripe seeds. The fixed oil is about 13%. Its oleoresin is brownish yellow with 3 lb. of oleoresin equivalent to 100 lb. freshly ground coriander seed.

Coriander leaf or cilantro has 0.1% to 0.2% essential oil, mostly aliphatic aldehydes (benzyl benzoate, cinnamaldehyde, *trans*-2-tridecenel, and decanal), caryophyllene, and eugenol. It has less linalool than the seed. Indian variety has lower linalool than the European variety but has higher ester, linalyl acetate. The *trans*-2-tridecenel and decanal are responsible for the fresh taste of cilantro. The pungent aroma of long coriander is due to the long chain aldehydes.

The fresh leaf contains vitamin A (5200 IU/100 gm), calcium, vitamin C (250 mg/100 gm), and potassium. The dried leaf contains higher amounts of vitamin C, potassium, phosphorus, calcium, magnesium, and iron.

How Prepared and Consumed: the Romans were the first Europeans to introduce coriander as a cooking spice. Coriander pairs well with lentils, beans, onions, potatoes, coconut milk, vegetables, and fish. In the United States, the ground form and oil are used in hot dogs, chilis, pastries, sausages, frankfurters, luncheon meats, confections, stews, cookies, and desserts.

When young, the entire plant is used to make chutneys and sauces in India. The mature, ripe seeds, whole or ground, are used in North American, Mediterranean, North African, Mexican, Indian, and Southeast Asian foods. Mexicans use the whole seeds inside a hard sugar candy called *colaciones*, which is used to make bread pudding. Coriander seed develops a more intense aroma when lightly dry roasted or toasted in oil. Indian and Southeast Asians roast the whole seed before using it in curries, pickles, or snacks.

Coriander seed is also coarsely crushed, like peppercorns, and added to foods. Turks flavor their strong coffee with ground coriander. Coriander is a popular ingredient in Tunisian spice blends, such as baharat and tabil, used in couscous, stews, and salads. Ethiopians use coriander as an ingredient in berbere to season sauces, stews, and dips. Yemenites mix coriander with dried fruits, green chilies, and other spices to make a seasoning called jhoug for dips and condiments. Moroccans use a rub of ground coriander combined with garlic, butter, and paprika on lamb before it is roasted.

Cilantro adds flavor and color to the cuisines of Southeast Asia, Mexico, India, the Caribbean, the United States (Tex-Mex), and North Africa. Cilantro is generally chopped or pureed to add flavor and color to green salsas, chutneys, samosa fillings, taco toppings, or bhel puris (North Indian snacks). It can be used whole or chopped to brighten dishes such as salads, dips, curries, salsas, or soups. It is commonly added to Middle Eastern and Northern African foods such as Egyptian bean puree (bissara), Moroccan tagines and soups (*harira*), and snacks such as *bisteeya* and falafel. Lamb dumplings are filled with ground lamb, pine nuts, and seasoned with coriander leaves, *smen* (clarified butter), and garlic. Whole cilantro is generally used fresh and is added toward the end of cooking to preserve its pungency and bright green color.

In India, the Marathis saute cilantro with mint and add it to chutneys or it is chopped and added as garnish to snacks or dips. It is an essential coloring and flavoring ingredient in a popular seasoning of North India called *chat masala*. The Thais add coriander root and stems to curries to give them an extra edge. In Malaysia, it is pureed with spearmint, green chilies, and other spices to make famous local kormas that accompany rice or local breads.

Cilantro pairs well with chile peppers (fresh or smoked), bell peppers, tomato, onion, nuts, lime or lemon, rice, corn, cumin, ginger, cinnamon, garlic, tortilla, couscous, chicken, and yogurt. Cilantro is also chopped and served with iced tea or coffee.

Culantro is commonly used by Mexicans, Puerto Ricans, and Central and South Americans as a garnish in cooked salsas, seviches, fillings, taco toppings, and soups. In Trinidad and Jamaica, it is a popular seasoning used with fish and meat marinades. Southeast Asians use it as a garnish for soups, noodles, and curries. Vietnamese use *culantro* as a wrap to give fragrance to meats, chicken, and seafood. The leaf has a great affinity to other herbs such as mint, parsley, basil, and lemongrass.

Coriander root and stem are chopped and used in Thai, Puerto Rican, and Asian Indian soup stocks, curries, or *recaitos*. They are indispensable ingredients in Thai green curries and in Vietnamese soups and salads.

Spice Blends: coriander seed—curry powders, chili powder, berbere, jhug/zhoug/sug, baharat, garam masala, and pickle blends.

Cilantro—chat masala/green masala, zhoug, Thai green curry blend, tabil, salsa verde blend, Malaysian kurma blend, ceviche blends, and black bean blend.

Culantro/recao—recados, salsa blend, ceviche blend, recaito, sofrito, chimmichurri, and fish marinade.

Therapeutic Uses and Folklore: coriander seed is an ancient spice eaten by the Chinese, Indians, and Arabs for longevity and as an aphrodisiac. Early Egyptians and European monks used coriander as a medicine for stomach ailments and to perfume cosmetics and flavor liqueurs. The Greeks flavored their wines and tonics with coriander seeds.

Coriander is traditionally used to treat migraines and indigestion, to help purify the blood and to relieve nausea, pain in joints, and rheumatism. It is also considered a mild sedative. The leaves are used to soothe skin burns.

CUMIN AND BLACK CUMIN

Cumin has been a popular spice in India since ancient times. Cumin got its name from the Sanskrit word *sughandan* meaning "good smelling." In Europe, cumin was a very valuable spice, and Romans and English used it to pay their taxes. They also used it to season food and as a digestion aid. The Romans introduced it to Germany, where it was named Roman caraway and is now called *kummel*, identifying it as a foreign variety of caraway. Then the Europeans confused it with caraway and called it Roman, Turkish, Egyptian, or Eastern caraway. Its popularity in Europe slowly disappeared after the Middle Ages, surpassed by caraway.

The ancient Egyptians used it to mummify royalty. It has now become a staple ingredient in South Asian and Latin American cooking.

Black cumin, a smaller variety of cumin with a different flavor, is a popular spice in North Indian, Pakistani, and Iranian foods. Also referred to as Kashmiri cumin in India, it was a popular ingredient in Muglai cooking and was given the name *shahi jeera*, meaning "imperial cumin," by the Moguls. The Moguls, who were Muslims, ruled North India during the sixteenth and seventeenth centuries. Muglai cooking combined Iranian and Middle Eastern ingredients with local preparation methods and influenced much of North Indian cuisine.

Scientific Name(s): cumin: *Cuminium cyminum;* black cumin: *Cuminum nigrum.* Family: Umbelliferae (carrot family) or Apiaceae (parsley family).

Origin and Varieties: it is indigenous to the eastern Mediterranean, Turkey, Egypt, Syria, Indonesia, China, and India. It is now cultivated in Iran, Mexico, the Middle East, and Argentina. Black cumin with origins in North India, Iran, and Egypt, is grown in Kashmir, Iran, and Pakistan.

Common Names: cumin—green cumin, white cumin, and cummin (English). It is also called kemun (Amharic), kamun (Arabic), ziya (Burmese), siew wui heung (Cantonese), spids kommen (Danish), zwarte komijn (Dutch), zireh kuhi (Farsi), cumin/cumin blanc (French), romischer kummel (German), kimino (Greek), jeeru (Gujerati), kamun (Hebrew), jeera (Hindi, Bengali), romai komeny (Hungarian), comino/comino bianco (Italian), kumin (Japanese), ma chin (Khmer), thien khao (Laotian), jintan putih (Malay, Indonesian), jeeragam (Malayalam), hsiao hui hsiang (Mandarin), spisskummen (Norwegian), cominho (Portuguese), kmin (Russian), suduru (Singalese), comino/comino blanco (Spanish), jira (Swahili), spiskummin (Swedish), jiragum (Tamil), met yeeraa (Thai), jeeraka (Telegu), kimyon (Turkish), and jirah (Urdu).

Black cumin—Kashmiri cumin, black caraway, and Roman nutmeg flower. It is also called kamun aswad (Arabic), cumin noir (French), schwarzer kreuzkummel (German), kala/shahi jeera (Hindi), comino nero (Italian), buraku kumin (Japanese), karun jeeragum (Malayalam, Tamil), svartkarve (Norwegian), kaluduru (Singhalese), comino negro (Spanish), svartkummin (Swedish), and kalajeera (Urdu).

Form: a dried, brown, ripe fruit or seed that is available whole or ground. Black cumin is dark brown, very thin, and much smaller fruit, about 3 mm long.

Properties: cumin seed is elongated and oval shaped with a ridged surface. It is light brown, grayish, or greenish yellow in color. Cumin has a nutty, earthy, spicy, and bitter taste with a penetrating, warm, and slightly lemony aroma. Ground, it has a dark brown to a brownish yellow color.

Black cumin seeds are smaller, thinner, and darker, almost black. They taste sweeter than cumin with intense lemony, caraway-like notes. When cumin is toasted in oil or dry roasted, it becomes nutty, with a richer and mellower flavor.

Chemical Components: cumin has 2.5% to 4.5% essential oil (depending on age and regional variations), which is pale to colorless, mainly containing monoterpene aldehydes. Chiefly, it contains cuminic aldehyde (33%), β-pinene (13%), terpinene (29.5%), ρ-cymene (8.5%), ρ-mentha-1,3-dien-7-al (5.6%), cuminyl alcohol (2.8%), and β-farnesene (1.1%). Cumin has 10% fixed oil.

Black cumin has 0.5% to 1.6% essential oil, mainly carvone (45% to 60%), limonene and ρ-cymene.

Oleoresin cumin is brownish to yellowish green in color; 5 lb. are equivalent to 100 lb. of freshly ground spice.

Cumin contains calcium, vitamin A, potassium, sodium, iron, magnesium, and phosphorus.

How Prepared and Consumed: in ancient Rome, cumin was ground and spread on breads and was used as a substitute for black peppercorns. Cumin pairs well with coriander, onions, garlic, ginger, turmeric, chile peppers, potatoes, and oregano. In Europe today, whole seeds are used mainly to flavor Swiss and Dutch cheeses or German sauerkraut, while in France and Germany, ground cumin is used in bread, cakes, pastries, and a liquor called kummel.

When dry roasted, whole cumin develops a more intense aroma and is added to Indian and Sri Lankan curries. It is a component of tandoori and garam masala to enhance chicken and meat dishes. In South India, whole cumin is roasted in oil or ghee, ground and mixed with curry leaves, lentils, and other roasted spices to make a spice blend called *sambar podi*. Cumin has become an essential part of Southeast Asian curries, especially where there is an Indian influence, such as Myanmar (Burma), Malaysia, and Singapore. It is also widely used in Iranian pickles and meat dishes.

Cumin is also a popular ingredient in Mexican, Southeast Asian, Cuban, and Tex-Mex cooking, especially in chili powder, which is used to flavor chili con carne, and enchilada and burrito fillings.

In Mexican and Caribbean cooking, cumin is a popular spice in achiote blends, adobos, black beans, and many tortilla fillings.

Black cumin is popular in North African and Middle Eastern cuisines, such as Yemeni *zhoug* and Saudi Arabian *baharat*. It is used in tagines (meat stews) of North Africa and in lamb roasts, couscous, sausages, vegetables, and ground meat dishes (*kibbeh*) of the Middle East. Iranians use black cumin to enhance many meat and lamb dishes.

In India, black cumin is used whole and roasted and is added to yogurts, chutneys, curries, biryanis, kormas, kebabs, garam masalas, lassis (yogurt drinks), and the breads of North India and Pakistan. It is an essential flavoring of meats and rices in Kashmiri cuisine, which is based on rich creamy sauces.

Spice Blends: garam masala, chili con carne panch phoron, ras el hanout, chili powder, sambar podi, jhoug, baharat, curry powder, and achiote blend.

Therapeutic Uses and Folklore: in ancient times, cumin was a popular flavoring that was also associated with superstition. In Europe during the Middle Ages, the Roman women drank tea spiked with cumin to give a paler complexion, and it was also used as a symbol of love. Greeks associated it with the symbol of greed, and German girls flavored bread with cumin to keep their lovers from straying.

In India, since ancient times, cumin was used for its medicinal properties to aid digestion and to treat dysentery. Egyptians and Indians took it to relieve stress and lower blood pressure. Cumin also stimulates circulation, dispels gas in the abdomen, and relieves cramping.

DILL AND DILLWEED

In the southern Mediterranean region, as early as 3000 BC, dill was popular for its medicinal properties. Dill promoted digestion and soothed the stomachs of Assyrians and later Egyptians, Romans, and Greeks, and eventually northern Europeans. Dill's name comes from the Old Norse word *dilla,* meaning "to soothe" because Norse parents used it to ease the stomach pains of crying babies and to lull them to sleep. Dill was also believed to have magical properties and was valued as an ingredient in love potions and aphrodisiacs.

Dill is still popular in northern Europe as a flavoring for breads, sauces, soups, potatoes, and pickles. In India it is used for treating gastric disorders.

Scientific Name(s): Anethum graveolens: European dill (true dill); *Anethum sowa:* Indian dill. Family: Umbelliferae (carrot famlly) or Apiaceae (parsley family). The East Indian dill is mellower and less intense than the European dill.

Origin and Varieties: dill is native to the southern Mediterranean and southern Russia. It is also cultivated in the United States, England, Poland, Scandinavia, Turkey, northern India, and Japan.

Common Names: European dill, American dill, and Indian dill. It also called insilal (Amharic), shabath (Arabic), samit (Armenian), samin (Burmese), see loh (Cantonese), dill (Danish), dille (Dutch), shivit and sheveed (Farsi), aneth odorant (French), dill (German), anitho (Greek), shamir (Hebrew), suwa/suwa patta (Hindi), adas manis/adas cina (Indonesian/Malaysian), aneto (Italian), diru (Japanese), phak si (Laotian), shih-lo (Mandarin), dill (Norwegian, Swedish), endro (Portuguese), ukrop (Russian), enduru (Sinhalese), eneldo (Spanish), satakuppi sompa (Tamil), pak chee lao (Thai), dereotu (Turkish), and tieu hoi hurong (Vietnamese).

Form: dill seeds are used whole or crushed and are sold as dried whole seeds or ground. The feathery leaves, called dillweed, are used fresh or dried and are sold whole, finely chopped, or ground.

Properties: dill seeds (or fruits) and the leaves possess different flavors. The seed is tiny, oval, and flat with a sweet, mild caraway-like odor and an aniselike taste. Its back notes are slightly sharp and pungent. The European dill seed is brown with a light rim and a curved, oval shape. Indian dill seed is tan or light brown in color with a yellowish edge and a pale ridge. It is longer and narrower than European dill. Its flavor is less intense than European dill, but it has a harsher, caraway-like taste.

The bluish green leaves of the European variety dill weed, taste anise-like and are fresher and milder than the seed. The dried leaves do not have much flavor.

Chemical Components: European dill seeds have 2.5% to 4% essential oil, which is pale yellow in color and consists mainly of carvone (30% to 60%), limonene (33%), and α-phellandrene (20.61%), including ρ-cymene, 3,9-epoxy-p-menth-1-ene, α-pinene, β-phellandrene, and dihydrocarvone.

Indian dill seeds have 1% to 4% essential oil, which is light brown in color. It has high dillapiole (52%) but has less carvone (21%) than the European type. It also has *trans*-dihydrocarvone (16.6%), limonene (6%), dihydrocarvone (17%), and α-pinene (1%).

The European dillweed has much less essential oil than the seed, about 0.35%. The essential oil is pale yellow and contains a high proportion of terpenes but contains less carvone (20% to 40%) than the seed. Principle components are limonene (30% to 40%), phellandrene (10% to 20%), myristicin, dillapiole, and other monoterpenes. Its typical fresh aroma is due to α-phellandrene.

Dill oleoresin is produced mainly from the seed with some from the European dillweed. It is greenish to pale amber in color, and 5 lb. of oleoresin are equivalent to 100 lb. of freshly ground spice.

Dill seeds contain calcium, magnesium, phosphorus, potassium, and vitamin A. Dill leaves contain calcium, iron, potassium, sodium, and folic acid.

How Prepared and Consumed: Europeans (Germans, Scandinavians, eastern and central Europeans) and North Americans use the seeds, leaves, or the oils in pickled cucumbers, dill pickles, sauerkraut, potato salad, breads, processed meats, sausages, seafood, salad dressings, soups, flavored vinegars, and stews. The seeds are ground and added as a seasoning to many dishes. Scandinavians love the flavor of dillseed with boiled fish and lamb and in accompanying creamy sauces.

North Indians use dill seed in braised beans, lentils, and other vegetable dishes. Dillweed is used to flavor green, leafy vegetables, especially spinach. It is an essential flavoring in *dhansak*, a popular meat, lentil, and vegetable dish of the Parsi community in Bombay. In the United States, Indian dill seed is used in sausages, cheeses, condiments, breads, pickles, and salad dressings. Dillweed is used in fish sauces and salad dressings.

Dillweed is a popular spice in the Middle East for rice, meat, bean, and vegetable dishes. Turks use dillweed with spinach, baby lamb, casseroles, bean dips, and yogurt. Pureed beans are mixed with onion, sour cream, lemon juice, black pepper, and dillweed and then eaten as a salad. Greeks stuff grape leaves with rice, garlic, parsley, pine nuts, and dillweed for their dolmadakis. Dillweed is enjoyed by Germans who add it to potato and fish soups with cucumber, onion, and white pepper, and by Russians who add it to pickled vegetables and to flavor sour cream sauce for stuffed cabbage. In Malaysia and Indonesia, dillweed is added with other leafy spices for rice dishes.

Dill seeds pair well with cabbage, onions, bread, vinegar, potatoes, cumin, chili powder, paprika, and turmeric. Dillweed goes well with rice, salads, fish, eggs, mayonnaise, mustard-based sauces, yogurt, sour cream, and mild cheeses. It is usually added toward the end of cooking to retain its flavor.

Spice Blends: dal curry blends, pickling spice, sauerkraut blend, salad dressing blend, pilaf blend, and dolmadakis blend.

Therapeutic Uses and Folklore: Egyptians and Greeks brewed dill seed into a magic potion used against witchcraft. They also used it for treating hiccups and many ailments. Asian Indians use dill seed to relieve stomach pains and hiccups because of its soothing effect on the digestive system. It is an important ingredient in a tonic that is given to babies in England, India, and Southeast Asia to relieve colic pains. Chewing on the seeds clears halitosis, stimulates appetite, and induces sleep.

EPAZOTE

Epazote characterizes the taste of Mayan cuisine in the Yucatan and Guatemala. The name "epazote" comes from the Nahuatl words, *epote,* meaning disagreeable or foul, and epa*tzotl,* meaning "sweat," reflecting its strong aroma. The Swedes call it *citronmalla,* because of its lemony undertones. Another variety is used in the southern United States as a cure for intestinal worms, thus, the origin of epazote's nickname, "wormseed."

Scientific Name(s): Chenopodium ambrosioides. Family: Chenopodiaceae (goosefoot family).

Origin and Varieties: epazote is native to central and southern Mexico and Central America. It is cultivated in the United States, Mexico, Asia, Europe, and Central America.

Common Names: pazote, bean herb, skunkwood, wormseed, sweet pigweed, West Indian goosefoot, Mexican tea, stinkweed, and Jerusalem parsley (English). It is also called mentruz (Brazilian), chau hang (cantonese), mexicanischer traubentee (German), mirhafu (Hungarian), ambrosia (Italian), amenka-xitasou (Japanese), katuayamodakam (Malayalam), chou xing (Mandarin), sitronmelde (Norwegian), erva formiqueira (Portuguese), epazot (Russian), epazote/yerba de Santa Maria (Spanish), citronmalla (Swedish), meksika cayi (Turkish), and cau dau hoi (Vietnamese).

Form: epazote has small serrated leaves which are tender and green when young and turning red and coarse when mature. The young leaves are milder and less bitter in flavor. It is used as fresh or dried forms. Both the leaf and the stem are used in cooking. Even the flowers and unripe seeds are used.

Properties: epazote has a herbaceous, strong, bitter taste with faint lemony notes. It has a strong pungent turpentine odor with camphor and mintlike overtones. The dried form has less flavor than the fresh form.

Chemical Components: epazote contains less than 1% essential oil and has mostly monoterpenes and its derivatives, such as ascaridol (up to 70%), limonene and ρ-cymene with camphor, α-pinene, myrcene, terpinene, thymol, and *trans-iso*-carveol.

Epazote contains calcium, iron, phosphorus, vitamin A, vitamin C, and niacin.

How Prepared and Consumed: Mexicans and Central Americans use epazote fresh in salads, soups, and meats and especially to enhance huitlacoche, mushrooms, bean- and chile-based foods such as refried beans (*frijoles refritos*), frijoles negros, moles, or rice and beans. It is usually added toward the end of cooking to prevent bitterness in the finished product. Spaniards flavor teas with epazote.

Epazote pairs well with cilantro, lime, chipotle peppers, cheeses, pork fat, black beans, pinto beans, cumin, garlic, onion, corn, and squash blossoms. It also goes well in tortilla soups, fillings and toppings, moles, quesadillas, huitlacoche, soups, and stews.

Spice Blends: black bean soup blend, refried bean blend, huitlacoche pate blend, tortilla soup blend, and mole blend.

Therapeutic Uses and Folklore: epazote helps prevent flatulence, and Mexicans use it in many bean dishes. Mexican mothers steep the herb in milk and sugar and serve this tonic to their children to rid them of intestinal parasites. It is also used to treat dysentery.

FENNEL SEED

Fennel seeds resemble anise, cumin, and caraway in flavor, so many cultures use them interchangeably. Called *madhurika* in Sanskrit, meaning sweet spice, fennel is thought to be a variety of anise (which is called *saunf* in Hindi) in India. In fact, the Hindi name for fennel, *moti saunf,* means big, thick anise seed. Indonesians think of fennel as a variety of cumin and call it *jintan manis,* which means "sweet cumin." Arabs confuse it with anise and many in Europe think it as a variety of dill. Native to the Mediterranean, fennel spread to the Far East and far north in Europe

To the ancient Greeks, fennel represented success, and so it was called "marathon" in reference to the battle where the Greeks defeated the Persians in 490 BC. Fennel was also a symbol of success to the Romans. They used all parts of this plant as ingredients in their foods and as medicine to fight diseases and to sharpen their vision. During the thirteenth century in England, fennel was considered a royal spice and was served to the king with fruits, bread, and pickled fish seasoned with fennel seeds.

Scientific Name(s): there are many varieties of fennel, but the two main types are sweet fennel and bitter fennel. The sweet fennel, called *Foeniculum vulgare* var. dulce, is also known as French or Roman fennel. The bitter fennel is called *Foeniculum (F) vulgare* Mill (or *F. officinalle* All). Family: Apiaceae (parsley family).

The Finocchio or Florence (Italian-type) sweet fennel is grown for its bulbous stalk, which is eaten as a vegetable.

Origin and Varieties: sweet fennel is native to the Mediterranean. It is cultivated in France, Italy, Morocco, India, Iran, and the United States. Bitter fennel is grown in central Europe, Russia, Germany, Hungary, Argentina, and the United States. Indian fennel is reported to be a distant variety. The most common variety in the United States is bitter fennel.

Common Names: called sweet cumin and anise, it is also called insilal (Amharic), shamar (Arabic), samit (Armenian), guamoori (Assamese), mouri (Bengali), samong saba (Burmese), wuih heung, hui-hsiang (Cantonese, Mandarin), fennikel (Danish), venkel (Dutch), razianeh (Farsi), fenouil (French), fenchel (German), finokio (Greek), wariari (Gujerati), shumar (Hebrew), moti saunf (Hindi), finocchio (Italian), uikyo (Japanese), phak si (Laotian), jintan manis (Malaysian, Indonesian), perun jeerakam (Malayalam, Tamil), badishep (Marathi), fennikel (Norwegian), fun-cho (Portuguese), saunf (Punjabi), fenkhel (Russian), maduru (Sinhalese), hinojo (Spanish), shamari (Swahili), fankal (Swedish), anis (Tagalog), yira (Thai), pedajirakaramu (Telegu), rezine (Turkish), sonf (Urdu), and cay thi la (Vietnamese).

Form: fennel is available as whole seeds or ground. The green leaf, stalk, and bulb of the Florence fennel are used as garnishes or vegetables in the Mediterranean region.

Properties: the fruits, also referred to as seeds, are oval and ridged. They are bright or pale green to yellowish brown in color. They can be slightly curved or straight. Fennel has a fresh, camphoraceous, aniselike aroma with a slightly bittersweet taste. The Indian fennel is smaller and straighter than the European fennel with a sweeter anise flavor. The Persian fennel is the smallest size with a strong anise taste. The leaves have an aniselike taste with a slight sweetness. The bulbous stalk of Florence fennel is sweet with aniselike notes. The seed becomes spicy and less sweet when it is toasted.

Chemical Components: fennel seed contains 1% to 6% volatile oil and 10% to 20% fixed oil. The bitter fennel contains 50% *trans*-anethole, 10% to 20% fenchone (which contributes to the pungent, camphorous, and bitter flavor), 10% to 30% limonene, 3% to 11% α-phellandrene, 12–16% α-pinene, with α-thujene, β-pinene, estragol (methyl chavicol), myrcene, and 1,8-cineole. The sweeter variety has 50% to 80% anethole, little (5%) or no fenchone, slightly higher levels of limonene with estragole, safrole, and pinene.

The oleoresin is brownish green in color, and 6.5 lb. are equivalent to 100 lb. of freshly ground fennel.

Fennel seeds contain calcium, potassium, magnesium, phosphorus, sodium, vitamin A, and niacin.

How Prepared and Consumed: fennel seed is a popular spice in European and Mediterranean regions. The English use the seeds in soups; the Germans enjoy them in breads, fish, and sauerkraut; the French toss them in fish soups and herbes Provence; Italians add to pasta sauces, sausages, pizzas, meatballs, salami, pepperoni, and sambuca; the Spanish flavor their cakes and other baked goods with it, and the Arabs use them in breads and salads.

In Asia, fennel seeds complement rich fish sauces, roast pork, mutton and lamb curries, sweet and sour dishes, roast duck, and cabbages. They are recommended for oily fish such as mackerel and salmon. Asian Indians roast the seeds in oil to intensify their flavor and then use them whole or ground in curries, spice blends, soups, lentils, vegetables, and breads. Fennel seeds are an essential ingredient in a Bengali spice blend called *panchporon*. Ground fennel is an essential flavoring in egg and fish curries of Kashmir and the stews and fiery curries of Sri Lanka. It is also a popular spice with the Mopla Muslims of Kerala and the Chettiars of South India.

In Southeast Asian curries, fennel adds sweetness and balances the spiciness of other ingredients. The Chinese include ground fennel in their five-spice blend for flavoring roast pork, roast duck, and other braised meat dishes. It balances the spicy smoked flavors.

Chopped, fresh fennel leaves garnish or flavor fish dishes, sauces, salads, stews, and curries.

In Europe, the bulbous root and stalk of Florence fennel are used in soups and sauces, deep-fried or baked. They are chopped and eaten raw drizzled with olive oil or sliced and mixed in salad and stuffing. They are braised or blanched and eaten as a vegetable. It goes well with cheese, butter, and black pepper.

Spice Blends: curry blends, Chinese five-spice, Bengali five-spice (panchporon), mirepoix blend, and herbes de Provence.

Therapeutic Uses and Folklore: ancient Europeans considered fennel a sacred herb. The seeds were taken to ease coughs, to stop hiccups, sharpen eyesight, and cure earaches. Egyptians, Chinese, and Asian Indians used it as a remedy for snakebites and scorpion stings. Asian Indians blend fennel into a spice mixture called *paan masala*, used in betel leaves, which is then chewed after meals to aid digestion and sweeten the breath. Fennel soothes stomach upsets and is an ingredient in "gripe water" that is given to infants to sooth colic. It is also used as a diuretic, to increase bile production, and to relieve gastrointestinal pains.

FENUGREEK SEED AND LEAF

Called in Sanskrit *methika,* fenugreek was cultivated in Egypt as early as 1000 BC. It was prized throughout the Middle East and in India as a flavoring, medicine, and fumigant, while in Europe mainly for medicinal use. The Egyptians and Indians soaked the seeds in water until they swelled and then used them to reduce fevers and aid digestion. By AD 1050, fenugreek had spread as far as China.

The name fenugreek is derived from the Latin word *foenum graecum,* which translates to "Greek hay" (strong haylike aroma of dried leaves); it was never used as a spice in Greek cooking. Today, North Indians and Middle Easterners enjoy fenugreek (both seeds and leaves) in their cooking, but many Westerners do not like its strong bitter taste.

Scientific Name(s): Trigonella foenum-graecum. Family: Leguminosae or Fabaceae (bean family).

Origin and Varieties: native to the eastern Mediterranean, fenugreek is now cultivated in India, Pakistan, France, Morocco, Greece, Lebanon, Germany, Argentina, and the United States.

Common Names: fenugreek was called Greek hay and goat's horn or cow's horn in ancient times because of its horn-shaped seed pods. The seeds and leaves are called by the same name in many countries where the leaves are eaten. Seed is also called abish (Amharic), hulba (Arabic), chaiman (Armenian), mithiguti (Assamese), methi (Bengali, Marathi), penantazi (Burmese), wuih lu bah, hu lu ba (Cantonese, Mandarin), bukkehorns klover (Danish), fenegriek (Dutch), shambelile (Farsi), fenugrek (French), bockshornklee (German), trigonella (Greek), menthro (Gujerati), methi (Hindi, Urdu), hilbeh (Hebrew), kelabat (Indonesian), fieno greco (Italian), koruha (Japanese), halba (Malaysian), venthiam (Malayalam), bukkehornklover (Norwegian), feno grego (Portuguese), menthri (Punjab), pazhitnik grecheskiy (Russian), uluhaa (Singhalese), alholva/fenogreco (Spanish), uwatu (Swahili), bockhornsklover (Swedish), vendayam, meti (Tamil), menti kura (Telegu), meeti (Tibetan), cemen (Turkish), and co cari (Vietnamese).

Fenugreek leaf: methini (Gujerati), saag methi (Hindi, Bengali, Marathi, Urdu), ho lo bah (Vietnamese).

Form: the seeds are enclosed in long sickle-shaped pods. They are light to yellow brown, smooth yet hard, and have a deep diagonal groove that divides each seed. The seeds come whole or ground. The green leaves, called *saag methi* in North India, are sold as fresh whole leaves and as dried whole or crushed leaves.

Properties: the seeds are bitter tasting, but when they are lightly dry roasted, their flavor mellows and some of the bitterness is removed. The aroma of roasted fenugreek is like that of burnt sugar, and its taste is comparable to maple syrup. Once roasted, the seeds are used whole, crushed, or ground. The leaves taste very bitter, slightly resembling lovage with a sweet haylike aroma.

Chemical Components: the seed has very little essential oil, less than 0.02%. Its aroma is due to a compound called 3-hydroxy-4,5-dimethyl-oxolane-2-furanone. It also includes *n*-alkanes, sesquiterpenes lactones and alkanoles. When the seed is toasted, pyrazines (major components) are formed. The fixed oil is 7% to 10%. The nonvolatile components contain trigonelline, choline, sterol and diosgenin derivatives, and furostanol glycosides, which are responsible for its bitter taste.

The oleoresin is dark brown to greenish yellow and is high in proteins, gums, mucilage, and saponin. The fixed oil has linoleic, oleic, and linolenic acids.

The seeds contain fiber (a high amount), folic acid, vitamin A (1040 IU/100 gm), calcium, potassium, phosphorus, iron, magnesium, and sodium. After roasting, the niacin content increases. The leaf contains calcium, iron, carotene, and vitamin C (43 mg/100 g) that decrease with frying, boiling, or steaming, with the last cooking method showing the least loss. Its mucilage content is 40%, which is a concentrated source of dietary fiber.

How Prepared and Consumed: fenugreek is an important spice in India, Egypt, Saudi Arabia, Iran, Armenia, and Turkey. The seeds pair well with fish and lentils and, thus, are commonly used in fish curries and dals in South India and Southeast Asia. Vegetarians in India dry roast the seeds or fry in hot oil and use them whole or ground to flavor curries, sambars, fermented breads, and chutneys. In India, fenugreek seed is an essential ingredient in many pickles and spice blends, such as *sambar podi* in the south, and *panchporon* in Bengal. In South India, the seeds are added to rice flour and lentil flour to make fermented flat breads such as *dosai* and *idli*. It is also roasted and ground and mixed with dried, ground chile peppers and other spices to be used as a dry-based seasoning dip for local breads.

In Iran, the seeds and the leaves are used. The seeds are toasted and added to salads to provide crunchiness. In the Middle East, fenugreek seeds are ground into a paste and rubbed on salted meat, which is then dried. This salted, spiced beef is called *pastourma* in Iraq and *aboukht* in Turkey. In Yemen, fenugreek is mixed with coriander leaf, tomato, chili paste, garlic, and other spices to make a hot dip called *hilbeh*, which is used as a spread for breads. Yemenites also use fenugreek in a seasoning called *zhug*, which is added as a topping to stews. Armenians use it with garlic and chile pepper in a spice called *chemen* to spice up a beef dish called *bastirma*. The Greeks boil fenugreek seeds and eat them with honey.

Ethiopians add fenugreek to a spice mixture, *berbere*, for seasoning meats, seafood, and vegetables, while Egyptians use them in sweets and flatbreads. In many regions of Africa, they are soaked until they swell up and then are used as legumes. In the United States, fenugreek seeds are used in soups, baked goods, icings, and meat seasonings. They are the main ingredients in artificial maple syrup. Because the seeds have high mucilage (40%), they can be used as a natural stabilizer in processed foods.

Fresh fenugreek leaves and young shoots are used as cooked vegetables by many cultures. The fresh leaves are enjoyed in many Iranian traditional dishes with parsley, mint, and other spices. In North India, the leaves are cooked with garlic, potatoes, and other root vegetables or used in breads such as methi naan. It goes well with fish, meats, and chicken curries of North India. The dried leaves are used as a seasoning in many dishes.

Spice Blends: fish curry blend, sambar podi, berbere, chemen, zhug, panch phoron, hilbeh, and aboukht blend.

Therapeutic Uses and Folklore: Egyptians used fenugreek paste on the body to reduce fever and to cure chapped lips and mouth ulcers. They also used fenugreek along with other spices in the embalming process. Romans ate fenugreek seeds because they were considered an aphrodisiac. The Greeks and Romans, not realizing its potential as a spice, used fenugreek as cattle fodder to increase cow's milk flow and to restore nitrogen to the soil.

In Indonesia, fenugreek is used to promote hair growth. In India, Southeast Asia, and Ethiopia, the seeds are soaked in water (the seeds swell because they are coated with mucilage) and are then drained and eaten to aid digestion, as a laxative, to treat bronchitis, and to cure sore throats. It is used to prevent sharp rises in blood sugar and to lower cholesterol.

In Ethiopia, nursing mothers take fenugreek to promote production of milk. In India, women take fenugreek to help with menstrual cramps and difficult childbirths and to promote lactation after the baby is born.

The leaves are believed to stimulate bile secretion and to aid in the digestion of fatty meat.

GALANGAL/GALANGALE/GALINGALE

Galangal is an aromatic, peppery, gingerlike spice. First used in Chinese and Southeast Asian cuisines, galangal has traveled around the world and is found in Egyptian, Indian, and a few European applications. There are three types of galangal: the greater galangal, the lesser galangal, and the kaempferia galangal. Greater galangal, called *liang xiang* in Mandarin (which means "mild ginger") or *lengkuas* in Malay, commonly used in many dishes of Southeast Asia, is more popular globally than the other varieties. Lesser galangal, called *kencur* in Malaysian, and kaempferia galangal, called *temu kunci* in Indonesian (referring to the many elongated fingers of the rhizome) are used as seasonings in many Southeast Asian pungent curry pastes and sambals.

Scientific Name(s): Alpinia galanga (greater galangal), *Alpinia officinarum* (lesser galangal), Alpinia rotunda (kaempferia galangal). Family: Zingiberaceae (ginger family).

Origin and Varieties: greater galangal is indigenous to Southeast Asia and southern China and is cultivated in Indochina, Thailand, Malaysia, and Indonesia. The lesser galangal is indigenous to Indonesia and South India and today cultivated in Southeast Asia and China. Kaempferia galangal has its origins in southern China and is cultivated in Thailand, Indonesia, Singapore, and other parts of Southeast Asia.

Common Names: greater galangal: galangale, Java galangal, galanga, Siamese or Thai ginger, mild ginger. Also called kholanjan (Arabic), kulinjan (Bengali), leung keung, liang chiang (Cantonese, Mandarin), galganga (Danish), galgant (Dutch, Russian), djus rishe (Farsi), souchet long (French), galangawurzel (German), galanki (Greek), kolinjan (Gujerati), galangal (Hebrew), kulinjan (Hindi), galanga (Hungarian), laos (Indonesian), galanga (Italian), nankyo (Japanese), rom deng (Khmer), kha tad eng (Laotian), lengkuas (Malay), aratta (Malayalam), koshtkulingan (Marathi), gengibre do Laos (Portuguese), galang (Spanish), galangarot (Swedish), palla

(Tagalog), arattai (Tamil), khaa (Thai), kachoramu (Telegu), sga skya (Tibetan), galangal (Turkish), kulanjam (Urdu), and rieng nep (Vietnamese).

Lesser galangal: Chinese ginger. Also called lille galangal (Danish), sian noih, shan nai (Cantonese, Mandarin), kentjoer (Dutch), galangal camphre (French), kleiner galgant (German), kineszike gao liang chiang (Mandarin), piperoriza (Greek), abuyu campa (Hindi), ban ukon (Japanese), van hom (Laotian), kachola (Malayalam), kencur, cekur Jawa (Malaysian, Indonesian), maraba (Russian), dusol (Tagalog), pro hom (Thai), and cam di la (Vietnamese).

Kaempferia galangal: aromatic ginger, finger root, Chinese ginger. Also called suo shi/lap seuh jeung, ao chian jiang (Cantonese, Mandarin), temoe koentji (Dutch), Chinesischer ingwer (German), temu kunci (Indonesian), kunci (Malaysian), gazutu (Japanese), kchiey (Khmer), kasai (Laotian), Chinese key (Singaporean), krachai (Thai), and ngai num kho (Vietnamese). In Indonesia, it is sometimes confused with lesser galangal and called kencur. Zedoary (called white turmeric), which has a very bitter taste, is sometimes confused with kaempferia galangal.

Form: the rhizome comes whole (fresh, frozen, or canned) and dried (whole, ground, sliced, or pickled). The slices of dried galangal need to be soaked in boiling water for 20 to 30 minutes before use. Its flowers are also eaten fresh. The lesser galangal is smaller in size, like ginger. Kaempferia galangal has a central smaller globular shape with ten or more slender, long tubes sprouting in the same direction from the central core, like fingers on a hand, and, hence, it is often called fingerroot.

Properties: fresh greater galangal has a knobby look with an inner yellowish brown to pale brownish skin that has reddish brown rings. The fresh form has different flavor profile from the dried form. The fresh galangal is aromatic, spicy, peppery, gingerlike and has a slightly sour note. The interior which has similar color as the exterior, is hard and woody in texture. Dried galangal is spicier with a cinnamon-like taste. Young rhizomes are pink in color and are more tender and flavorful.

The lesser galangal has a reddish brown skin with a soft white interior. It is crunchy and more pungent, with a gingery, cardamom-like taste and hints of eucalyptus notes. The dried version does not possess the same flavor intensity as the fresh form.

Kaempferia galangal has a pale reddish brown or yellowish brown skin, and the inside is watery and soft. Its taste is similar to lesser galangal, but it has a more camphorous and medicinal-like taste.

Chemical Components: greater galangal has 0.5% to 1.5% essential oil, with fresh galangal having mainly 1,8-cineol, α-pinene, eugenol, methyl cinnamate, farnesene, bisabolene, camphor, and d-pinene. The dried version has a different composition than the fresh, with lesser aroma components (farnesene and 1,8-cineol). Its pungent taste is due to galangol or alpinol, which are phenyl alkanones, and diaryl heptanoids.

Lesser galangal has 2.5% to 4% essential oil, with ethyl cinnamate, 1-para-methoxy cinnamate, 1,8 cineol, and β-phellandrene, the levels depending on its origins.

Kaempferia galangal has about 1% to 3% essential oil, mainly ethyl cinnamate, 1,8 cineol, camphor, ethyl-*p*-methoxycinnmate, *p*-methoxycinnamic acid, and 3-carene-5-one.

How Prepared and Consumed: greater galangal is used abundantly in Malaysia, Thailand, Vietnam, Singapore, Cambodia, and Indonesia. By itself, galangal's texture is woody and its flavor is undesirable, but when it is added and cooked with other ingredients, it enhances the overall flavor profiles of applications.

Greater galangal pairs well with coconut, garlic, chile peppers, kaffir lime leaves, turmeric, fish sauce, tamarind, and shallots. In Malaysia and Singapore, the Malays and Nonyas (female descendants of Chinese traders and local Malay women) use it abundantly with lemongrass, garlic, scallions, and tamarind for laksas, soups, and curries. In Thailand, it is used in pungent curry pastes, meat marinades, soups, and stir-fries. Indonesians use it in their fiery hot *rendangs* and popular rice dishes, such as *nasi goreng* and *nasi padang*. Thais and Indonesians use more galangal than ginger and enjoy its flowers when they are fried or pickled.

Lesser galangal is generally less popular than greater galangal. It is freshly grated or sliced and added to satay sauces and fiery, pungent spice pastes of Indonesia that season meats, vegetables, and fish. It is also combined with chile peppers, ginger, turmeric, and other spices to create bumbu, a seasoning used to spike up sauces and soups in Indonesia. In Bali, lesser galangal is used in a spice paste called *jangkap* with lemongrass, chile peppers, ginger, nuts, and other spices. This spice paste is rubbed on duck that is then wrapped in banana leaf and baked or roasted. It is also commonly used by the Indonesians in Netherlands.

Kaempferia galangal is used in liqueurs, bitters, and beers of Russia and Scandinavia. It is popular in Thailand, where it is grated for use in meat curry pastes, soups, and fish curries with chile peppers, kaffir lime leaf, and coconut. It is also used in Indonesia and in pungent Malay-style curries, sambals, and condiments.

Spice Blends: Thai red curry paste blend, bumbu blend, rendang blend, laksa blends, Thai stir-fry blend, Thai five-spice blend, Nonya curry blend, tom kha blend, and jangkap blend.

Therapeutic Uses and Folklore: Asians use galangal for indigestion, respiratory problems and to stop nausea.

GARLIC

One of the world's most popular spices, garlic is used extensively from China to the Americas, in French aioli, Turkish cacik, Vietnamese *pho bo*, Indian korma, or Greek *skordalia*. Garlic's name is derived from the Anglo-Saxon word *garleac,* meaning "spear plant." Since ancient times, garlic has been used as a cure as well as a food. Egyptians used garlic since 3700 BC to provide strength and prevent disease. Jews ate it on their long journeys, Romans honored garlic for providing strength and courage, and Greeks used it to treat colds and coughs. Many cultures call it "white onion," including Malaysians, Indonesians, Sri Lankans and Ethiopians.

Scientific Name(s): Allium sativum (softneck); *Allium ophioscorodon* (hardneck). Family: Alliaceae (onion family).

Origin and Varieties: garlic is indigenous to central Asia, and it was brought to the Mediterranean regions, Egypt, France, Spain, and Italy. It is now cultivated in the United States, Asia, Europe, the U.K., Taiwan, South America, Mexico, and Hungary. There are about 200 varieties of garlic, with different colors, sizes, shapes, and flavors, but only two cultivated varieties exist—hardneck and softneck varieties. The hardneck variety (Siberian, Spanish roja, Prussian white, Italian rocombole, German white, German red, Marino Xian, Persian star, and Romanian red) is superior in flavor but is harder to grow and has smaller yields and a shorter shelf life. Softneck varieties have smaller cloves that are harder to peel. Approximately 90% of all garlic sold in the United States is of the softneck variety from California. Other softneck varieties are French white, Creole Red, Burgundy, Inchelium, and Chinese.

Elephant garlic is not really garlic, but is a member of the leek family. It is mild and does not have the pungency of regular garlic.

Common Names: garlic is also called netch shinkurt (Amharic), thum (Arabic), sekhdor (Armenian), naharu (Assamese), rasun (Bengali), chyethonphew (Burmese), suen tauh, suan tou (Cantonese, Mandarin), huidlog (Danish), knoflook (Dutch), sir (Farsi), ail (French), knoblauch (German), skordo (Greek), shum (Hebrew), lasun (Hindi, Marathi), fokhogima (Hungarian), aglio (Italian), ninniku (Japanese), phak thien (Laotian), bawang putih (Malaysian, Indonesian), velluthulli (Malayalam), alho (Portuguese), chesnok (Russian), sudulunu (Singhalese), ajo (Spanish), kitunga saum (Swahili), vitlok (Swedish, Norwegian), velai pundoo (Tamil), kra tiem (Thai), vellulli (Telegu), goghpa (Tibetan), sarmisak (Turkish), lehsun (Urdu), and cay tai (Vietnamese).

Form: each garlic bulb contains plump and succulent egg-shaped bulblets called cloves, within an outer skin that is white, buff, rose, or purple, depending on the variety. Garlic comes fresh or dried. Fresh, it is a firm whole clove that is also available crushed, sliced, minced, chopped, roasted, or as juice. Dried garlic comes powdered, granulated, flaked, diced, ground, minced, chopped, sliced, or added to salt.

Properties: the flavor of freshly cut garlic ranges from mild and sweet to strongly pungent, depending on the variety. Some can be pungent at first but become milder during cooking. Hardneck varieties are pungent and strong flavored, that usually lingers after cooking. They have a wider range of tastes than softneck varieties. German White has a strong pungency with heat that is experienced at the back of the throat. Prussian White has a pleasant aftertaste and Spanish Roja has a lingering taste with a pungency on the tip of the tongue. Softneck varieties can be harsh to mildly sweet to pungent and sometimes contain sharp bites. Mild softneck garlic can become more pungent during storage.

Whole garlic is odorless when intact but gives a strong aroma when cut or bruised. When fresh garlic is cut or bruised, the enzyme allinase acts on alliin in garlic to produce allicin. This breaks down to diallyl disulfide which gives the penetrating sulfur-type aroma. The sharp bite typically subsides after cooking or roasting. Cooking softens the flavor, while roasting gives garlic a well-balanced, delicate, nutty flavor. Dried garlic (mainly from the white-skinned variety) has a very strong, persistent aroma and taste. Roasted garlic has a slightly sweet and delicate flavor.

Chemical Components: garlic has 0.1% to 0.25% essential oil, formed enzymatically when cloves are crushed, cut, or rehydrated. It consists of sulfur compounds, 60% diallyl disulfide, 20% diallyl trisulfide, 6% allyl propyl disulfide, and diallyl sulfide. Allicin (diallyl disulfide) contributes to its strong aroma.

Oil of garlic (undiluted form) has 200 times the strength of dried garlic and 900 times the strength of fresh garlic. The oils come in many dilutions for easier blending and handling in applications. Oleoresin garlic is brownish yellow and contains 5% of the volatile oil of garlic; 8 lb. of oleoresin are equivalent to 100 lb. of dehydrated garlic, and 2 lb. are equivalent to 100 lb. of fresh garlic. Garlic salt consists of garlic powder, salt, and an anticaking agent (tricalcium phosphate or starch).

Raw garlic contains calcium, phosphorus, vitamin C, and potassium. Dried garlic has higher potassium, phosphorus, and magnesium levels but has a lower calcium level and negligible or no vitamin C.

How Prepared and Consumed: garlic is widely used in French, Italian, Spanish, Middle Eastern, Asian Latin American, and American cooking. It is fried in oil or gently simmered before onions or other spices are added to create a wonderful base for stir-fries, curries, soups, or sauces. Garlic pairs well with onions, ginger, basil, turmeric, greens, beans, spinach, chicken, pork, and seafood. It rounds up and modifies other flavors, such as tomato, chilis, onions, and ginger.

Garlic is an essential component of many spice blends that enhance the flavor of soups, stews, and curries in Asian cooking. For Chinese and Indians, garlic is a must with onions and or ginger in curries, stir-fries, pickles, or BBQs. Chinese and Thais pickle garlic in vinegar for use in noodle dishes, roast pork, and chicken. Thais fry whole garlic in oil and add this to create enhanced flavors and textures in many dishes. Vietnamese add to spring rolls, marinades, and noodle soups, or phos.

Mediterraneans roast garlic for many dishes, such as cassoulet, beef bourguignonne, sofritos and *gambas alajillas*. Raw garlic is added to Greek *skordalia* (with potatoes) and *tzatziki* with souvlaki or hummus with sesame paste. For Italians, it is an integral flavoring along with olive oil, basil, and tomatoes in their pasta sauces (pestos, vegetables, soups and sprinkled over pizzas for added flavor). Latin Americans and Caribbeans add dried garlic to meat marinades, many salsas, condiments, and seasonings such as adobos, mojos, rouille, sofritos, jerk paste, chimmichurri, and soups and stews. Cuban mojos or Puerto Rican adobos will not exist without garlic. Mexicans use garlic in their famous mole sauces and condiments.

Spice Blends: tabil, rouille, aioli, refogado blend, adobos, sofritos, moles, hummus, ketchup, mojo, tzatziki, and chimmichurri.

Therapeutic Uses and Folklore: called the "stinking rose" by the English, garlic has been used since ancient times to drive away evil spirits, kill intestinal parasites, and stimulate blood flow for a healthy face and body. Known in Russia as "Russian penicillin," it is one of the world's oldest medicines. In the sixteenth century BC, the Egyptians listed twenty-two remedies containing garlic, using garlic to treat everything from heart disease and tumors to insect bites. The Greeks gave it to the condemned to purify them of their acts. While wealthy Romans did not eat garlic, they gave it to their soldiers to make them strong.

In Asia, Hindu priests and the strict Buddhists eliminated garlic from their diets because of its reputation as a sexual stimulant.

Garlic's curative powers have been recorded since the eighteenth century. It has been reported to stimulate the digestive system, thereby helping poor digestion, lower blood pressure and help blood circulation by reducing the amount of fat in the blood and thinning it, thereby enabling platelets to move more freely.

Recent research has focused on garlic's role in preventing heart disease, decreasing cholesterol and blood pressure, enhancing the immune system, preventing cancer, and enhancing memory.

Since ancient times, garlic has been used as an antibiotic and as a fungicide.

GINGER

Asia's most treasured spice, ginger, derives its name from the Sanskrit word *shringavera,* meaning "shaped like a deer's antlers." Used by Indians since 5000 years ago and by Chinese since the sixth century BC, it is highly esteemed throughout Asia for its therapeutic effects as well as for its flavor. Ginger is most noted for its soothing effect on the stomach, as it has been a popular traditional cure in Ayurveda and Chinese Traditional Medicine for common stomach ailments, nausea, and motion sickness. Indispensable in Indian, Chinese, and Southeast Asian cooking, ginger's spicy-sweet flavors complement and enhance many of their curries, stir-fries, marinades, and soups.

Arab traders brought ginger to the Mediterranean region from Asia before the first century AD for use by Romans. The Greeks made gingerbread, which later gave rise to gingersnaps, cakes, and biscuits. In the sixteenth century, Spanish explorers grew ginger in Jamaica and exported it to Europe.

Scientific Name(s): Zingiber officinale. Family: Zingiberaceae (ginger family).

Origin and Varieties: there are many varieties of ginger—Indian (Cochin, Calicut), Chinese, Jamaican and African. Indigenous to southern India and Southeast Asia, ginger is also grown in Hawaii, Fiji, Sierra Leone, Nigeria, Jamaica, Japan, Mexico, Sri Lanka, and Australia. The names for fresh and dried ginger differ in many regions of the world.

Common Names: Fresh and dried ginger are called by different names in Asia. Root ginger, green ginger, gingerroot, stem ginger, and black ginger. It is also called zinzibil (Amharic), zanjabeel (Arabic), gojabheg (Armenian), ada-fresh (Bengali, Assamese), ginh-fresh (Burmese), sang geung-fresh, geung-dried (Cantonese), ingefer (Danish, Norwegian), gember (Dutch), jamveel (Farsi), gingembre (French), ingwer (German), piperoriza (Greek), adhu-fresh, sunth-dried (Gujerati), zang'vil (Hebrew), kai (Hmong), adrak (fresh), sauth (dried) (Hindi), gyomber (Hungarian), zen-zero (Italian), shoga (Japanese), khnthey (Khmer), saenggang (Korean), khing (Laotian), halia (Malaysian, Indonesian), inchi (Malayalam), sheng jiang-fresh, gan jiang dried (Mandarin), ahle-fresh, sunth-dried (Marathi), gengibre (Portuguese), imbir (Russian), inguru (Singhalese), jengibre (Spanish), tanga wizi (Swahili), inge-fara (Swedish), luya (Tagalog), ingi/ellam-fresh, sunthi-dried (Tamil), khing (Thai), elamu-fresh, sonthi-dried (Telegu), gamug (Tibetan), zenefel (Turkish), adraka-fresh (Urdu), and sin gung-fresh, gung-dried (Vietnamese).

Form: ginger is a rhizome (an underground stem that looks like a thick root) that is available as a spice in many forms: fresh, dried, black, or white. Fresh ginger is preserved and crystallized. The fresh form is knobby and branched, firm and tan colored, and comes whole (unpeeled), sliced, julienned, chopped, crushed, or grated. Young or immature fresh ginger is juicy, less pungent, and more delicate in flavor with pink shoots, a thin skin, and a crispy texture. The mature fresh ginger has a tough, shriveled skin with a desirable flesh underneath. Some types of ginger are peeled before sale.

Dried ginger is used bruised, sliced, or powdered. Black ginger is created by scalding or steaming the rhizome and then drying it. White ginger is made by peeling the outer layers before washing and drying it. Preserved ginger is made from fresh young or green rhizomes that are peeled and sliced and then cooked in heavy syrup or brine. Crystallized ginger is preserved ginger, which is fresh ginger that is peeled, cooked in sugar syrup, dried, and rolled in sugar. Pickled ginger is fresh, peeled gingerroot that is thinly sliced and pickled in a vinegar solution, which gives it a pink color.

The ginger leaves are used to flavor some Asian dishes. The pink ginger bud (or torch ginger bud), called *bunga kantan* in Malaysian lends a floral and aromatic sensation to laksas or noodles in fragrant spicy broths and curries. This torch ginger bud is from the Etlingera elatior family.

Properties: ginger's flavor and color vary with its origin and harvesting, storage, and processing conditions. Fresh ginger has a juicy, spicy, refreshing and slightly sweet, lemonlike aroma, along with a strong bite. It is more aromatic than dried ginger. Drying conditions change ginger's flavor and pungency. Fresh ginger is firm and plump, not shriveled, and its color varies from tan to pale brown. Dried ginger is fibrous and has less pungency than fresh ginger. Aged or older ginger is fibrous, tough, and harsher in taste, whereas young ginger is tender and delicate in taste. Longer cooking time tends to increase fresh ginger's pungency and decrease its aroma.

Jamaican ginger is light tan in color and has a delicate aroma. It is more pungent than African ginger, which is darker in color and has a weak aroma and a harsh flavor. Cochin and Calicut (South Indian) gingers are strongly aromatic and pungent with a lemon aroma. They are considered the best gingers. Japanese and Chinese ginger are weak in pungency and aroma. Chinese ginger is whiter in color, more fibrous, slightly bitter and is usually used in the preserved form (in sugar syrup or candied).

Chemical Components: the essential oil (mainly sesquiterpenes) from ginger is pale yellow in color and ranges from 1% to 4%, depending upon the variety. The oil gives ginger its characteristic aroma but not its bite. The chief aromatic constituents are zingiberene (70% in fresh and 20% to 30% in dried), curcumene, α-pinene, sabinene, limonene, borneol, linalool, farnesene, and citral. Ginger's bite or pungency is due to nonvolatiles—gingerol, zingerone, shogaol, and paradol. Fresh ginger (depending on the variety) is the most pungent because of high levels of gingerol. Dried ginger or fresh ginger stored for a while has less pungency because it has less gingerol, because of its conversion to milder shogaol, zingerone, and paradol contents.

Oleoresin ginger is dark brownish green in color and is extracted from dried ginger. It contains 20% to 30% volatile oil, 10% fixed oil, and 50% to 70% pungent resinous constituents (mainly gingerol) that are responsible for its pungency.

Fresh ginger contains magnesium and potassium, whereas ground ginger contains calcium, potassium, phosphorus, magnesium, sodium, niacin, vitamin A, and manganese.

How Prepared and Consumed: ginger can add flavor, round off flavor, and enhance particular flavors. While ginger is used throughout the world, its primary popularity as a savory spice is in Asian countries such as India, Korea, China, and Southeast Asia. In Asia, it is eaten fresh, dried, pickled, preserved in syrup, and crystallized. Fresh ginger is peeled before use. Many Asians, especially the Chinese and Indians, like to freshly grate it into their meat, poultry, vegetable, and fish dishes.

A must in Southeast Asian cooking, fresh ginger is used raw and grated or finely chopped for flavoring curries, condiments, and marinades, as well as tenderizing and preserving meats. East Asians and Southeast Asians like young ginger for its intense flavor. The Chinese use the fresh, pickled, and preserved gingers to create spicy, sweet flavors in rice porridges (congee), soups, and stir-fried meats and vegetables. Dried ginger is used in seasoning blends such as five-spice and for flavoring sauces and soups that require long simmering times.

In India and Sri Lanka, dried and fresh gingers are used frequently. Fresh ginger is fried or boiled for sauces and curries. Dried ginger is part of curry blends, stir-fries, snacks, teas, and desserts. It has an affinity with garlic, spring onions, fish, soy sauce, chile peppers, oranges, and vinegar. Ground ginger is popular in hot and cold beverages such as tea and ginger beer as well as baked goods and confections. Indian vegetarians use ginger abundantly to flavor many dishes.

The young green ginger, either pickled or crystallized, is popular in the Caribbean, Asia, Australia, and Europe. Pickled ginger, sweet or hot, is commonly eaten as a snack in Southeast Asia and China. The Japanese serve thinly sliced pink or red pickled ginger (young ginger) as a garnish or as a side dish called *gari* to refresh the palate in between meals. It also accompanies their sushi dishes.

Scandinavians still enjoy their gingerbread, ginger cakes, and confections made with ginger. Dried ginger is commonly found in Mediterranean spice blends such as quatre epices, *berbere*, *la kamut*, and *ras-el hanout*. Arabs use great amounts of the dried ginger in many of their seasonings. Caribbeans use ginger in marinades, curries, stews, ginger beer, and ginger ale.

Europeans and North Americans traditionally prefer the dried, crystallized, or preserved form of ginger as a table seasoning in cakes, puddings, cookies, gingerbread, pies, preserves, pickles, fish soups, marmalades, ice creams, sausages, frankfurters, puddings, and teas. Today, fresh ginger has become more popular in the United States and Europe to flavor teas, sodas (ginger ale), bakery products, and confectionary, and for Asian style applications.

The fragrant ginger flower called *bunga kantan* in Malaysia provides a unique flavor to Malay and Nonya (term given to female ancestry from Chinese traders and local Malay women) curries and laksas.

Spice Blends: curry blends, berbere, five-spice, quatre epices, ras-el-hanout, stir-fry blends, congee blend, laksa blend, ginger beer, and ginger ale.

Therapeutic Uses and Folklore: ginger is one of the oldest and more popular medicinal spices. It has a warming effect on stomachs, soothes digestion, and is traditionally used to neutralize toxins, decrease acidity, and increase blood circulation in the gastrointestinal tract. The Greeks prescribed ginger for stomach ailments and as an antidote to poisons.

The Chinese and Indians use it in a variety of remedies and consider it to be effective against motion sickness and nausea, to discharge phlegm, and even to relieve pains during childbirth. The Chinese chewed on it to prevent motion sickness during their sea travels. Indians used ginger as a kidney and bowel cleanser and for blood circulation. In Asia, it is also brewed in teas to ease menstrual cramps, bloating, and morning sickness.

Today, ginger ale is still a common home remedy for an upset stomach. Recent studies have shown ginger to be a strong antioxidant to prevent cancers, to prevent or enhance gastrointestinal disorders, as an antitumor (skin cancer) agent, and to decrease cholesterol. It has also been shown to be an antioxidant for lipid-based foods, and its gingerol content is an effective antimicrobial agent against E. coli and B. subtilis.

GRAINS OF PARADISE

Also called Melegueta or Malagueta pepper, guinea grain, or guinea pepper, grains of paradise were prized as a spice and as a substitute for black pepper in Europe during the Middle Ages. Grains of paradise were combined with ginger and cinnamon to flavor wine called *hippocras*. Currently, they are not widely used around the world, except in northern Africa and in the western regions of Africa, such as Ghana, Benin, and Nigeria. The Yorubas in West Africa call them *atare*.

Scientific Name(s): Aframomum (A) melegueta or *A. grana paradisi.* Family: Zingiberaceae (ginger family).

Origin and Varieties: grains of paradise are indigenous to the West African coast, along the Gulf of Guinea, from Congo to Sierra Leone. The grains of paradise were traded along the West African coastal area and thus this region came to be known as the "pepper coast" or "Grain Coast." Grains of paradise are cultivated in the Ivory Coast, Sierra Leone, Guinea, and Ghana, but mostly in Ghana. Grains of paradise are related to cardamom. They are grown on a small scale in India and the Caribbean, where they were brought in the 1800s on slave ships from West Africa. Sometimes this spice is confused with mbonga spice.

Common Names: Guinea grains, Malagueta pepper, Alligator pepper. They are also called kewrerima (Amharic), hubub al janel, gawz as Sudan (Arabic), paradijs korrels (dutch), graines de paradis/poivre de Guinee/malaguette (French), paradieskorner (German), piperi melenketa (greek), grani de Meleguetta (Italian), rajskiye zyorna (Russian) and malagueta, grano de parai (Spanish), and idrifil (Turkish).

Form: grains of paradise are small, dried, reddish brown, pyramid-shaped seeds with a shiny exterior, enclosed in a large pod. They are sold with or without the enclosed seedpod or fruit (60–100 seeds in a pod). The seeds are sold whole or ground. The ground seeds are greyish in color.

Properties: the seeds have a whitish interior and a reddish brown exterior. They have a pungent, peppery taste with bitter notes and a faint odor that is similar to cardamom. They have less pungency than black pepper.

Chemical Components: grains of paradise contain 0.5% to 1% essential oil, which is yellow in color and contains mostly humulene and caryophyllene. Their mild pungency is due to paradol, shogoal, and gingerol.

How Prepared and Consumed: grains of paradise go well with black pepper, ginger, cinnamon, clove, coriander, and cumin. The seeds are usually ground and added toward the end of cooking in order to retain their aroma. Northern Africans use them to create aromatic stews and also add them with other spices to flavor their coffees. They are an important ingredient in the Moroccan spice blend *ras-el-hanout*, and are added to tagines, lamb dishes, couscous, and rice. In Tunisia, grains of paradise are used in a blend called *galat dagga* (or Tunisian five-spice), to flavor lamb and vegetable dishes, especially eggplant dishes. In West Africa, they are used to spice many vegetables such as pumpkin, potatoes, eggplant, or okra.

In Europe, grains of paradise were used traditionally to spice wine and beer but are now used in vinegars and liqueurs.

Spice Blends: ras-el-hanout, galat dagga, couscous blend, and Nigerian vegetable stew blend.

Therapeutic Uses and Folklore: in West Africa, grains of paradise were used as aphrodisiacs and to relieve pain and cure fevers. They were also chewed on cold days to warm up the body.

HORSERADISH

A relative of turnip, cabbage, and mustard, horseradish, called *peberrod* or *pepperot*, meaning "pepper root" in northern European languages, is a popular seasoning in central and northern European cuisines. The German name for horseradish, *meerettich,* means "more radish" and reflects horseradish's much stronger pungency than the ordinary radish. Japanese refer to it as *seiyo* wasabi or western wasabi, and wasabi daikon or radish wasabi.

Scientific Name(s): Armoracia rusticana. Family: Brassicaceae (cabbage family).

Origin and Varieties: horseradish is indigenous to eastern and northern Europe and the Mediterranean and is also cultivated in central Europe.

Wasabi is Japanese horseradish, which has a slightly stronger taste than horseradish.

Common Names: it also called fejel har (Arabic), laht gahn, la ghen (Cantonese, Mandarin), peberrod (Danish), mierikwortel (Dutch), torob (Farsi), piparjuri (Finnish), raifort (French), meerettich, kren (German), hazeret (Hebrew), torma (Hungarian), barbaforte, rafano (Italian), seiyo wasabi, wasabi daikon, hosuradishu (Japanese), pepperrot (Norwegian), raiz-forte (Portuguese), khrjen (Russian), ra'bano picante (Spanish), mronge (Swahili), pepparrot (Swedish), kamungay (Tagalog), and yaban turbu (Turkish).

Form: horseradish is a root with a brown outer skin and a fleshy white interior. Fresh forms are sold sliced, grated, or shredded. Dried horseradish comes flaked, granulated, or powdered. It is also pickled and canned. The leaves are broad and green.

Properties: the root by itself does not give much aroma, but releases a pungent, burning, mustardlike aroma upon grating, cutting, or shredding, which quickly disappears. Horseradish has a sharp, acrid, burning, bitter taste.

Chemical Components: horseradish has about 0.2% to 1.0% essential oil, mainly sinigrin, sinigrin-derived allyl isothiocyanate, diallyl sulfide, phenyl propyl thiocyanate, and phenethyl. Myrosinase enzyme acts on sinigrin to give allyl isothiocyanate, which gives horseradish its burning taste.

Horseradish has high vitamin C (302 mg/100 gm) content.

How Prepared and Consumed: in Europe, the U.K., and the United States, horseradish is popular in cocktail sauces, prepared mustards, relishes, cured ham, baked beans, salad dressings, roast meats, and seafood. The leaves are used in salads and stews. The British use horseradish as a condiment that enhances the flavor of roast beef. In Austria, it is used with ham and grated apples as a relish. When mixed with ketchup, horseradish becomes a cocktail sauce or dip for seafood, such as shrimp and oysters. It pairs well with tomato, eggs, vinegar, mustard, cheese, soy sauce, scallions, and garlic. It accompanies gefilte at Passover meals.

Since the allyl isothiocyanate is not heat resistant, horseradish is usually best served with cold dishes or added after cooking and before serving. Vinegar and other forms of acids are added with horseradish to prevent the pungency from disappearing quickly.

Spice Blends: relish blend, cocktail sauce blend, salad dressing blend, and mustard horseradish blend.

Therapeutic Uses and Folklore: horseradish was traditionally used by Europeans to treat gout, kidney stones, asthma, and bladder infections. It was used in Europe to prevent scurvy before vitamin C was discovered. Grated horseradish mixed into a paste is a home remedy for chest congestion and stiff muscles because it brings blood to the surface of the skin and warms the skin.

Horseradish has antioxidant and antimicrobial properties.

JUNIPER

Juniper characterizes gin flavor and comes from the colder regions of Europe and North India. In ancient times, the berries, bark, root, and other parts of the plant were used for medicinal purposes. It is a popular seasoning for venison dishes and German sauerkraut.

Scientific Name(s): Juniperus communis. Family: Cupressaceae (cupress family).

Origin and Varieties: juniper is indigenous to the U.K. and also cultivated in Italy, the United States, and India.

Common Names: araar (Arabic), ardug (Armenian), junipero (Danish), jeneverbes (Dutch), sarv kuhi (Farsi), genièvre (French), wacholder (German), arkevthos (Greek), ar'ar (Hebrew), borokha (Hungarian), seiyo suzu (Japanese), junipero (Portuguese), junipero, enebro (Spanish), enbar (Swedish), araar (Hindi), ginepro (Italian), einer (Norwegian), mozhzhevelnik (Russian), mrteni (Swahili), abhal (Urdu), ardic (Turkish), and cay boch xu (Vietnamese).

Form: juniper is a bluish black, small, pea-shaped, ripe berry. The berry has three seeds inside. It is dried and used whole, but generally crushed lightly or coarsely ground before use.

Properties: juniper's flavor is woody and astringent with sweet, lemon and pinelike overtones. It has a ginlike aroma with a slightly bitter aftertaste. Its flavor is released when it is lightly crushed.

Chemical Components: juniper generally has 0.5% to 2% essential oil, with sometimes as much as 2.5%. It is mainly made up of monoterpenes such as 80% α- and β-pinene, thujene, and sabinene, 5% α-terpineol, borneol, geraniol, and camphene with traces of sesquiterpenes such as caryophyllene.

It contains calcium, vitamin A, potassium, phosphorus, and magnesium.

How Prepared and Consumed: juniper is commonly used in England and Europe to flavor pork and remove the strong, gamey flavor from roasted or barbecued wild meat, such as venison, squab, and wild boar. It is also used in liver pates, stuffing, and German sauerkraut, with caraway and bay leaves. Juniper is added to cheeses, goose, pot roasts, pickled meats, and seafood. It pairs well with black pepper, bay leaf, vinegar, cabbages, and alcohol for marinades.

Spice Blends: bouquet garni, barbeque blends, sauerkraut blend, and pickling spice blend.

Therapeutic Uses and Folklore: in ancient Europe, juniper was used to treat tapeworms, keep away evil spirits, and clean the air. Long known as a diuretic, it was used to treat urinogenital tract infections and gonorrhea. Juniper was also used for gastrointestinal infections, gout, and rheumatic pains. It is a home remedy for hangovers.

KAFFIR LIME (LEAF AND FRUIT)

Kaffir lime is a very popular spice used in Southeast Asian foods, especially in Thai, Malaysian, Cambodian, and Balinese cooking. The leaves and the rind of the fruit are similar in taste except the leaves are more intense with the zesty pungent citrusy notes which Southeast Asian cultures enjoy.

Scientific Name(s): Citrus hystrix. Family: Rutaceae (citrus family).

Origin and Varieties: indigenous to Southeast Asia.

Common Names: Indonesian lime and wild lime. Also called lemo (Balinese), shaunk nhu (Burmese), Danish (kaffir lime), kaffir limon (Dutch), syun kahm, suan gan (Cantonese, Mandarin), limettier herisse (French), kaffernlimette (German), aley kafir (Hebrew), mauw nas (Hmong), kaffer citrom (Hungarian), kobumikan (Japanese), kraunch soeuth (Khmer), kok mak khi hout (Laotian), limau purut (Malaysian, Indonesian), lima Kaffir (Spanish), kafir lime (Swahili), nardanga (Tamil), makrut (Thai), and truc (Vietnamese).

Kaffir lime leaf is also called daun lemo (Balinese), daun limau perut (Malay), daun jeruk purut (Indonesian), hojas de lima Kaffir (Spanish), bai makrut (Thai), and chanh sac truc (Vietnamese).

Form: Kaffir lime fruit is a dark green, knobby, pear-shaped, wrinkled fruit. The fresh rind of the fruit is sliced or grated for use in cooking. It can be used as whole pieces, chopped, or julienned. Unlike the lemon, Kaffir lime fruit has virtually no juice. Kaffir lime leaves have a shiny green surface and an appearance of two leaves attached together on a stem. They are sold fresh, frozen, or dried.

Properties: both the fresh and dried leaves have a strong, pungent, floral, and lemonlike aroma. The fruit also has a lemonlike flavor with intense bitter overtones. The dried fruit pieces are reconstituted and added to foods.

Chemical Components: S-citronellal is responsible for its distinct aroma. The fruit has limonene, β-pinene, and citral. The leaf has higher essential oil (about 80%) with mainly citronellol. It also contains citral, nerol, and limonene.

How Prepared and Consumed: Kaffir lime leaf characterizes Thai cooking and is used abundantly in soups, such as *tom yom goong* (hot and sour soup) and *tom kha gai* (coconut soup). It is also used with stir-fries, grilled fish or chicken, spicy salads, sauces, and pork curries. The leaves are finely shredded and added to salads, while the rind is grated and added to sauces and curries. Kaffir lime pairs well with coconut, fish broth, squid, galangal, garlic, ginger, chile peppers, basil, and mango. In Indonesia, Malaysia, and western Cambodia, the grated fruit is used in fish and chicken dishes, while the leaf is finely shredded or added whole to minced fish and cooked sauces.

Spice Blends: tom yom goong blend, tom kha gai blend, yellow curry, and spicy Thai salad blend.

KARI (OR CURRY) LEAF

Kari (or curry) leaf (called *kariveppilai* in Tamil) is grown all over India and has been used for centuries in South India and Sri Lanka as a flavoring for curries, chutneys, vegetables, and beverages. South Indian traders introduced it into Malaysia, Burma, and Singapore. When the British were in India, they called it curry leaf, naming it after the seasoned sauce (called kari in Tamil) that it was added to.

Scientific Name(s): Murraya koenigii. Family: Rutaceae (citrus family).

Origin and Varieties: indigenous to India and cultivated all over India, including the Himalayas, Sri Lanka, Southeast Asia, and the United States (California and Florida).

Common Names: curry leaf. It is also called barsunga (Bengali), pindosin (Burmese), gai leu yiph (Cantonese), karry blad (Danish), kerriebladerer (Dutch), feuilles de curry (French), curryblatter (German), kari patta, meetha neem (Hindi), aley kari (Hebrew), curry levelek (Hungarian), fogli di cari (Italian), daun kari (Indonesian/Malaysian), kore rihu (Japanese), karibue (Kannada), khibe (Laotinan), kareapela (Malayalam), kadhi limbu (Marathi), karriblader (Norwegian), folhas de caril (Portuguese), bowala (Punjabi), listya karri (Russian), karapincha (Singhalese), hojas de curry (Spanish), bizari (Swahili), bignay (Tagalog), kariveppilai (Tamil), karepeku (Telegu), bai karee (Thai), and la cari (Vietnamese).

Form: kari leaf is very fragrant when used fresh, but it loses its flavor intensity when dried. The fresh or dried leaf is used whole, crushed, and chopped.

Properties: the fresh leaf has a spicy, strong piney-lemony aroma, and a slightly tangerine peel-like taste.

Chemical Components: the essential oils vary based on different varieties. The fresh leaf has about 0.5% to 2.5% essential oil, mostly monoterpenes. There is a gradual decrease in volatile content with advancing maturity so fresh leaves have more volatiles than the older leaves. The essential oils mainly consist of sabinene (9% to 34%), α-pinene (5% to 27%) and dipentene (6% to 16%) with β-caryophyllene (8% to 20%), β-gurjunene, β-elemene, β-phellandrene, limonene, β thujene, and bisbolene.

Curry leaf has a good amount of vitamin A (beta-carotene is 12,600 IU/100 gm), with calcium (810 mg/100 gm), phosphorus (600 mg/100 gm), iron (3.1 mg/100gm), vitamin C (4 mg/100gm), and fiber (6.1%). It also has high levels of oxalates (1.35%).

How Prepared and Consumed: it is an essential spice in South Indian, Sri Lankan, and Malaysian curries, dals, samosas, dosai fillings, chutneys, snacks, sambars, soups, breads, and vegetables. Kari leaf is popularly used in South Indian vegetarian and fish dishes and Sri Lankan meat and chicken curries. Kari leaves par well with mustard seeds, turmeric, ghee, cumin, coriander, fenugreek, dals, ginger, garlic, tomatoes, and yogurt. It provides a certain zest to yogurt-based salad dressings and vegetable dishes, such as fried cabbage, lentils, beans, okra, or eggplant. It is usually removed before the food is eaten.

Kari leaf also provides a distinct spicy flavor to cold dishes and buttermilk. It gives a more intense flavor and crunchiness when it is toasted in oil or ghee, and this mixture is then added to many vegetarian foods. Sometimes it is toasted, ground, or crushed to season or garnish soups, sambars, and curries.

Kari leaf can be kept frozen or refrigerated in a plastic bag for about two weeks. Freezing better retains its flavor, but its color changes to black. To retain its fresh flavor, it is best not to remove the leaves from its branches until ready to use.

Spice Blends: curry blends, sambar podi, rasam podi, chutney blends, and fish curry blends.

Therapeutic Uses and Folklore: the leaves, root, and bark are used as medicinal aids in India. The leaves are used to help blood circulation and menstrual problems. The fresh leaves are taken to cure dysentery, and an infusion made of roasted leaves stops vomiting. It is also recommended for relieving kidney pains. Recent studies have shown that it has a hypoglycemic action, thereby a possible treatment for diabetes, as well as found to prevent formation of free radicals. It is shown to prevent rancidity of ghee (or clarified butter).

LAVENDER

Used in ancient times by the Phoenicians, Egyptians, and Arabs as a perfume and for mummification, and later, by Romans and Greeks to scent bath water. Lavender got its name from the Latin word '*lavare*,' meaning 'to wash.' As part of wedding bouquets, it sends a message of good luck to the bride and groom. It is a favorite with the French who not only use its oil for perfumery and cures but for culinary purposes as well. A member of the mint family, its dried flowers make up the potpourri used to mask unpleasant odors and for perfuming many household items. Though it is used mainly in the pharmaceutical industry and in a few culinary applications such as in sweet-based products and herbes Provence. There is great potential in using its flavor in many savory and sweet applications including soups, sauces, dressings, desserts, confectionary, and ice creams.

Scientific Name(s): Lavendula (L) augustifolia, L. stoechas (French lavender), L.latifolia (spike lavender), L. dentata, L viridis, L. multifada (fern leafed lavender), L. vera, L. officinalis, and many hybrid lavenders such as Lavandin. *Family: Lamiaceae (mint family).*

Origin and varieties: originated in western Mediterranean region and cultivated today in France and Mediterranean regions, Hungary, Bulgaria, North Africa, India, and Saudi Arabia. About thirty other varieties are present and names can become confusing because of climatic and growing conditions.

Common Names: khuzama (Arabic), hoosam (Armenian), lavandula (Bulgaria), fan yoi chou (Cantonese), xeun yi cao (Mandarin), lavendel (Danish), lavendel (Dutch), ostukhudus (Farsi), tupsupaallaventeli (Finnish), lavande, lavando (French, Provencal), lavendel (German), levanta (Greek), lavender (Hebrew), levendula (Hungarian), lavanda (Italian), rabenda (Japanese), rabandin (Korean), lavandel (Norwegian), alfazema (Portuguese), lavanda (Russian), lavanda (Spanish), lavendel (Swedish), lawendeot (Thai), lavanta cicegi (Turkish), and hoa oi huong (Vietnamese).

Form: The plant has grey-green spikes of leaves and purplish flowers which are sold dried or as oils. The leaves are long, narrow, and spiky while the flowers are tiny and tubular and surround the stem in clusters. The dried flowers are used more often in cooking than the dried leaves. Fresh flowers are used to create a visual appeal for many dishes.

Properties: The flowers have a strong perfumy odor with crusty woody undertones acamphorlike notes. The leaves have herbaceous and more pronounced bitter notes.

Chemical Components: the flowers are rich in flavones and anthocyanins, with 1% to 3% essential oil, mainly linalyl acetate (about 35% to 55%), linalool, β-ocimene, 1,8 cineol, terpineol, coumarin, β-caryophyllene, and camphor. Rosmarinic acid and chlorogenic acid are common in the leaves. Lavandin has higher camphor and 1,8 cineol which provides its harsh notes while its content of geraniol, nerol, citronellal, neryl acetate, and geranyl acetate gives a sweet rose-like odor. Angustifolia, which has mainly linalool (25% to 37%) and linalyl acetate (25% to 46%), has a perfumy odor with less harsh notes (again subspecies differ here).

How Prepared and Consumed: lavender is not commonly used as a spice but the French enjoy it in many dishes such as aioli, bouillabiase, ratatouille, meat dishes, and soups. Flowers are candied and also flavors fruit salads, sweets, and desserts. Lavender pairs well with strawberries, mango, guava, blueberries, vanilla, coocnut, white wine, citrus, rosemary, mints, basil, clove, cinnamon, shallots, chives, fennel, star anise, rose, or saffron. It is used to flavor cakes, biscotti, ice cream, syrups, liquers, preserves, lamb, cookies, and steamed or broiled fish. Lavandin and lavender oil is used to flavor baked products, frozen ice cream, confectionary, puddings, beverages, and teas.

Spice Blends: herbes Provence, curry powder, ice cream blend, milk shake blend, kesari blend, and vinaigrette blend.

Therapeutic and Folklore: believed to ward off insects, its essential oil has been since ancient times used externally as a sedative and to relieve stress and headaches. The French used lavender oil to dispel worms and heal wounds. They let lamb graze in lavender fields so the meat becomes tender and fragrant. The

English take lavender tea to ease migraines. Recent research has led to its use in relieving rheumatic pain and for treating skin irritations.

LEMON BALM

Lemon balm, called *bosem* in Hebrew or *citronelle* in French, originated in the Middle East. While it is a popular spice in Mediterranean cooking, lemon balm is valued more for its medicinal properties, for stomach ailments or nervous disorders, than as a seasoning.

The Latin word for lemon balm, *melisa* (which means bee) came from the Greek practice of planting this spice to feed bees.

Scientific Name(s): Melissa officinalis. Family: Lamiaceae (mint family).

Origin and Varieties: lemon balm is indigenous to Greece and the Middle East and is cultivated in Middle Asia, North Africa, Hungary, Egypt, Italy, and the United States.

Bergamot (Monarda didyma) or bee balm, a relative of lemon balm, and which has harsher notes than lemon balm, is used to flavor teas.

Common Names: balm mint, sweet balm, common balm, and true balm. It is also called habak, turijan (Arabic), heung fung chao, xiang feng cao (Cantonese, Mandarin), citronemellise (Danish), bijenkruid (Dutch), badrang buye (Farsi), citronelle (French), zitronenmelisse (German), melissa (Greek), bosem (Hebrew), mezfu (Hungarian), melissa, citronella (Italian), seiyo yamahakka (Japanese), sitronmelisse (Norwegian), erva cidreira (Portuguese), limonnik (Russian), balsamita maior, balsamo limon, toronjil (Spanish), citronmeliss (Swedish), and ogulota (Turkish).

Form: the fresh and dried lemon balm leaves are used whole or chopped.

Properties: lemon balm leaf is light green with jagged edges. It has a sweet delicate lemon aroma and a slightly minty taste.

Chemical Components: it has about 0.1% essential oil, mainly citronellal, β-caryophyllene, geranial, including neral, citronellol, linalool, and limonene.

How Prepared and Consumed: lemon balm is used as a garnish or is added to salads to give a lemony flavor. It goes well in teas, vinegars, stewed fruits, jellies, puddings, and custards. It is also added to fish, poultry, eggs, salads, and soups. Lemon balm pairs well with fresh fruits, so it is a common flavoring in fruit drinks, iced teas, and fruit-based desserts. It can be made into a pesto or a sauce with olive oil, garlic, fruits, and almonds.

Lemon balm is sometimes used to substitute for lemongrass and to intensify a lemony aroma in a dish that contains lemon juice.

Spice Blends: pesto blend, lamb stew blend, and barbecue blends.

Therapeutic Uses and Folklore: since ancient times, lemon balm was used to heal wounds, sores, and bee and wasp stings. Called the "elixir of life" in Europe, it was mixed into a drink to ensure longevity and to treat asthma, stomach ailments, indigestion, menstrual cramps, and fevers.

Lemon balm has also been used traditionally to promote emotional well-being by relieving tension, helping depressive illnesses, calming nerves, and preventing headaches.

LEMON VERBENA

Lemon verbena is a popular spice for European teas, fruit drinks, and desserts. It is indigenous to South America and was brought to Europe by the Spaniards. Verbena, a Latin word meaning leafy branch, was originally used as a perfume. Though not widely used in culinary applications, it adds a refreshing taste to many cold dishes.

Scientific Name(s): formerly *Lippia citriodora,* now called *Aloysia triphylla.* Family: Verbenaceae (verbena family).

Origin and Varieties: it is indigenous to Central and South America and is cultivated in Latin America, France, China, and Algeria. There are many species of Lippia in Latin America and Africa such as Mexican oregano and Ethiopian *koseret.*

Common Names: lippia, cedronella, erba, ning mang mabinchou, meng ma bincao (Cantonese, Mandarin), jernut (Danish), citroen verbena (Dutch), limou (Farsi), verbena odorosacitronelle, verveine odorante (French), zitronen verbena (German), verbena (Greek), verbena, lipia limonit (Hebrew), citro verbena (Hungarian), cedrina (Italian), verbena limonnay (Russian), and yerba de la princesa, vervena, verbena limon (Spanish).

Form: lemon verbena is a long, pointed, green leaf available fresh or dried. It is used whole or chopped. The dried form retains its flavor well.

Properties: fresh lemon verbena has a strong, lemon-lime-like flavor, with a fruity and penetrating aroma.

Chemical Components: lemon verbena contains 0.1% to 0.2% essential oil, which is yellowish green. It is chiefly composed of aldehydes-citral, neral, and geranial. Others are monoterpenes-cineol, dipentene, limonene, linalool, borneol, geraniol, and nerol.

How Prepared and Consumed: it is commonly used by Europeans to flavor fruit-based drinks, fruit salad dressings, fish soups, marinades, puddings, jams, and desserts. It does not tend to lose its flavor during cooking. It pairs well with fruits, vanilla, and seafood dishes.

Spice Blends: fish marinade blend, soup blend, and pickle blend.

Therapeutic Uses and Folklore: lemon verbena has been used traditionally by Europeans as a diuretic and a gout remedy, to treat inflammation of the liver or spleen, and even to aid depression. It is also brewed in tea as a home remedy to relieve colds and fevers. Lemon verbena is a natural insect repellent.

LEMONGRASS

Lemongrass derives its name from its characteristic scent and appearance. It is widely used in Southeast Asian (including Thai, Malaysian, Singaporean, Vietnamese, Cambodian, Indonesian as well as Sri Lankan) cooking – soups, curries, sauces, and desserts.

Scientific Name(s): Cymbopogon citrates: Southeast Asian, West Indian; *Cymbopogon flexuosus*: Sri Lankan, Thai, Burmese/Myanmar. Family: Poaceae (grass family).

Origin and Varieties: lemongrass is indigenous to Southeast Asia and South Asia. It is cultivated in Southeast Asia, Sri Lanka, India, Australia, the Caribbean, Central America, Africa, and the United States. There are many species, but the two main varieties are the South Asian and Southeast Asian types. The latter is also cultivated in the West Indies.

Common Names: Indian verbena, ginger grass, Java citronella, and Cochin grass. It is also called hashisha al limon (Arabic), zabalin (Burmese), chou geung, ng mao chao (Cantonese), citrongras (Danish), citroengras, sereh (Dutch), sitruunaruoho (Finnish), lemongrass, verveinedes Indes (French), zitronengras (German), lemonochorto (Greek), essep limon (Hebrew), sera (Hindi), tawng dubh (Hmong), sereh (Indonesian), citronella, erbe de limon (Italian), remonsou (Japanese), bai mak nao (Khmer), bai mak nao, si khai (Laotian), serai (Malaysian), xiang mao cao, chao jiang (Mandarin), erva cedeira, capim-santo (Portuguese), limonnoe sorgo (Russian), sera (Singhalese), zacate de limon, herbe de citron, hierbe de limon (Spanish), tanglad (Tagalog), servi-pillu, karpurapul (Tamil), ta khrai (Thai), limon ortu (Turkish), and xa, sa chanch (Vietnamese).

Form: lemongrass has a long stalk (or stem), about 4 to 6 inches long, with pale green, slender, long leaves at the top and a bulbous base that is woody and fibrous. The hard, fibrous outer sheaths of the lemongrass stalk are peeled and discarded, and the delicate inner stalk is used. It comes fresh, frozen, or dried. The fresh form is whole, sliced, finely chopped, or coarsely pureed, while the dried form is whole, shredded, or ground. The fresh stem is bruised or macerated to release its flavor. The dried forms need to be soaked in hot water before use; otherwise, they are woody and fibrous. The ground form is added directly into cooking.

Properties: its flavor is found in the lower tender part of the stem, about 4 inches from root end to 3 inches from the upper leaves. The root-end stalk gives the most flavor. The whole fresh stalk becomes aromatic when it is crushed or cut. It gives a refreshing lemon-lime-like taste with a tinge of mint and ginger. It has a delicate citral flavor with floral-like (rose) and a delicate fresh, grassy aroma. The dried form has very little aroma or taste.

Chemical Components: the essential oils (terpenes, alcohols, and aldehydes) in lemongrass vary with the variety, from 0.2% to 0.5%. *C. citratus* has 0.3% to 0.4% essential oil with 70% citral, others being citronellol, geraniol, limonene, linalool, geranial, and neral. *C. flexuosus* has 0.2% to 0.5% essential oil with 80% to 85% citral. The other components are nerol, limonene, β-caryophyllene, myrcene (14%), geranyl acetate (3%), methyl heptenone (2%), and linalool (1%).

The essential oil has a pale yellowish to a brownish yellow color, depending on the variety.

Lemongrass contains calcium, vitamin A, iron, potassium, magnesium, phosphorus, sodium, and manganese.

How Prepared and Consumed: lemongrass enhances many ingredients and does not dominate the flavor profile of a dish. It pairs well with garlic, galangal, shallots, cilantro, turmeric, kaffir lime leaves, candlenuts, ginger, chicken, pork, fish, and chile peppers. Lemongrass is used whole in soups or is chopped and pounded for use in soups, stews, curries, *laksas*, *rendangs*, and condiments of Thailand, Indonesia, Vietnam, and Malaysia. The ubiquitous *sambals* of Southeast Asia contain blended lemongrass as an essential ingredient with chilies, garlic, ginger, and shallots.

Lemongrass is an important flavoring ingredient in Southeast Asian spice blends, especially in Nonya-style dishes of Malaysia and Singapore as well as in Thai,

Vietnamese, and Indonesian cooking. Malay and Thai curry pastes and Nonya laksas frequently contain lemongrass. It provides a wonderful aroma to *nasi lemak*, a coconut-milk-based rice dish from Malaysia, enhances Vietnamese soups and meat marinades, and balances the fiery chilies and galangal in Indonesian *bumbu* pastes. In Southeast Asia, lemongrass is added to many local desserts that have rice flour, *gula melaka* (dark palm sugar), and coconut milk.

The stalk is sliced into small rings and added to sauces and curries, or the stalk is cut into pieces, crushed with the back of a knife or spoon, and added to flavor soups or stews. The whole lemongrass stalk can be tied into a knot and added to soups or stews to give them an aromatic, fresh, citral flavor. It is discarded before serving. When cooked, lemongrass imparts a fresh floral, citrus, taste to foods. In Malaysia, whole lemongrass is made into a skewer or a brush to add marinade to meat and chicken satays (barbecues). In Indonesia, it is combined with turmeric, chile peppers, and other spices to make *bumbu*, a seasoning used to flavor the local sauces and soups.

Spice Blends: sambal blends, nasi lemak blend, rendang blend, satay sauce blend, bumbu blend, assam fish blend, and Malay chicken curry blend.

Therapeutic Uses and Folklore: in Southeast Asia, lemongrass is traditionally used to improve blood circulation and ease muscle pain. It is also used in teas to cure loss of appetite and reduce fever. In Thailand, lemongrass is used as a smelling salt and to relieve headaches.

LOVAGE

Lovage belongs to the parsley family, and its seeds, leaves, and roots are commonly used in Europe for flavoring foods and beverages and for their medicinal properties. The Romans, who introduced lovage to Europe, used it widely in their cooking as well as to reduce fevers and treat stomach ailments. Germans called it *maggikraut* because its aroma reminded them of maggi cubes (meaty yeast extracts). Today it is popular in South and Central European cuisines.

Scientific Name(s): Levisticum officinale. Family: Umbelliferae or Apiaceae (parsley family).

Origin and Varieties: true lovage is native to Southern Europe but cultivated in western Asia, Germany, Italy, France, Czechoslovakia, Hungary, and the United States. There are two other types of lovage that grow wild. One variety, called sea lovage, Scottish lovage, or shunis, grows in northern Britain and along the north Atlantic coast of the United States. The other type, called black lovage or alexanders, grows in Britain and around the Mediterranean.

Common Names: love parsley, garden lovage, Italian lovage, true lovage, maggi herb, and old English lovage. It is also called habak (Arabic), yuhn yih dong gwai, yahn ye dang gui (Cantonese, Mandarin), lovstikke (Danish), lavas, magi plant (Dutch), anjodan romi (Farsi), liveche (French), maggikraut/liebstockl (German), levistiko (Greek), levistico (Italian), robezzi (Japanese), monari (Korean), haulopstikke (Norwegian), levistico (Portuguese), ljubistok (Russian), ligustico (Spanish), and libsticka (Swedish).

Form: lovage leaves, seeds, stems, and rhizomes are used in foods, but the leaves are the most commonly used flavoring. Lovage has green, serrated leaves and hollow stems that are sold fresh, dried, frozen, or crystallized. The leaves, which resemble celery leaves, can be used whole or chopped. The younger leaves are smaller in size.

The seeds (which resemble ajowan seeds) are tiny, ridged, crescent-shaped, brown, and aromatic. The roots are slightly thick and fleshy with a greyish brown color.

Properties: the fresh leaves have a sharp, yeast-like and musky taste with a lemon and celery-like aroma. The dried leaves have a stronger flavor than the fresh leaf.

Chemical Components: the fresh leaf has 0.5% to 1% essential oil, while the dried leaf has 0.2% to 0.5% essential oil, which is yellow amber to greenish color. It consists mainly of phthalides (ligustilide, butylphthalide, sedanolide) with lesser amounts of α-terpineol, eugenol, and carvacrol.

How Prepared and Consumed: ancient Greeks and Romans commonly used the seeds, leaves, and roots in their cooking. Today, lovage is a favorite flavoring in Britain and southeastern Europe. It is eaten cooked or raw. The leaves are used in soups, stocks, flavored vinegars, pickles, stews, and salads. In Italy, lovage is used with oregano and garlic for tomato sauces. The seeds are sprinkled over salads and mashed potatoes and are crushed for breads, pastries, biscuits, and cheeses. The stems and stalks are chopped for use in sauces and stews, while the crystallized leaves and stems are used for decorating cakes. The roots are peeled to remove the bitter skin and are then used as a vegetable or are pickled.

Spice Blends: tomato sauce blend, soup blend, stew blend, and stock blend.

Therapeutic Uses and Folklore: Europeans traditionally use lovage as a digestive stimulant, for stomach upsets, water retention, and skin problems. It was also taken to treat poor circulation and menstrual irregularities.

MACE

Called "flower of nutmeg" in France, Sweden, or Germany, mace was first introduced into Europe during the eleventh century by Arab traders to flavor beer. Mace's popularity as a spice reached its height during the seventeenth century when the Dutch monopolized the nutmeg trade in the Spice Islands. Mace and nutmeg are from the same fruit. Mace is the lacy covering of the seed, which is the nutmeg (with a ratio by weight of 1:25). They have different color and flavor profiles.

Scientific Name(s): Myristica fragrans. Family: Myristicaceae (nutmeg family).

Origin and Varieties: mace is native to the Banda Islands in eastern Indonesia (Moluccas) and is cultivated in the Banda Islands, Grenada, the Caribbean, South India, Sri Lanka, Malaysia, Sumatra, and Brazil.

Common Names: fuljan, basbasa (Arabic), meshgengouz (Armeinan), dao kao syuh, dau kuo shu (Cantonese, Mandarin), muskat blome, (Danish, Norwegian), foelie (Dutch), jouz hendi (Farsi), fleur de muscade (French), muskatblute (German), moschokarido (Greek), jaipatri (Gujerati), meyz (Hebrew), javitri (Hindi), szerecsendio (Hungarian), sekar pala (Indonesian), macis (Italian), mesu (Japanese), kembang pala (Malaysian), jathipathri (Malayalam), jaypatri (Marathi), muskatnot (Norwegian), macis (Portuguese), sushonaya shelukha (Russian), wasa vas

(Singhalese), macia/ maci (Spanish), muskotblomma (Swedish), jattikai, athipalam (Tamil), jatikayi (Telegu), dok chand (Thai), besbasi (Turkish), and dau khau (Vietnamese).

Form: mace is the deep red, lacy, netlike, leathery covering of the nutmeg seed called aril or mace blades. These blades are removed and dried in the sun, which then become brittle and assume a red color. Mace is used whole, broken, or ground. It retains its flavor better than nutmeg when it is ground. There are different grades of mace based on regional variations.

Properties: mace blades are smooth and shiny and can be up to about 4 cm long. Mace can be pale orange, yellowish brown, orange, or reddish brown in color. The color and flavor of mace varies depending on its origins—reddish orange from Indonesia and brownish yellow from the West Indies. It is spicy and bitter with clovelike and piney overtones. Its aroma is terpeny. Mace is more aromatic than nutmeg but has more bitter notes. Ground mace is lighter in color than ground nutmeg.

Chemical Components: depending on its origins, mace has 7% to 14% essential oil and about 30% fixed oil. It contains the same aroma compounds as nutmeg but in different amounts, mainly monoterpenes (87.5%), monoterpene alcohols (5.5%), and other aromatics (7%). Like nutmeg essential oil, the main constituents of mace essential oil are sabinene, α-pinene, myrcene, limonene, 1,8-cineole, terpinen-4-ol, myristicin, γ-terpinene, and safrole. Mace oil is more expensive than nutmeg oil.

The mace oleoresin is amber to dark red in color; 7 lb. of mace oleoresin are equivalent to 100 lb. of freshly ground spice. Mace butter, which has fixed oils and volatiles, has 60% unsaturated fats and 40% saturated fats.

Ground mace contains vitamin A, phosphorus, potassium, magnesium, sodium, and calcium.

How Prepared and Consumed: in Europe and the United States, mace is usually used in cakes, confections, and light-colored products, such as cream soups, cream sauces, chowders, crackers, pie fillings, and cakes. Mace is also used with fish and in vegetable purees, meat stews, and meat pies. Commercially, mace flavors frankfurters, doughnuts, pickles, preserves, ice cream, confections, icings, sausages, knockwurst, ham, soup mixes, and poultry.

Mace pairs well with fruit, sugar, chocolate, and milk-based products—cakes, cookies, and doughnuts. It provides an intense aroma to Middle Eastern and Asian foods. In Middle Eastern, Iranian, and northern Indian recipes, mace is often combined with nutmeg. Arabs add mace to mutton and lamb dishes and many spice blends such as *ras-el-hanout, baharat,* and *galat dagga.* It is ground and sprinkled over North Indian pulaos, lamb, and other meat dishes to add aroma. Ground mace is commonly used in Southeast Asian, Chinese, and Indian foods such as garam masalas, curries, sauces, puddings, cakes, pies, and cookies.

Spice Blends: curry blends, garam masala, ras-el-hanout, baharat, galat dagga, and rendang blend.

Therapeutic Uses and Folklore: Asian Indians have traditionally treated stomach pains, dysentery, vomiting, and the symptoms of malaria with mace. It is also chewed to prevent foul breath.

MARJORAM

In Europe, marjoram was a traditional symbol of youth and romantic love. Used by Romans as an aphrodisiac, it was used to cast love spells and was worn at weddings as a sign of happiness during the Middle Ages. Greeks who wore marjoram wreaths at weddings called it "joy of the mountains." It was used to brew beer before hops was discovered, and flavored a wine called *hippocras*. A cousin of the oregano family, marjoram originated in Mediterranean regions and is now a commonly used spice in many parts of Europe. Called *za'tar* in the Middle East and often mistaken for oregano, it is also a popular spicing in eastern Europe.

Scientific Name(s): sweet marjoram: *Origanum (O) hortensis* (or *Majorana hortensis*). There are a few varieties of sweet marjoram. There are also many wild varieties, the main one being pot marjoram: *O. onites* and wild majoram: *O. vulgare*. Syrian majoram is called *za'tar*. Family: Labiatae or Lamiaceae (mint family).

Origin and Varieties: marjoram is indigenous to northern Africa and southwest Asia. It is cultivated around the Mediterranean, in England, Central and Eastern Europe, South America, the United States, and India.

Common Names: sweet marjoram, knotted marjoram, and annual marjoram. It is also called marzanjush, za'tar (Arabic), marzanan (Armenian), mah yeh lah fah, ma yuek lan fa (Cantonese, Mandarin), merjan (Danish), marjolein (Dutch), avishan (Farsi), marjolaine (French), majoran (German), matzourana (Greek), mayoran, za'tar (Hebrew), mirzan josh (Hindi), majorama (Hungarian), maggiorana (Italian), mayorana (Japanese), maruvammu (Malayalam), merian (Norwegian), manjerona (Portuguese), majoran (Russian), mejorana (Spanish), mejram (Swedish), maruvu (Tamil), mercankosk, kekik out (Turkish), and marva kusha (Urdu).

Pot marjoram: rigani, common marjoram, dictamo, oregano, French marjoram, golden marjoram, curly marjoram, gold splash marjoram, and al maraco.

Form: marjoram leaf is used fresh, as whole or chopped, and dried whole or broken, and ground. The flowering tops and seeds, which are not as strong as the leaves, are also used as flavorings.

Properties: sweet marjoram is a small and oval-shaped leaf. It is light green with a greyish tint. Marjoram is fresh, spicy, bitter, and slightly pungent with camphorlike notes. It has the fragrant herbaceous and delicate, sweet aroma of thyme and sweet basil. Pot marjoram is bitter and less sweet.

Chemical Components: sweet marjoram has 0.3% to 1% essential oil, mostly monoterpenes. It is yellowish to dark greenish brown in color. It mainly consists of *cis*-sabinene hydrate (8% to 40%), γ-terpinene (10%), α-terpinene (7.6%), linalyl acetate (2.2%), terpinen 4-ol (18% to 48%), myrcene (1.0%), linalool (9% to 39%), ρ-cymene (3.2%), caryophyllene (2.6%), and α-terpineol (7.6%). Its flavor varies widely depending on its origins. The Indian and Turkish sweet marjorams have more *d*-linalool, caryophyllene, carvacrol, and eugenol.

Its oleoresin is dark green, and 2.5 lb. are equivalent to 100 lb. of freshly ground marjoram.

Marjoram contains calcium, iron, magnesium, phosphorus, potassium, sodium, vitamin A, vitamin C, and niacin.

How Prepared and Consumed: marjoram is typically used in European cooking and is added to fish sauces, clam chowder, butter-based sauces, salads, tomato-based sauces, vinegar, mushroom sauces, and eggplant. In Germany, marjoram is called the "sausage herb" and is used with thyme and other spices in different types of sausages. It is usually added at the end of cooking to retain its delicate flavor or as a garnish. It goes well with vegetables including cabbages, potatoes, and beans. The seeds are used to flavor confectionary and meat products. The French add marjoram to bouquet garni and herbes fines for flavoring pork, fish, and lamb dishes.

It is popular in Greek cooking, for grilled lamb and meats and to complement onions, garlic, and wine. Italians use it in tomato sauces, pizzas, fish dishes, and vegetables. In Eastern Europe, it is added to grilled meats and stews with paprika, chilies, fruits, nuts, and other dried spices.

North Africans and Middle Easterners use marjoram in lamb, mutton, barbecues, vegetables, and seafood. In the United States, it is used commercially in poultry seasonings, liverwurst, bologna, cheeses, sausages, soups, and salad dressings.

Spice Blends: bouquet garni, fines herbes, khmeli suneli, sausage blend, and pickle blends.

Therapeutic Uses and Folklore: Greeks used marjoram extensively to treat dropsy, convulsions, and poisons. Traditionally, it was used in tea to cure headaches, head colds, calm nervous disorders, and to clear sinuses. Marjoram has also been used to comfort stomachaches and muscular pains and improve circulation. It is found to have good antioxidant properties with fats and helps to retain color of carotenoid pigments

MINTS: SPEARMINT AND PEPPERMINT

Peppermint was discovered in the seventeenth century in England as a wild plant and was mainly used as a medicinal infusion. From here it spread to Europe, North Africa, and to the United States. The familiar "after dinner mints" originated in England to soothe the stomach after a meal. Nowadays, mint is widely used in confectionery, liqueurs, teas, and chewing gums.

Spearmint was the symbol of hospitality in traditional Europe, where it was crushed and used in baths and other leisure places. It was called Herba Santa Maria in Italy and Our Lady's mint in France. The Greeks used it as an aphrodisiac, while the Romans used its aroma as an appetite stimulant. Today, spearmint is commonly used in chutneys, curries, and sauces in India and Southeast Asia. It is also a popular flavor for teas around the world.

Many other varieties include Japanese mint, orange mint, or Eau de Cologne, apple mint, and ginger mint mainly used in teas. Vietnamese mint or rau ram is not a true mint botanically but belongs to the coriander family.

Scientific Name(s): peppermint: *Mentha (M) piperita;* spearmint: *M. spicata;* cornmint or Japanese mint: *M. arvensis;* orange mint: *M. citrata; orange mint: M. rotundifolia; ginger mint: M. gentilis.* Family: Labiatae or Lamiacea (mint family).

Origin and Varieties: peppermint is native to southern and central Europe and is now cultivated in northern Africa, Russia, the United States, India, and England. Spearmint is cultivated in China, Southeast Asia, India, and Japan. There are many varieties and cultivars, and some plants that are called mints are not real mints but tend to have the flavor of mints, like Vietnamese mint or lemon mint. Spearmint and peppermint (whitish and black) are the two most commonly used mints. Other mints (and so-called mints) are lemon mint, apple mint, basil mint, Vietnamese mint (or Vietnamese coriander), shiso (or Japanese basil), perilla (*la tia to* in Vietnamese is sometimes referred to as shiso in the United States because of its similar flavor), *huacatyl* or South American mint, pineapple mint, ginger mint, orange mint, or lavender mint.

Common Names: mint (general): naana (Arabic), ananookh (Armenian), bokh hoh, boh he (Cantonese, Mandarin), menthe (French), menta (Greek), menta, nana (Hebrew), pudina (Hindi, Bengali, Marathi, Punjabi), phum hub (Hmong), menta (Hungarian), cheep poho (Khmer), menta (Italian, Portuguese), daun pudina (Malaysian, Indonesian), pothina, mutthina (Malayalam), mente (Portuguese), myata (Russian) hierba buena, mentas (Spanish), pereminde (Swahili), poliyas (Tagalog), pothina (Tamil, Telegu), bai saranae (Thai), nane (Turkish), and rau thom (Vietnamese).

Peppermint: naana (Amharic, Arabic), pebermynte (Danish), pepermunt (Dutch), nanah (Farsi), naana (Hebrew), piparminta (Finnish), menthe poivre, menthe anglaise (French), pfefferminze (German), menta pepe (Italian), seiyo hakka, pepaminto (Japanese), hortela pimenta (Portuguese), myata pjerechnaya (Russian), meenchi (Singhalese), menta, piperita (Spanish), and pepparmynta (Swedish).

Spearmint: naana (Amharic, Arabic), menthe verte (French), grune minz (German), menta meshubelet (Hebrew), podina (Hindi, Marathi, Punjabi), daun pudina (Indonesian, Malaysian), mutthina (Malayalam), mentastro verde (Italian), menta, hierba buena (Spanish), pothina (Tamil), bai saranae (Thai), and rau thom, hang que (Vietnamese).

Form: mint is available fresh or dried. The dried form is sold whole, as flakes, chopped, and fine or coarse. Fresh, it is eaten raw, pureed, or cooked. The leaves can also be crystallized.

Properties: peppermint has bright green leaves, with a fresh, slightly sweet, tangy, peppery, and strong menthol notes. Menthol gives peppermint its cooling sensation. Spearmint has a slightly pungent, warm, fresh and herbaceous taste, with lemony and sweetish notes. Spearmint does not have menthol and so does not have the cooling sensation of peppermint.

Other so-called mints: Vietnamese mint has aromatic coriander lemony flavor; shiso has a decorative mint- and basil-like flavor, and perilla, which is similar to the shiso, has a lemony mint-like flavor. When bruised, these mints lose their flavor easily.

Chemical Components: peppermint has 0.5% to 5% essential oil that is pale yellow. It mainly consists of menthol (26% to 46%), menthone (16% to 36%), menthyl acetate (3.8% to 7%), menthofurane (2% to 8%), isomenthone (2% to 8%), limonene (2.5%), pulegone (1.4% to 4%), and β-pinene (1.5% to 2%). Menthol and menthyl acetate are responsible for the refreshing and cooling pungent odor and are mostly found in the older leaves.

Spearmint has 0.5% essential oil, mainly 50% to 70% carvone and dihydrocarvone, including dihydrocuminyl acetate, dihydrocuminyl valerate, phellandrene, limonene, menthone, menthol, and 1,8-cineol.

Oil of corn mint or Japanese mint is referred to as mint oil in the United States and is blended with peppermint oil because it is less expensive. Corn mint has about 1% to 2% oil, mainly 28% to 34% menthol, 16% to 31% menthone, 6% to 13% isomenthone, 5% to 10% limonene, and a higher content of α- and β-pinenes.

Mints contain calcium, potassium, vitamin A, niacin, sodium, magnesium, phosphorus, vitamin C, and iron. They also have vitamin A (2700 IU/100 gm), iron, phosphorus, and other minerals.

How Prepared and Consumed: Greeks and Romans used peppermint in condiments, cordials, and fruits. Europeans use it in sweet products such as desserts, candy, jams, jellies, chocolates, cordials, liqueurs, and cigarettes. In Europe, the crystallized leaves are also used as decorations in cakes and pastries. Today in the United States, peppermint is mainly used in bakery products, teas, and confectionery.

In England, peppermint is used in savory products, such as sauces for roast lamb, boiled mutton, peas, potatoes, and in teas, chocolate, and vinegar. Middle Easterners use chopped peppermint in many dishes—yogurt dressings, dips, salads, vegetables, grilled lamb, poultry, fish, and in teas. Dried leaves are popular in many North African and Middle Eastern cuisines including Turkish and Iranian cooking. They are used in dry blends, meat and fish marinades, and beverages. Mint tea is widely consumed by Arabic cultures.

In India, spearmint is ground with coconut, green chilies, onion, and green mango to flavor green or chat masalas for chutneys and curries, and is added with yogurt and cucumber in raita as dips and drinks. It enhances the Mogul-style biryanis of North India. In Southeast Asia, Vietnamese, Malaysians, and Thais use fresh spearmint as garnishes in salads (along with basil and cilantro), in fish and poultry curries, seasoned rice, sauces, biryanis, and soups. The Vietnamese mint is very popular in Vietnam and is always used fresh as a garnish for most southern Vietnamese dishes, for soups, salads, spring rolls, stir-fries, dips, and as food wrappers.

Spearmint pairs well with cardamom, shallots, basil, cilantro, lemongrass, green chilies, lime, green papaya, yogurt, and green mango. Normally, mints are added at the end of cooking because heat rapidly dissipates their flavor. In Mexico, mint is popular in fruit-based beverages and teas.

Spice Blends: chat masala, Malaysian kurma blend, podina chutney blend, tabbouleh blend, mint sauce blend, and Thai green curry blend.

Therapeutic Uses and Folklore: mints were used as a symbol of hospitality in many cultures. They are also used as a general body cleanser and strengthener. Mint has been used to relieve cold and flu symptoms, fevers, headaches, muscular aches, sore throats, and toothaches. Peppermint tea helps relieve nausea and is taken to relieve seasickness. It is also used to promote digestion and relieve stomach upsets. Mint has antibacterial properties.

Vietnamese mint: although called mint, it does not belong to the mint family. In Malaysia and Indonesia, it is called *daun laksa* (or laksa leaf) because it is an essential flavoring ingredient in the stew-like, pungent, hot noodle dishes called laksas created by the Nonyas of Malaysia and Singapore.

Scientific Name(s): Polygonum odoratum. Family: Polygonaceae (buckwheat family).

Common Names: it is also called Vietnamese coriander, Cambodian mint, smartweed, or hot mint. Its other names are pakarmul (Bengali), laak sah yip, la sha ye (Cantonese, Mandarin), Vietnamesisk coriander (Danish), renouee odorante (French), Vietnamesischer coriander (German), luam laus (Hmong), Vietnami menta (Hungarian), chee krassan tomhom (Khmer), phak phew (Laotian), daun kesum, daun laksa (Malaysian), kupiena lekarstvennaya (Russian), pak pai (Thai), and rau ram (Vietnamese).

Properties: it has a clean, coriander-like lemony taste.

Chemical Components: its essential oil contains aldehydes, mainly dodecanal (45%), decanal (30%), and decanol (10%), and sesquiterpenes such as α-humulene and β-caryophyllene.

How Prepared and Consumed: native to Southeast Asia, Vietnamese mint is commonly used in Vietnam, Singapore, Cambodia, and Malaysia. The Nonyas of Singapore and Malaysia add chopped fresh mint with fermented shrimp paste, turmeric, galangal, and lemongrass to lend a unique flavor to their spicy soups, curries, laksas (or spicy noodle broths), and condiments. It is an important flavoring in Vietnam, where whole fresh leafy spices are a significant part of preparing or garnishing a meal or dish. Vietnamese mint leaves are sometimes used to wrap beef, seafood, or vegetables that are cooked or served as salads.

MUSTARD

Mustard seed was a symbol of fertility for the ancient Indians and has been used by Greeks, Chinese, Indians, and Africans since ancient times. Mustard paste making was introduced to central and northern Europe by the Romans. The Romans ground mustard seeds, which were called *sinapis*, with grape must to make the first table mustard, which they called *mustum ardens* meaning "burning or hot must." Eventually, *mustum ardens* became mustard, and before long, the seed took this name as well. The prepared mustard condiment was first made by the French in Dijon during the thirteenth century, and mustard flour was first made during the eighteenth century in England. Mustard oil is a popular cooking oil in North India.

Scientific Name(s): there are three types of mustard seeds: pale yellow or white mustard (*Brassica (B)/Sinapis alba* or *B. hirta* Moench); Indian brown and Oriental mustard; (*Brassica juncea*), a hybrid of B. nigra and B. campestris); and the black or dark brown mustard (*Brassica nigra*). Family: Brassicaceae (cabbage family).

Origin and Varieties: yellow or white mustard is indigenous to southern Europe and western Asia; brown mustard is indigenous to northern India, China, Iran, Afghanistan, and Africa; and black mustard is indigenous to South Mediterranean. These mustards are also cultivated in Nepal, Russia, Canada, southern Italy, northern Africa, and Central and South America. Black mustard seed is not as popular in the United States or Europe because it is difficult to harvest. Brown and white mustards are the most popular types used in the United States, but black and brown are the most popular in Asia and Europe.

Common Names: mustard (in general), khardal (Arabic), senafich (Armenian), gai chay, chieh kai (Cantonese, Mandarin), sennap (Danish, Norwegian, Swedish), mosterd (Dutch), khardel (Farsi), moutarde (French), senf (German), moustarda

(Greek), hardal (Hebrew), rai (Hindi, Bengali, Urdu), mustar (Hungarian), senape (Italian), garashi (Japanese), sien (Laotian), biji sawi (Malaysian, Indonesian), molari (Marathi), mostarda (Portuguese), gorchitsa (Russian), mostaza (Spanish), mostaza (Tagalog). kadugu (Tamil, Malayalam), hardal (Turkish), and cai denh (Vietnamese).

Yellow/white mustard seeds: netch senafitch (Amharic), kardhal abyed (Arabic), bankh gai choy, bai chieh gai (Cantonese, Mandarin), hved sennap (Danish), witte mosterd (Dutch), khardel sefid (Farsi), sufed rai (Hindi), moutarde blanche (French), weiber senf (German), sinapi agrio (Greek), hardal levan (Hebrew), angse mustar (Hungarian), senape biancha (Italian), shiro garashi (Japanese), som sien (Laotian), biji sawi putih (Malaysian, Indonesian), huit sennap (Norwegian), mostarda branca (Portuguese), gorchitsa belaya (Russian), mostaza blanca, mostaza silvestre (Spanish), vit sennap (Swedish), byaz hardal (Turkish), and banh cai denh (Vietnamese).

Brown mustard includes two varieties: one called Oriental used mostly by Chinese, and the other a darker, stronger brown mustard that is used by Indians. There are many cultivars of these different brown types, which are hybrids of the black mustard: also called gai choy, chieh gai (Cantonese, Mandarin), rai (Hindi), moutarde de Chine (French), indischer sonf (German), senape India (Italian), and mostazo India (Spanish).

Black mustard seeds, also called brown mustard or true mustard, also called tikur senafich (Amharic), kardhal asuja (Arabic), mananekh (Armenian), zwarte mosterd (Dutch), moutarde noir (French), schwarzer senf (German), sinapi nauro (Greek), hardal shaor (Hebrew), rai kala (Hindi), fekete mustar (Hungarian), senape nera (Italian), biji sawi hitam (Malaysian, Indonesian), svart sennap (Norwegian), mostarda preta (Portuguese), gorchitsa chyornaya (Russian), abba (Singalese), haradali (Swahili), brun sennap (Swedish), and mostaza negra (Spanish).

Form: mustard seeds range anywhere from large yellow to yellowish brown seeds, medium-sized brown to dark brown seeds or small black seeds, depending on the variety. The seeds are used whole, crushed, ground, or as flour. Mustard also comes in wet or prepared paste forms with water, vinegar, sugar, oil, and spices called prepared mustards. Ground mustard is made from yellow or brown whole seeds.

Mustard oil is the fixed oil in mustard seed and has little pungency. It is pale yellow in color and has a raw, pungent, and bitter taste with an unpleasant aroma, but it becomes pleasant and sweeter during cooking. Mustard meal (that includes bran) is made mostly from yellow mustard and is sometimes blended with brown mustard. The seed is ground with the bran. Mustard flour/powdered mustard is usually a blend of yellow or brown mustard seeds. Their proportions depend on the region. Mustard seeds are milled to remove the bran. A wide variety of flours are available based on volatile oil and particle size. Prepared mustard is a smooth condiment made from mustard seeds, salt, spices, and vinegar.

Mustard leaves, also called mustard greens, have a radishy taste and are used as a prepared vegetable or are put into salads. The most common mustard greens are from the brown mustard variety and are commonly eaten in the southern United States, China, India, and Southeast Asia.

Properties: the smaller and darker the seed, the more intense and hot the flavor is. Black mustard is oblong and is the smallest in size. It has the sharpest flavor along with a nutty aftertaste. The brown mustard seed is spherical and medium in size, and has a nutty, sweeter and mellower burning note than black mustard. It has a taste sensation like horseradish. The white or yellow mustard is largest and has a delicate flavor that is the least burning.

The whole seed has no flavor, but can provide a pungent taste after chewing for some time. The heat experienced in yellow mustard is on the tongue, but in brown and black mustard, the heat is also felt in the nose and eyes. This latter pungency is more intense and lasts longer than the former, though bite does not build up.

The pungent aroma varies among the different mustards. The white or yellow type has a less pungent aroma than the brown mustard seeds that have a very pungent aroma. The black mustard seeds have the most pungency.

Ground mustard has no aroma. Its flavor and pungency is experienced by triggering an enzyme action that releases mustard's flavor or pungency.

Mustard pungency is due to a variety of isothiocyanate compounds that exist in mustard tissue as glycosides. The major pungent compound of black and brown mustard is allyl isothiocyanate. The release of sensation, especially in brown and black mustard, is delayed and begins at the back of the mouth, with a shooting sensation to the sinuses. This is due to the activation of an enzyme, myrosinase, that, in the presence of water, breaks down the glycoside (sinalbin) in yellow mustard or (sinigrin) in black or brown mustard to para-hydroxybenzyl-isothiocyanate or allyl isothiocyanate, which gives the characteristic pungent aroma. The odor lasts until all enzymatic activity has ceased.

Acids are poor triggers of mustard's overall flavor, but they extend mustard's penetrating odor. Heat stops the flavor release, so it is important to let the mixture of mustard and water stand for ten minutes before it is added to cooking. When water, vinegar, milk, wine, or beer is added to mustard, mixed and left to stand for a few minutes, different degrees of flavor sensations are produced. With water, a very sharp and hot taste is produced; with vinegar, a milder flavor occurs; with milk, a milder, spicier, and pungent flavor is created; and with beer, a very hot flavor is produced.

Whole mustard seeds are toasted in heated oil or are "popped" to give a nutty, sweetish, aromatic flavor in South India cooking.

Mustard is a natural physical emulsifier and binder for hot dogs, salad dressings, and many sauces.

Chemical Components: mustard is high in fat (35%) and protein (28%). Mustard seed is cold pressed to extract the fixed oil from the press cake. The press cake is hydrolyzed by enzymes and is steam distilled or solvent extracted to give the essential oil. The white/yellow seed has 2.5% sinalbin and very little volatile oil (below 0.2%). When the cells are damaged, sinalbin is hydrolyzed by enzyme myrosinase to produce the para–hydroxybenzyl-isothiocyanate. The brown and black seeds have 1% sinigrin and 0.5% to 1.2% volatile oil, mostly 95% to 99% allyl isothiocyanate. Certain types of brown mustards contain up to 2.9% essential oil.

Mustard seeds have 28% to 35% fixed oil, called mustard oil that is pale yellow in color and contains linoleic, linolenic, and 20% to 30% erucuc acids. Oleoresin mustard is yellow to light brown in color, and 4.5 lb. are equivalent to 100 lb. of ground yellow mustard.

Generally, mustard seeds contain calcium, vitamin C, potassium, magnesium, phosphorus, and niacin. Also, when mustard is ground with its bran, it results in mustard meal, which is high in protein.

How Prepared and Consumed: white/yellow mustard is mostly used for preparing mustard pastes or condiments with wine, vinegar, and other spices to accompany boiled or roast meats, or added to stews and sauces in Europe. The vinegar in mustard paste stabilizes and mellows its pungency, while wine extends and enhances its pungency. Only a small amount of black mustard is used in preparing condiments because its pungent principle is very volatile. Therefore, white/yellow mustard is preferable for condiments.

Mustard seeds were sprinkled over foods by the ancient Greeks and Romans. The early French mixed it with grape must or honey and vinegar. The yellow prepared mustards have sharp tongue tastes but not pungent aromas because of their lesser pungent principles. German, Dijon, English, and Chinese prepared mustards contain black, brown, or Oriental ground mustard that have more pungent aromas and bites.

Dry Coleman's mustard and the North American hot dog mustards contain the white/yellow type of mustard. For English mustard, brown mustard is mixed with yellow mustard; In Germany, black mustard is made into Lowensenf, while sweet Bavarian has coarsely ground white mustard, honey, and spices to accompany sausages. French Dijon has a sharp taste and contains mainly black mustard with sour grape juice and spices for flavoring roasts and boiled meats. The milder Bordeaux has white mustard with its bran layer making it dark in color. Many emerging new types contain exotic spices, wine, and chilies. Chinese mustard is powdered ground brown mustard mixed with water and other ingredients. It is used as a table condiment in Chinese restaurants and for other ethnic foods.

Mustard goes well with cold meats, sausages, grilled steak, poultry, fish, herbs, wine, garlic, sauerkraut and fruits. Today, there are many customized prepared mustards that use different spices, fruits, wine, and chile peppers.

Ground yellow mustard acts as a physical emulsifier and stabilizes mayonnaise and salad dressings. It is also used as a flavor enhancer and a water binder in processed meats. In the United States, yellow mustard is combined with sugar, vinegar, and turmeric and commonly used as a spread for hot dogs, hamburgers, and sandwiches and luncheon meats, while the seeds are used as toppings on cooked vegetables. It is an important flavoring in baked beans, combined with brown sugar, ketchup, beans, onions, and bacon.

Whole yellow mustard seeds are used in pickled condiments, not only for flavor but also for their preservative function. The paste form is popular in salad dressings and hot sauces.

The English enjoy brown mustard with roast beef and ham. The Japanese use the oriental brown variety as a dip for raw fish. The Barbadians and other populations in the Caribbean use yellow or brown mustard with fruits and chile peppers for great tasting sauces, marinades, and stews. In Indian cooking, especially in southern India's vegetarian meals, whole brown or black mustard seeds are "popped" in heated ghee or oil to bring out their nuttiness, and they are then added to sauces, chutneys, pickles, curries, *sambars*, and dals. Black mustard is sometimes used to flavor ghee in South India. Ground mustard seeds provide flavor and consistency in Bengali fish curries.

Fixed oil from the brown and black seeds is used in many northern Indian dishes and pickles, especially in Bengali, Kashmiri, and Pakistani cooking. The raw, pungent notes of this oil dissipate with cooking, and it becomes sweeter. It is becoming a popular drizzling oil in many Indian-inspired dishes in the United States.

Sprouts from the mustard seeds are used in salads in many Asian recipes.

Spice Blends: Dijon mustard blend, baked bean blend, Bajan sauce blend, pickle blend, chutney blend, sambar podi, panchphoron, and Chinese mustard blend.

Therapeutic Uses and Folklore: in ancient times, the Greeks used mustard for scorpion and snake bites. Mustard plasters were used to stimulate blood circulation and to warm up cold feet, to relax stiff muscles, and to treat arthritis and rheumatism. It treats skin diseases because of its high sulfur content.

Mustard also stimulates the flow of salivary and gastric juices and promotes appetite. It has been used as a laxative, as a remedy for asthma, and to induce vomiting or relieve coughs.

Asian Indians use mustard oil in cooking to increase blood flow and cleanse the blood of any toxic materials.

Mustard has preservative properties.

MYRTLE

Myrtle leaves, branches, and fruits have been used since Biblical times to flavor smoked or roasted meats in the Eastern Mediterranean islands of Sardinia, Corsica, and Crete and western Asia. The dried berries were used in the Mediterranean as a substitute for black pepper.

Scientific Name(s): Myrtus communis. Family: Myrtaceae (myrtle family).

Origin and Varieties: myrtle is indigenous to Mediterranean regions of northern Africa, southern Europe, and the Middle East. Lemon myrtle: Backhousia citriodora is native to Australia.

Common Names: myrtle pepper: Corsican pepper, allspice. It is also called adus (Amharic), hadass, raihan (Arabic), mrdeni (Armenian), bilati mehendi (Bengali), heung tu much (Cantonese), myrte (Danish), mirte (Dutch), moord (Farsi), mirtia (Greek), hadas (Hebrew), vilayati mehndi (Hindi, Punjabi), mirto (Italian), myrtti (Finish), myrte (French, German), jinbaika (Japanese), tao jian niang (Mandarin), murta mirto (Portuguese, Brazilian), myrt (Russian), arraya'n/mirto (Spanish), myrten (Swedish), kulinaval (Tamil), mersin (Turkish), and habulas (Urdu).

Form: myrtle seeds are purple-black berries that are used whole or coarsely ground. Its leaf is used whole or chopped.

Properties: myrtle berries are sweet, with juniper and rosemary-like flavors. The leaves have spicy, astringent, and bitter taste with a refreshing, fragrant, and orange-like aroma. Lemon myrtle has a refreshing aroma and taste.

Chemical Components: leaves have 0.2% to 0.8% essential oil, with myrtenol, myrtenol acetate, α-pinene, limonene, linalool, camphene, cineol, geraniol, and nerol. Lemon myrtle leaf has 4% to 5% oil, mostly terpenoid aldehydes, citral (90%), neral, and geranial.

How Prepared and Consumed: in the Mediterranean, the berries were initially used to flavor wine but are now more commonly used in desserts, liqueurs, and sweet dishes. The leaves are used in stews, roast meats, stuffings, salads, and meat ragouts. The leaves are used to wrap wild game or roast pork before cooking.

Italians, especially Sardinians, use myrtle branches in the same way allspice branches are used in the Caribbean. They wrap meat, other game meats, birds, and poultry with myrtle branches and then roast, broil, or smoke them. The leaves are also stuffed in the meats and are removed before serving. The burning myrtle wood and leaves provide a fragrant note to the meat.

Australians add it to roast poultry, seafood, salad dressings, many sauces, and curries, and infuse it in vinegar.

Spice Blends: smoked meat blends, roast blends, and meat ragout blend.

Therapeutic Uses and Folklore: the berries are used as an antiseptic for bruises, and the leaves are used to relieve gingivitis and sinusitis.

NIGELLA

Nigella, called *kalonji* in Hindi, is a popular spice for breads and vegetables in North India, Lebanon, Turkey, and Iran. It was a culinary spice and a medicine for Romans, ancient Hebrews, and Greeks. Nigella was introduced by the Armenians into the United States as *charnushka.*

From the Latin word *"niger'* or *"nigellus"* meaning black, it is often confused with black cumin because of its similar size and color. Thus, it is called black cumin in many cultures, including *kala jeera, jintan hitam, or grano negro.* It grows wildly in India and has been used as a flavoring for pickles since ancient times.

Scientific Name(s): Nigella sativa. Family: Ranunculaceae (buttercup family).

Origin and Varieties: nigella is indigenous to western Asia and the Middle East. It is cultivated in northern India, Sri Lanka, Pakistan, Nepal, western Asia, Middle East, Egypt, and other regions of northern Africa and central Europe.

Common Names: black caraway, black onion, love-in-a-mist, devil in a bush, fennel flower, and onion seed. It is also called tikur azmud (Amharic), habat albaraka, habit assuda (Arabic), chernuska (Armenian), hak jung chao, hei jhong cao (Cantonese, Mandarin), zortkommen (Danish), nigelle (Dutch), siah daneh (Farsi), nigelle, poivrette (French), schwarzkummel, nigella (German), melanthion ninkela (Greek), ketzah (Hebrew), kalonji, kala jeera (Hindi), feketekomeny (Hungarian), nigella, grani neri (Italian), nigera (Japanese), jintan hitam (Malaysian, Indonesian), svartkarve (Norwegian), nigela (Portuguese), chernushka (Russian), kaluduru (Sinhalese), neguilla, pasinara (Spanish), svartkummin (Swedish), karun jeeragum (Tamil), nellajira kaira (Telegu), kolongi, core kotu (Turkish), and charnushka (United States).

Form: nigella is a small, irregular-shaped black seed that is used whole or ground. The pods are harvested before they are fully ripe in order to prevent the pods from bursting open and discarding the seeds.

Properties: whole nigella has little flavor but when ground, has a warm, slightly fruity and oregano-like aroma with a sharp, peppery, and nutty taste. The essential oil is yellowish brown in color while fixed oil is reddish brown.

Chemical Components: it has 0.5% to 1.5% essential oil, mainly thymoquinone (54%), ρ-cymene (30%), α-pinene, ethyl linoleate, and dithymoquinone with smaller amounts of carvacrol, carvone, limonene, citronellol, and 4-terpineol. Other components include glucosides, melanthin, melathingenin, and nigellone. Its fat content or fixed oil is about 32% to 40%. It is rich in unsaturated fatty acids, mainly linoleic (55%), oleic (20%), and dihomolinoleic acid (20%).

It contains iron, niacin, and calcium.

How Prepared and Consumed: Indians, Middle Easterners, Turks, and Egyptians commonly sprinkle whole nigella seeds on breads before baking to provide flavors and textures. In North Indian cooking, whole seeds are dry roasted or fried in oil to give a more intense aroma to kormas and garam masalas. They are used in curries, naan, dals, yogurts, vegetables, and chutneys. In Punjab and Iran, nigella is used mostly to enhance vegetable dishes. It can also be used in salads, cottage cheese, lamb, poultry, and pickles.

Nigella is an essential ingredient in a spice mixture of Bengal, Bangladesh, and Sikkim, called *panchphoron* (five-spice seasoning). It is blended with cumin, mustard seed, ajwain, and black pepper and is then fried in mustard oil to flavor eggplant, cabbage, squash, and ground meats. In the Middle East, it is added to bread dough and is important in *choereg* rolls. Indians add nigella to preserve pickles.

Spice Blends: panchphoron, dal podi, chutney blend, and lamb curry blend.

Therapeutic Uses and Folklore: Indians eat nigella to reduce flatulence, treat nervous and stomach disorders, and induce sweating. It is also taken by lactating women to induce milk flow. Today's research shows it has strong antimicrobial activity and good immunological property. Ancient Greeks used it to treat headaches, toothaches, and intestinal parasites. Its melanthin content shows good emulsifying properties. Egyptians drink tea made with ground nigella, fenugreek, and other ingredients to treat diabetes.

NUTMEG

Nutmeg has its origins in the Spice Islands of Indonesia. Arab traders brought nutmeg to India (thus called "Indian nut" by many European languages), then to Europe, and eventually it was taken to the Caribbean by the Spanish. It was widely popular in Europe and in India for its flavoring and medicinal properties. During the Middle Ages, fashionable Europeans carried their own nutmegs and graters to eating establishments as a status symbol. Nutmeg's name is derived from the Latin word *nux muscatus,* meaning musky nut. The East Indian (Indonesian) nutmeg is superior in flavor to the West Indian nutmeg (Caribbean). Nutmeg and mace come from the same fruit. Nutmeg is the seed that is covered by the lacy mace (see mace).

Scientific Name(s): Myristica fragrans. Family: Myristicaceae (nutmeg family).

Origin and Varieties: nutmeg is indigenous to the Banda Islands (Moluccas) in Indonesia. It is also cultivated in the Caribbean, South India, Sri Lanka, Sumatra, and Malaysia. There are many "adulterant" species found in South India and Indonesia with flavors that sometimes lack the aroma of true nutmeg. Nutmeg's flavor varies depending on its origins. The United States buys Indonesian nutmeg, which has higher volatile oils than Caribbean nutmeg.

Common Names: gewz (Amharic), jouz atib (Arabic), meshgengouz (Armenian), mutwinda (Burmese), muskatmod (Danish), notemuskaat (Dutch), djus hendi (Farsi), noix de muscade (French), muskat (German), moschokarido (Greek), jaiphal (Gujerati, Marathi), ego musket (Hebrew), jaiphal (Hindi, Urdu), szerecsendio (Hungarian), noce moscata (Italian), natumegu (Japanese), pok kak (Khmer), chan thed (Laotian), buah pala (Malaysian, Indonesian), jo tou kuo (Mandarin), jathikka (Malayalam), noz moscada (Portuguese), muskatniy orekh (Russian), sadikka (Singalese), nuez moscada (Spanish), muskotnot (Swedish), duguan (Tagalog), atipalam, jatikkai (Tamil), and luk jan (Thai).

Form: nutmeg is a light brown or grayish wrinkled seed inside a smooth, hard, blackish brown nut. The nut is dried in the sun until the inner seed rattles when shaken.

Properties: nutmeg is used ground, grated, or crushed. It loses its flavor easily when ground, so it should be bought whole and then ground when desired. Depending on the type, its flavor can vary from a sweetly spicy to a heavier taste. It has a clovelike, spicy, sweet, bitter taste with a terpeny, camphorlike aroma. It is sweeter in flavor than mace.

Chemical Components: nutmeg has 6.5% to 16% essential oil, which is pale yellow in color and is called oil of myristica. Depending on the source, the essential oil has mainly sabinene (15% to 50%), α-pinene (10% to 22%) and β-pinene (7% to 18%), with myrcene (0.7% to 3%), 1,8-cineole (1.5% to 3.5%), myristicin (0.5% to 13.5%), limonene (2.7% to 4.1%), safrole (0.1% to 3.2%), and terpinen 4-ol (0% to 11%). Their amounts depend on whether the oil is of West Indian, Indian, or Sri Lankan origin. The fixed oil is a pale yellow to golden yellow viscous oil, and 6 lb. of oleoresin is equivalent to 100 lb. of freshly ground nutmeg.

Nutmeg butter, which consists of fixed oil and volatile oil, is orange red to reddish brown and has the consistency of butter at room temperature. It contains trimyristin (70%), fats (4%), resins (13%), and other constituents (2%). The fats are mainly saturated (90%) with 10% unsaturated fats.

Nutmeg contains potassium, magnesium, and phosphorus.

How Prepared and Consumed: nutmeg loses its flavor quickly when ground, so generally it is grated just before cooking or baking. Also, to retain its flavor, nutmeg is added toward the end of cooking. It complements chocolate, fruits, custards, vanilla, coconut milk, lemongrass, and kari leaves. Europeans use it in mashed potatoes, rice dishes, pastas, soups, rice puddings, pies, eggnog, biscuits, and milk-based drinks.

Nutmeg provides an intense, sweet, spicy aroma to pastries, cakes, sweet rolls, banana bread, pumpkin pies and apple pies, ice cream, chocolate, and lemon desserts. Nutmeg is also used in cheese fondues, and it enhances savory products such as vegetable stews, bechamel sauce, tomato sauce, processed meats, and pork patties.

Italians flavor spinach with nutmeg for stuffed pastas. It is also a favorite spice of the Dutch, who use it in potatoes and other vegetables. Nutmeg is an important ingredient in the French spice blend, quatre epices, is used to flavor meats that are braised or cooked for a long time, such as ragouts or stews.

Along with mace, it is an important ingredient in spice blends of India, the Middle East, and North Africa. The pungent garam masalas of North India commonly use nutmeg. The aromatic spice blends of the Middle East and Asia contain it as well. It is generally used sparingly during the cooling process. In Indonesia, nutmeg is used in sauces and curries, and nutmeg pulp is made into a local jam.

Nutmeg has become a popular spice in Caribbean cooking and is added to jerk seasoning, pastries, ice creams, fruit punches, eggnogs, breads, and cakes.

Spice Blends: garam masala, bechamel sauce blend, quatre epices, baharat, ras el hanout, curry blends, jerk seasoning, and Indonesian ikan pedas curry blend.

Therapeutic Uses and Folklore: traditionally, nutmeg has been used to treat digestive disorders, such as nausea and diarrhea, and kidney ailments. Southeast Asians also treat fevers, headaches, and bronchial problems with nutmeg. The Chinese consider it to be an aphrodisiac.

ONION

Onion is an ancient spice, native to Asia, and it is long noted for its flavoring and pickling properties. From the Latin word *"cep"'* meaning onion, Greeks valued it for its curative powers, while Egyptians ate it raw. Today, onion is an indispensable ingredient for flavoring many ethnic cuisines, whether sauteed, roasted, or pickled. Indians savor it not only for its flavor but also for the texture and consistency it provides to curries. There are many types of onions that vary in color, size, and flavor, such as yellow, red, purple, Italian, sweet, and pearl onions, and shallots. Shallot, whose name is derived from Askalon, West Asia, is favored in Southeast Asian and Mediterranean cuisines. Scallions are salad onions or green onions that are chopped and added as garnish to Asian cooking.

Scientific Name(s): onion: *Allium cepa;* shallot: *Allium ascalonicum.* Family: Alliaceae (onion family).

Origin and Varieties: onion is indigenous to central Asia and is now cultivated in Egypt, Japan, North America, South America, Europe, Southeast Asia, France, and Mexico. The yellow or white is the most common onion, and includes the Spanish and Bermuda. Sweet onions include Vidalia, Texas, and Maui. Others are the Italian cipollini and purple onions.

Common Names: common onion and garden onion. In some cultures, onion and garlic have similar names. For example, in Indonesia and Malaysia, shallots are referred to as "red" onions (bawang merah) because of their red-colored sheath, while garlic is called "white onion" (bawang putih) because of its white sheath. It is also known as key shinkurt (Amharic), basal (Arabic), sok (Armenian), poneru (Assamese), pianj (Bengali), kesuni (Burmese), yeung chung, yan tsong (Cantonese, Mandarin), loeg (Danish), oi (Dutch), pias (Farsi), oignon (French), zwiebel (German), kremidi (Greek), batzal (Hebrew), pyiaz (Hindi, Urdu), hagyma (Hungarian), cipolla (Italian), atasuki, tamanegi (Japanese), khtim slek (Khmer), pak bhou (Laotian), ulli, vengayan (Malayalam), bawang merah, bawang besar (Malaysian, Indonesian), kepalok (Norwegian), cebola (Portuguese), peyaz (Punjabi), luk (Russian), lunu (Singhalese), cebolla (Spanish), kitungu (Swahili), lok (Swedish), sibuya (Tagalog), vengayam (Tamil), nirulli (Telegu), hua hom (Thai), btsong (Tibetan), and hanh (Vietnamese).

Shallot: eschalot, schalotte and Spanish garlic. It is also called key shinkurt (Amharic), skalot (Danish), sjabt (Dutch), ciboule (French), ascalonia (Spanish), schalotte (German), askalonia (Greek), kanda lasum (Hindi), mogorohagyma (Hungarian), scalogna (Italian), eshirroto (Japanese), bawang merah (Indonesian, Malaysian), hong tsong (Mandarin), sjalott-lok (Norwegian), cebolha roxa (Portuguese), shallot (Russian), chalota (Spanish), kitungu kidogo (Swahili), schalottenlok (Swedish), vengayam (Tamil), hom daeng/lek (Thai), and btsong gog (Tibetan).

Scallion: spring onion, green onion, chung tao, jiao tou (Cantonese, Mandarin), piaz che (Farsi), ujhagyma (Hungarian), khtim kraham (Khmer), daun bawang (Malaysian, Indonesian), ton hom (Thai), and hanh tay (Vietnamese).

Form: the common onion is a white, yellowish white, or brownish red bulb, depending on the variety and source. Onion comes fresh or dried. The fresh form is chopped, sliced, or diced. The fresh onions can be roasted, grilled, or pickled. As dried, it comes granulated, powdered, ground, minced, chopped, or toasted. Pearl onions are tiny, red onions that are larger and round to oblong, and shallots are a cluster of small- to medium-sized bulbs with a reddish brown to orange brown outer skin. Scallions, also called spring onions or green onions, are long, slender onions that are immature yellow onion bulbs. Both their bulbs and stems are consumed.

Properties: onions are mild or pungent depending upon the variety. Onions do not give any flavor until they are cut or bruised. These actions cause an enzymatic action and give rise to a mixture of sulfides. The regular white or yellow onions have a strong, pungent, and penetrating odor when raw but become sweet when cooked. The red onions are pleasant and sweet. Pearl onions do not have much aroma but taste sweet when cooked. Vidalia onions are sweet with high sugar content, and have less sulfur than regular onions. The shallot is more delicate and has a slightly pungent, sweet flavor. Scallions are mild and slightly sweet.

Chemical Components: regular onions contain 0.01% to 0.015% essential oil that has a dark brown color. They consist mainly of sulfur compounds, such as *d-n*-propyl disulfide, methyl-*n*-propyl, disulfide, vinyl sulfide, and thiols. The flavor-contributing chemicals are methyl propyl disulfide, methyl propyl trisulfide, and dipropyl trisulfide that are released through enzymatic action when the cells are damaged. The oleoresin is brownish in color, and 1 lb. is equivalent to 100 lb. of dehydrated onions or 400 lb. of fresh onions.

How Prepared and Consumed: onions add flavor, color, and texture to foods. They are added whole and glazed for stews; chopped or minced for soups and sauces; chopped or sliced raw as garnish for salads, omelettes, and dips; and deep-fried, simmered, stir-fried, roasted, or pickled.

Yellow onions are good for stews, soups, or sauces that require long cooking times. Sweet onions are tasty when baked, battered, and fried (as onion rings), sliced and fried as toppings, or eaten with broiled or roasted meats. Red onions are used raw in sandwiches and salads and can be pickled or curried. Pearl onions are glazed, pickled, or added whole to soups and stews.

Onion-filled dumplings called *pierogi* are popular Polish food in the United States which was brought to central Canada by the Ukranian immigrants. In classic French cooking, pearl onions and shallots are glazed, pickled, or roasted and used in salads, roasted meats, poultry, and seafood. In France, shallots are very popular and are used with red wine and are mostly braised or boiled. They are an important ingredient with tarragon in béarnaise sauce and many other French sauces. Fried onion rings are used as appetizers in the United States and in Germany as garnish with mashed potatoes.

In India, onions are the base of many sauces and curries, providing not only flavor but also body. They are chopped and fried in oil or ghee until golden brown, before other ingredients, such as garlic, ginger, and spices, are added. When fried, onions become sweet and aromatic. Onions pair well with garlic, ginger, tomatoes, yogurt, coconut, and many other spices.

Shallots are an essential ingredient that adds flavor and consistency to Southeast Asian cooking. They are sliced and mixed raw into sambals or are served alongside satay. They are pounded and added to spice blends and condiments and cooked in sauces or *rendangs*. They are also fried in oil and added as toppings to stir-fried noodles and *nasi goreng* (fried rice) to give a crunchy texture and caramelized taste.

In Malaysia, Singapore, and Indonesia, shallots are pureed with spices and chile peppers and are "*tumised*," or fried in oil, until a fragrant aroma develops to create cili boh. Similarly, in Burma, onions are blended with vinegar, chilies, and ginger into a smooth paste and then fried in sesame oil until the oil separates and the sauce becomes fragrant.

Scallions are chopped or sliced raw and are commonly used in soups, stir-fries, or pickles in East Asian cooking. In Chinese and Southeast Asian cooking, they are chopped or sliced and added raw as toppings over cooked noodles, *laksas*, and rices, or mixed into condiments and dumpling sauces. The Afghans have ravioli that is stuffed with scallions and served with yogurt sauce and onions.

Spice Blends: cili boh blends, rojak blend, curry blends, pickle blend, pierogi filling blend, and onion soup blend.

Therapeutic Uses and Folklore: onions have been used traditionally as an expectorant, a diuretic, and to lower blood pressure. Raw onion has been used to fight colds and flus and prevent edema. Shallots are traditionally used to cure earaches and fevers.

OREGANO

Oregano's name comes from the Greek word *oros ganos,* meaning joy of the mountain. There are many varieties of oregano, including Greek, Mexican, African, Italian, Spanish, Syrian, and Moroccan. The word "oregano" in Spanish, German, French, and Swedish means wild marjoram and, thus, oregano is sometimes confused with sweet marjoram. Originally used to flavor ale and beer in Europe, oregano is an essential spice in Italian pizzas, Tex-Mex chili con carne, and Mexican beans. The European, generally Greek, oregano is used fresh or dried in North American cuisine while the Mexican oregano is popular in Mexican-style cuisine.

Scientific Name(s): common oregano (wild European): *Origanum (O) vulgare;* Italian oregano: *O. viride* and *O. virens; Thymus capitatus:* Spanish oregano; O. Syriacum: Syrian oregano. Family: Labiatae or Lamiaceae (mint family). Mexican oregano (or called Mexican sage): *Lippia graveolens.* Family: Verbenaceae (verbena family).

Origin and Varieties: there are many species of oregano indigenous to the Mediterranean (Italy, Greece, Turkey, France, Israel, Morocco), India, Syria, Mexico, and South America.

The United States uses Mediterranean oreganos (Turkish, Greek, Italian, and Spanish) and Mexican oregano. Mexican oregano is not botanically related to the common oregano and has a more intense taste than the common oregano.

Common Names: common oregano—wild marjoram, winter sweet, origanum, wild European oregano, origany, pizza herb, common marjoram, and Spanish marjoram. It is also called mardakosh, anrar (Arabic), oregano (Armenian), ngau lakh gong, hao le ganh (Cantonese, Mandarin), wilde marjolein (Dutch), avishan kuhi (Farsi), origan, marjoleine batarde (French), dosten, oregano (German), rigani (Greek), oregano (Hebrew), sathra (Hindi), oregano (Japanese), origano (Italian), oregao (Portuguese), dushitsa (Russian), oregano (Spanish), oregano (Swedish), kekik out (Turkish), and mirzan josha (Urdu).

Form: oregano's dark green fresh leaves are larger than marjoram leaves. Fresh oregano is available whole, chopped, or minced. The dried light green leaves are available whole, flaked, or ground.

Properties: Oregano can be mild or strong depending on its origins. Typically, it has a phenolic, herbaceous, slightly floral, and bitter taste, with slight lemony and pungent notes. It has an aroma resembling that of marjoram. European or Mediterranean oregano is milder and sweeter in flavor than Mexican oregano that is darker in color and stronger and more robust in flavor. Oregano has a more potent flavor when it is dried.

Chemical Components: fresh oregano has up to 4% essential oils, mainly phenols and monoterpene hydrocarbons. European or Greek oregano (used mostly in the United States) has about 1.0% to 2% essential oil, which is yellowish red to dark brown, and contains 60% to 70% phenols, mainly thymol and carvacrol. Greek oregano has terpinen-4-ol (46%), α-terpinene (7%), α-terpineol (7.6%), sabinene (2.8%), linalool (1%), *cis*-sabinene (7.6%), ρ-cymene z (3.2%), linalyl acetate and β-caryophyllene (2.6%). Mexican oregano has 3% to 4% essential oil, which is higher than the Mediterranean types.

The amounts of the two types of phenols, carvacrol and thymol, vary depending upon the origin of the oregano, thus giving differences in flavor profiles. The common or Greek oregano has 5% thymol and 7.5% carvacrol; Mexican oregano 50% thymol and 12% carvacrol; Turkish oregano 83% thymol and 0.8% carvacrol; Italian oregano 60% thymol and 9% carvacrol, and Spanish oregano 18% thymol and 4.5% carvacrol.

Oleoresin oregano (made from Spanish oregano) is a dark, brownish green liquid, and 4 lb. of oleoresin are equivalent to 100 lb. of freshly ground oregano.

Dried oregano contains vitamins A, E, C, niacin, B6, riboflavin, potassium, iron, magnesium, calcium, and phosphorus. Oregano inhibits iron absorption.

It is an effective antioxidant for mayonnaise, other fat-based dressings, and meat products. It also inhibits molds, S. aureus, E. coli, B. subtilis, and other bacteria.

How Prepared and Consumed: the ancient Greeks and Egyptians used oregano to flavor fish, meats, vegetables, and wine. Today, oregano is still a popular spice for the Greeks, who use it in salads, chicken, and seafood dishes. It is an essential flavoring in Italian pizzas, pasta sauces, and roast or grilled meats. It is also used in Italian, Spanish, and Latin American dishes—meat stews, lamb and pork roasts, veal dishes, stuffings, salad dressings, soups, taco fillings, chili, corn, omelettes, and bean dishes. Oregano is a popular flavoring in Middle Eastern dishes. It has great affinity for garlic, cumin, cilantro, sweet basil, lemon, chile peppers, tomatoes, eggplant, peppers, zucchini, capers, olives, vinegar, and cheeses.

Mexican oregano is used in authentic Mexican cuisines, in sauces, bean dishes, empanada fillings, burritos, or tacos. It is full bodied and pairs well with ground meat, garlic, onion, cumin, and chili powder. It is used with paprika, cumin, and chile peppers in chili powder, a common flavoring for Tex-Mex style beans, meat stews, sauces, roasts, soups, and in chile con carne. Oregano is also popular in Southwestern foods, such as green chile blends, soups, chowders, black beans, and sauces.

Spice Blends: chili powder, pizza sauce blend, pasta sauce blend, zatar, chili con carne blend, green chile-blend, frijole blend, Sante Fe seasoning, and Southwestern blend.

Therapeutic Uses and Folklore: ancient Egyptians used oregano for healing and as a disinfectant; Greeks used it to treat spider and scorpion bites, and Romans used it to stimulate hair growth and give a clear complexion. It is added to tea and taken to treat exhaustion, nervous disorders, and tension. Oregano is used as a traditional remedy for stomach upsets, low blood pressure, whooping cough, skin irritations, throat inflammations, toothaches, headaches, and asthma. Recent studies have found it to be effective against respiratory tract and urinary tract disorders.

PAPRIKA

It comes from the Hungarian word *paparka*, a variation from the Bulgarian *piperka,* which in turn comes from the Latin *piper*, meaning pepper. In the United States and the rest of the world, paprika is the term for the dried ground or powdered form of the dried peppers of the nonpungent to slightly pungent red varieties of Capsicum annum. Brought by the Ottoman Turks to Hungary during the sixteenth century, the Hungarian term paprika includes all of the fresh varieties of peppers, including green bell peppers, as well as the ground form of all capsicum annum fruits. Hungarians have different grades of paprika, including sweet, semisweet, and pungent, based on the varieties and their degree of ripening.

Paprika was introduced from the New World into Spain and Turkey. Then, it was taken into Hungary in the sixteenth century when the Ottoman Turks invaded Hungary. The Hungarians came to call it *paparka*, which means Turkish pepper. Through different climates, soil conditions, and crossbreeding, milder capsicum annums were obtained from earlier, hotter American relatives.

Scientific Name(s): Capsicum annum. Family: Solanaceae.

Origin and Varieties: paprika peppers are indigenous to Central and South America but are now cultivated in Hungary, Morocco, Mexico, Spain, Turkey, and the United States. There are four types: American, Spanish, Mexican, and Hungarian, all varying in pungency and sweetness. Climate and environment affect the color of paprika. In the United States, the paprika used is the brilliant red, sweet type.

Common Names: pimiento, sweet red peppers, pod pepper, Hungarian paprika, tomato pepper, pimienton, sweet paprika, noble paprika, and Spanish paprika. It is also called filfil hila (Arabic), yafranj (Amharic), gamir bhbheg (Armenian), rod peber (Danish), timh jiu, tian jiao (Cantonese, Mandarin), paprika (Dutch), piment doux (French), piperia (Greek), paprika metuka (Hebrew), deghi mirch (Hindi), paprika-edeo/piros (Hungarian), paprica (Italian), papurika (Japanese), pimentao doce (Portuguese), peret krasnij (Russian), apapr (Serbian), pimiento dulce, paprika, pimenton (Spanish), pilipili hoho (Swahili), paprika (Swedish, German, Dutch), siling pangsigang (Tagalog), siper ngonpo (Tibetan), and kirmizi biber (Turkish).

Form: paprika is the brilliant red, ground powder from the thick flesh of dried red capsicum annum pods, with most of the seeds and veins removed. The ripe fruits are dried, ground, and sometimes cured. Paprika has a wide range of colors and pungencies. Depending on its origin and grade, it can be sweet to pungent.

Properties: paprika has a deep red or russet to a reddish orange and brownish red color depending on its grade. As it ripens, it becomes darker in color, from a purplish to almost blackish purple due to formation of anthocyanins. Spanish paprika is slightly more pungent than the paprika consumed in the United States. The U.S. paprikas have more pungent flavors. Hungarian paprika has a deep, round flavor and can range from a delicate and mild, very sweet with caramelized notes, to more pungent flavors. Hungarians classify their paprika as fiery (*eros*), semisweet (*feledes*) and premium sweet (*edesnemes*).

There are many grades based on degree of ripeness–pungency, color, and flavor and include *kulonleges* (special) which is bright red, delicate, aromatic, and mild pungency; *csemege* (delicatesse), with stronger flavor but no pungency; *edesnemes* (sweet), a dark red and slightly pungent grade and *rozsa* (rose), a dull yellowish red color, with no sweetness but most pungency, and which is exported around the world. Paprika loses its aroma and develops bitterness when exposed to high heat. It loses its color quickly (from a russet red to tan red and finally to a brownish red) when exposed to light and air. The paler reds and the brownish reds which are the poorer quality are the most pungent paprikas.

Chemical Components: paprika has very little essential oil, from 0.001% to 0.005%. Its oleoresin is dark red with 8 lb. equivalent to 100 lb. of freshly ground spice. Its reddish coloring matter is due to carotenoids (0.3% to 0.8% in fruit) and consists of 35% to 60% capsanthin, 18% capsorbin, 8% to 23% alpha- and beta-carotene, 8% to 10% zeaxanthin, 3% to 5% cryptoxanthin, and 8% to 10% lutein. Its aroma is due to 3-isobutyl-2 methoxy pyrazine and overall flavor due to long chain aliphatic hydrocarbons and fatty acids. When fresh paprika ripens further, its color ranges from a dark purple to almost black color due to presence of anthocyanins.

Paprika contains good amounts of vitamin A and vitamin C, with niacin, potassium, magnesium, sodium, phosphorus, and calcium. Sugar is high in the ripe paprika.

How Prepared and Consumed: paprika is an indispensable spice in Hungarian cooking and many eastern European countries. It is used in Hungarian and Serbian soups, stews, barbecues, omelets, and sausages. In Hungarian cuisine, it is an essential flavor in *porkolt* (thick stew of pork, goose, beef, duck, and game with onions), *tokany, szekelgulyas* (sauerkraut and goulash), chicken *paprikas* (creamy chicken stew), and beef *gulyas*/goulash (beef stews with potatoes, carrots, and onion). In *gulyas*, paprika is fried in lard to bring out its flavor. It also gives the red coloring in Hungarian salami and is used as a tabletop condiment, combined with other spices and lard.

Paprika is used in cream-based sauces in European cooking, mainly to add color. Spanish paprika, called pimento, is a favorite ingredient in Andalusian cooking. It has a rich, deep, and smoky flavor. It is used in fish dishes, sofritos, romesco (red pepper almond sauce), and in *berza*, a thick bean soup with vegetables, black sausage (*morcilla*), chorizos, ham hocks, clove, nutmeg, and cumin. Smoked paprika comes as dulce (sweet), *agridulce* (medium hot), and picante (hot).

In the United States, paprika is valued for its coloring and mild flavor. It provides visual appeal for lighter colored foods. Paprika is used to color and garnish potatoes, cheeses, eggs (deviled eggs), salad dressings, fish, soups, and vegetables, and drizzled over meats and chicken. It is used commercially in sausages, condiments, salad dressings, and other meat products. It flavors stuffings containing olives, fish, and rice. Southwestern and Texas cuisines enjoy paprika in their chilis, soups, stews, and sauces. It goes well with the added cumin, oregano, coriander, and cilantro.

Paprika is a popular spice in North Indian, Turkish, Moroccon, and Middle Eastern stews, sauces, dips, and roasts. In North India, paprika is fried in oil (not overdone to cause bitterness) with other spices to provide color and taste for curries. In the Arab regions of the Middle East, paprika forms an important ingredient in many spice mixtures, such as *ras-el-hanout*, *berbere*, *galat dagga*, and *baharat*. *Baharat* is a spice mixture that is fried in oil or clarified butter for use in mutton and other dishes. *Chermoula*, a puree of paprika, garlic, saffron, cayenne, lemon juice, and cilantro, is a marinade or sauce used for meats and fish in Morocco, Yemen, and other neighboring regions.

Paprika also pairs well with onions, sour cream, chile peppers, black pepper, corn, black beans, tomato, turmeric, and garlic and pork. It develops a brilliant reddish orange color and a distinct sweet, pungent flavor when fried in oil. Due to its high sugar content, it should not be overheated, otherwise it will turn bitter.

Spice Blends: goulash blend, chermoula blend, baharat, porkolt blend, curry blends, paprikash sauce blend, sofrito, romesco, ras-el-hanout, berbere, chili blend, and galat dagga.

Therapeutic Uses and Folklore: paprika is traditionally used to help prevent colds and influenza and aid circulation problems.

PARSLEY

Parsley is popularly used as a garnish all over Europe. Derived from the Greek word *petroselinon*, meaning rock celery, it is an important component of bouquet garni and fines herbes in French cooking. It is also used by Europeans as an after-dinner breath cleanser. Throughout the centuries, parsley was not always appreciated as a culinary ingredient. At one time, it was associated with ill luck by the English and the Greeks. But later, the Greeks and Romans associated it with speed and strength. It has also been used since antiquity in Jewish Passover meals as a symbol of new beginnings.

Scientific Name(s): flat leaf parsley: *Petroselinum (P) satiuum;* curly leaf parsley: *P. crispum;* root parsley (Hamburg): *P. crispum tuberosum.* Family: Apiaceae (parsley family).

Origin and Varieties: parsley is indigenous to the eastern Mediterranean regions, Greece, Turkey, and Sardinia. It is now grown throughout Europe, especially in Hungary, Germany, France, Holland, the Middle East, and the United States. There are many varieties, but the most commonly used types are flat leaf (or Italian parsley), curly leaf, and Hamburg parsley. In Asian regions, parsley is referred to as a foreign or western coriander.

Common Names: garden parsley, common parsley, western coriander and Chinese coriander. It is also peterzili (Amharic), called makdunis (Arabic), azadkegh (Armenian), heung choi, xiang cai (Cantonese, Mandarin), peterselie (Dutch), zaafari (Farsi), persil (French), petersilie (German), maintano (Greek), petrozil'ya (Hebrew), ajmood (Hindi), seledri (Indonesian), prezzemolo (Italian), paseri (Japanese), vhans bairang (Khmer), kothambelari (Malayalam), pasli (Malaysian), persille (Norwegian), salsinha (Portuguese), petrushka (Russian), perejil (Spanish), persilja (Swedish), kintsaj (Tagalog), pak chee farang (Thai), maydanoz (Turkish), and rau mui tay (Vietnamese).

Form: the leaves come fresh or dried. They are used whole, flaked, chopped, minced, or pureed. The seeds, root (Hamburg parsley), and stems of the parsley plant (Neapolitan parsley) are also used as flavorings.

Properties: flat leaf parsley has a less harsh flavor than the seed. Curly leaf parsley has no flavor, so it is used as a garnish. Flat leaf parsley has a grassy, green taste with a herbaceous and slight lemony aroma. Dry parsley has no flavor. The greyish brown parsley seed has a bitter, harsh, and terpeny taste. The root has a strong aroma.

Chemical Components: commercial parsley oil is manufactured from the whole herb or seed. Parsley leaf has 0.06% to 0.1% essential oil and has a more desired aroma than the seed oil. The leaf has mainly myristicin (20.6%), apiole (18.3%), α-pinene (5.1%), β-phellandrene (12.1%), myrcene (4.3%), limonene (3.6%), *p*-mentha-1,8-triene (9.2%), α-*p*-dimenthylstyrene (7.2%), aldehydes, ketones, and phenols.

The seed has 1.5% to 3.5% essential oil, with mature seeds containing up to 6%. Its apiole content is 36.2%, and its myristicin content is 13.3%. Apiole gives its characteristic odor and taste. It is toxic at higher levels, so only small quantities of fresh parsley should be eaten at any one time.

Oleoresin parsley is prepared from the seed and blended with the leaf oil. The oleoresin from the leaf is deep green; 1/3 lb. is equivalent to 100 lb. of fresh parsley, and 3 lb. are equivalent to 100 lb. of dried parsley.

Fresh parsley has high levels of vitamin A (8230 IU/100 gm) and vitamin C (125–300 mg/100 gm), with folic acid, iron, potassium, sodium, magnesium, and calcium. The dried form has higher levels of vitamin A (23,340).

IU/100 gm), calcium, iron, sodium, phosphorus with vitamin C, niacin, magnesium, and manganese.

How Prepared and Consumed: curly leafed parsley is used primarily as a great garnish to add color and visual appeal to many foods. Flat leafed parsley has a mild flavor that can harmonize mild leafy spices, such as chives and chervil, tame more assertive ones such as cilantro, bay leaf, or tarragon, and extend pungent types such as basil or mint. Flat leafed parsley is used as a garnish (for salads, soups, boiled potatoes, and egg dishes) and is chopped and blended in dips, condiments, quick cooked sauces, and stews. Fresh parsley is typically added toward the end of cooking, because prolonged heat tends to destroy its flavor.

Parsley is chopped and used as a popular garnish in Central Europe. It is commonly found in many European foods such as soups, sauces, fish, meats, pies, and poultry dishes. Parsley is a popular addition to Turkish, Iranian, Lebanese, and other Middle Eastern dishes. It is used in appetizers such as hummus or *baba ghanoush*, bean soups, *taboulleh*, rice dishes, couscous, meats, and stews. The French finely chop parsley and garlic, then saute this mixture and add it to broiled lamb, chicken, beef, or fried fish.

Parsley pairs well with garlic, shallots, chicken, butter, lentils, beans, vinegar, potatoes, and legumes. It is finely chopped and mixed with garlic or shallots to make *persillade* which is added toward the end of cooking to flavor French stews, vegetables, and meat dishes. Parsley root is used as a vegetable to enhance soups, stews, and condiments of Europe. It retains its flavor after long cooking times. In Latin America, it is chopped and added with garlic and olive oil for chimmichurri and other sauces and condiments.

Spice Blends: bouquet garni, fines herbes, persillade blend, gremolada blend, pesto blend, and chimmichurri blend.

Therapeutic Uses and Folklore: traditionally, parsley has been used as a diuretic and to treat stomach ailments, menstrual problems, arthritis, and colic. It is believed to maintain the elasticity of blood vessels, clean the blood, and speed oxygen metabolism. With its high chlorophyll content, it absorbs odors and is used to remove halitosis.

PEPPERS: BLACK, WHITE, GREEN, PINK, LONG/PIPPALI, CUBEB, NEGRO, TASMANIAN

Black pepper has long been known as the "King of Spices." The word "pepper" comes from the Sanskrit word *pippali,* which is Indian long pepper, a relative of black pepper. Native to the Malabar Coast of southwest India, black pepper was introduced to Egypt and Europe through Arab traders.

Black pepper was a precious commodity in ancient times. Measured like gold, it played a vital role as a medium of exchange, whether to pay taxes, rents, dowries, or ransoms. In ancient times, the Romans lavishly used it to spice their foods, and it became a status symbol of fine cuisine in Europe. Black pepper became so highly prized that it became a quest for European nations to discover the fastest routes to use to obtain the spice.

Long pepper or *pippali* has origins in India and Indonesia. Long before the use of black pepper, this pepper was highly regarded in Europe and, thus, became the source of the word pepper. Today, black pepper is used abundantly all over the world, while long pepper's use has disappeared, except in India.

When pepper became a monopoly by Arab and other traders, and price of black pepper was high, "substitute" peppers, such as grains of Paradise, Negro pepper, and cubeb pepper were used abundantly in ancient times. Nowadays, they are rarely used, except in West and North Africa and Indonesia. Many kinds of fruit peppers exist that have varying flavors and colors, based on their origins and how they are processed.

The mild and sweet-tasting pink peppercorns or rose peppers, from Brazil, South America, that are found in mixed peppercorns, are not true peppers. They are from the cashew family.

Cubeb pepper, also called Javanese pepper, derives its name from the Indonesian word "cabe" meaning pepper. It is indigenous to Indonesia and West Africa. These are small berries of a pepper relative, and are similar in appearance to black pepper except that they have a "tail." They are also sometimes confused with cassia buds. Cubeb pepper was first brought to Africa and the West by Arab traders who used it to season their meat dishes. Later, it became a popular spice in North Africa. In Benin, West Africa, it is referred to as "piment pays," meaning "pepper of the country." Ashanti peppers are false cubeb peppers.

Negro pepper, also called Moor pepper or "kien" in West African regions, is a popular spice used from Ethiopia, north east Africa, to Ghana, southwest Africa. It was popular in Europe before black pepper. Belonging to the custard apple family, these are long dark brown pods that are smoked before use. It has a mixture of cubeb and macelike notes.

Szechwan pepper or *fagara* (discussed earlier in this chapter) from China, and *sansho* (Japanese pepper) and the pink and green peppercorns from South America are also popular peppers emerging in the United States.

Tasmanian pepper or mountain pepper, popular in Australian cuisine, and which resembles black pepper in color and size, is becoming better known in the United States. It has a sweet, pungent, and numbing taste. It is popular with meat marinades, stews, and many other local applications.

Scientific Name(s): black/white/green pepper: *Piper (P) nigrum;* long pepper: *P. longum* (India) and *P. retrofractum* (Indonesia); cubeb pepper: *Piper cubeba* (Indonesia, North Africa). Family: Piperaceae (pepper family). Negro pepper: Xylopia (X) aethiopica or X. aromatica, Family: Annononaceae (custard apple family). Tasmanian pepper: Tasmannia (T) lanceolata or T. aromatica, Family: Winteraceae, are not from the pepper family but will be included here.

Origin and Varieties: black and white peppers are native to the southwestern Malabar Coast of India. There are many varieties of black pepper, depending on its origins. It is now cultivated in Thailand, Sarawak (in Malaysia), Sumatra (in Indonesia), Madagascar, Vietnam, Singapore, Sri Lanka, and Brazil. Long pepper is native to India and Indonesia. Grades of pepper are identified by their origins, with Tellicherry (or Thalassery) and Malabar being the higher grades of peppers. Cubeb pepper comes from west Africa while Negro pepper is found in Africa and Brazil, and Tasmanian pepper originates from Australia, especially grown in Tasmania.

Common Names: black pepper is known as black gold. It is also called kundo berbere (Amharic), fulfol aswad (Arabic), beghbegh (Armenian), kalomirich (Bengali), nayukhon (Burmese), hak wuh jiu, hay hu chiao (Cantonese, Mandarin), sort peber (Danish), zwarte peper (Dutch), felfel siah (Farsi), poivre noir (French), schwarzer pfeffer (German), piperi mauro (Greek), pilpel shahhor (Hebrew), kali mirchi (Hindi, Urdu), feketebors (Hungarian), merica hitam (Indonesian), pepe nero (Italian), kosho (Japanese), lada hitam (Malaysian), kuru mulagu (Malayalam), hay hu chiao (Mandarin), kala mire (Marathi), pimienta preta (Portuguese), chyorny pjerets (Russian), pimienta negra (Spanish), svartpeppar (Swedish), karu mulagu (Tamil), prik (Thai), kara biber (Turkish), and tieu den (Vietnamese).

White pepper: fulful abyad (Arabic), bhak wuh jiu, bai hu chiao (Cantonese, Mandarin), hvid peber (Danish), witte peper (Dutch), felfel sefid (Farsi), poivre blanc (French), weiber pfeffer (German), piperi aspro (Greek), pilpel lavan (Hebrew), mirch (Hindi), feherbors (Hungarian), hvit pepper (Norwegian), merica putih (Indonesian), pepe bianco (Italian), lada putih (Malaysian), bai hu chiao (Mandarin), pimento branca (Portuguese), byely pjerets (Russian), pimienta blanca (Spanish), vitpeppar (Swedish), paminta (Tagalog), vella mulagu (Tamil/Malayalam), prik kao (Thai), and beyaz biber (Turkish).

Green pepper: poivre vert (French), gruner pfeffer (German), pilpel yarok (Hebrew), merica hijau (Indonesian), pepe verde (Italian), lada hijau (Malaysian), pimiente verde (Portuguese), zelyony pjerets (Russian), and gronpeppar (Swedish).

Long pepper: Balinese pepper and Bengal pepper. It is also called timiz (Amharic), fleyfla taweela (Arabic), pipoli (Assamese), pipool (Bengali), yeung jiu, xiang jiao (Cantonese, Mandarin), langwer piber (Danish), poivre long (French), langer pfeffer (German), macro piperi (Greek), pipara (Gujerati), pipli, pipal (Hindi), Bengali bors (Hungarian), cabe bali (Indonesian), Indonaga koshou (Japanese), morech ansai (Khmer), salipi (Laotian), thippali (Malayalam), cabai Jawa (Malaysian), magha (Punjabi), tipli (Singhalese), langpeppar (Swedish), litlit (Tagalog), tippali (Tamil), phrik hang (Thai), uzun biber (Turkish), pipul (Urdu), and tieu hoi (Vietnamese).

Cubeb pepper: Java peppercorn, Javanese pepper, and tailed pepper. It is also called hab alaru (Arabic), kabab chini (Bengali), kubeba (Dutch), kubabah (Farsi), poivre du Java, cubebe (French), kubeben pfeffer (German), koubeba (Greek), thada miri (Gujerati), kabab chini (Hindi), Javai bors (Hungarian), cabe java (Indonesian), lada berekor (Malaysian), kubeba (Japanese), cubeba (Portuguese), kubeba (Russian) and kubeba peppar (Swedish), china milagu (Tamil, Malayalam), hint biberi (Turkish), and tieu that (Vietnamese).

Negro pepper: also known as grains of Selim, Moor pepper, Senegal pepper, Kani pepper. Called fulful as Sudan (Arabic), piment noir de Guinee, poivre de Senegal (French), Senegalpfeffer (German), Afrikaniko piperi (Greek), arabbors (Hungary), kusillipia (Korean), pimento da Africa (Portuguese), and kumba perets (Russian).

Tasmanian pepper: mountain pepper, native pepper. Also called shan hu jiao (Mandarin), bergpeper (Dutch), poivre indigene (French), Tasmanischer pfeffer (German), Tasman bors (Hungary), and Tasmanijskij perets (Russian).

Form: pepper is a dried berry called peppercorn and comes whole, cracked, or ground. Ground pepper is available coarse or fine and is a mixture of light and dark particles. The black, white, green, and red peppercorns are berries picked at different stages of growth and processed using different methods.

Black peppercorn is a globular, wrinkled, dried, unripe berry or fruit with its outer hull or skin intact. This green or greenish yellow peppercorn (the immature berry) is picked, fermented for a few days, and then dried in kilns. During drying, it shrivels and becomes wrinkled and black. The most popular black peppers are as follows: Tellicherry pepper, from the coast of southwest India, is the most aromatic black pepper, with a fruity clean bouquet with little pungency; Malabar is the regular grade black pepper also with good aroma and little pungency; Lampong pepper, from Sumatra, Java, and Borneo in Indonesia, is small and grayish in color, is very pungent but not as aromatic as Tellicherry pepper or Malabar pepper. Sarawak pepper, from Malaysia, is milder than the Lampong pepper. Other varieties include Chinese black pepper (light in color with a mild flavor), and from Brazil, Madagascar, Vietnam, Ceylon, and Singapore, which are all mild with little aroma.

White peppercorns are berries stripped of the outer hull and picked when near ripe, when they are yellowish red or red in color. The berries are then soaked in water or steamed to soften and loosen their skin. The outer hulls or skins are removed by rubbing, leaving a smooth, light-colored berry. They are then bleached, rinsed, and sun dried. White peppercorns can also be prepared from black peppercorn by mechanically removing the outer hulls, a process called decortication. This latter type of white pepper tastes more like black pepper. So, white pepper is actually the inside seed and not the whole fruit like black pepper. The more popular varieties of white peppers are Muntock (from Banda Island near Sumatra, Indonesia), Sarawak (Malaysia), Brazilian, and Chinese.

Green peppercorns, from Amazonas, Brazil, are the unripe tender berries that are picked and air dried or freeze-dried and then packed in brine or wine vinegar to retain their color.

Red peppercorns are the ripe matured berries that are dried at high temperature. They are also pickled.

Long pepper/*pippali* are tiny berries that merge into a single rodlike structure that is about 1.5 inches long and is slightly tapered. Its taste is similar to black pepper, but it has a slight sweetness and a stronger taste. In India, it is an essential ingredient of the spice mixture used with betel nuts in paan leaf, which is chewed after a meal.

Cubeb pepper is a berry that is slightly larger than the peppercorn and has a furrowed surface. It comes with the stem attached, and thus, it is also called tailed pepper. Most cubeb berries are hollow. Cubeb pepper is sold whole, crushed, or ground.

Negro peppers are kidney-shaped seeds encased in long slender bean pods that are dark brown in color and come in clusters. Tasmanian peppers are sold dried and look like shriveled whole black peppercorns.

Properties: pepper provides aroma as well as a "bite" that is different from chile peppers, ginger, mustard, or other pungent spices. The essential oil gives the aroma, whereas the nonvolatiles, such as piperine, give the pungency. Each pepper has different flavor characteristics.

Black pepper has a less biting taste than white pepper but has a wonderful bouquet—a penetrating pungent and woody aroma when freshly ground. It has slight lemony and clove tones. White pepper has a sharp, winey, and biting taste with less harsh notes. It has little aroma and lacks the bouquet of black pepper. Both peppers lose flavor easily when ground. Green peppercorn is aromatic, with less of a sharp flavor than black or white pepper. Red peppercorns have a delicate sweet taste. Long pepper is hot and more pungent but with sweet tones. Cubebs are pungent and tealike with slightly musky and bitter notes. They have a terpeny aroma. When ground, cubebs release cumin-like notes and, when cooked, they release an aroma that combines allspice with curry. Negro peppers are aromatic with slightly bitter and pungent notes that are quite similar to cubeb and nutmeg or mace. Tasmanian peppers resemble black peppercorns in regard to color and size. The fruits have an initial sweet taste, followed by pungent notes and a later numbness on the tongue, like Sichuan pepper. Its leaves are also used sometimes.

Chemical Components: it has a colorless to pale greenish colored essential oil. Black pepper generally contains 1% to 2.6% essential oil, going up to 5%, while white pepper and long pepper contain less than 1.0%. Fixed oil is 2% to 9% in black and white peppers. Essential oil consists primarily of monoterpenes (80%), such as sabinene, α-pinene, β-pinene, limonene, and 1,8-cineol, all of which are responsible for the pungent aroma. During storage, ground black pepper can lose these monoterpenes. Sesquiterpenes make up the other 20% which include β-caryophyllene and humulene.

Black pepper has 10% to 15% oleoresin, which is dark green to olive green. In white pepper, the essential oil is about 8%.

The bite and pungency in black and white peppers is primarily due to the nonvolatile alkaloids, piperine and chavicin. The ratio of these two alkaloids varies in different peppers, thus giving rise to the different pungencies. White pepper lacks chavicin (which is present in the epicarp) and most of the essential oil (present in the outer mesocarp), but contains piperine, which gives the bite but not the aroma. Piperine accounts for about 98% of the total alkaloid compounds in both of these peppers, others being piperittine and piperyline.

Black pepper has aroma and bite. Its piperine content varies with its origins. Indian peppers are very aromatic, whereas Malaysian and Indonesian peppers are less aromatic but have more bite; Brazilian peppers have a milder bite than other black peppers because of their low piperine content.

Pipalli or long pepper has 1% essential oil, mainly phellandrene and limonene. It has a slightly higher piperine level than black pepper.

The dried fruit of the cubeb pepper has up to 10% essential oil, mostly sabinene, 1,4-cineol, and carene. Pungency is mainly due to cubebin with other related compounds.

Negro pepper has 2% to 4.2% essential oil, mainly β-pinene, 1,8 cineol, α-terpineol, paradol, linalool, β-ocimene, β-pinene, and other terpenes with traces of vanillin, depending on the fruit. Tasmanian pepper has polygodial which is responsible for its pungency, and its essential oil contains monotrpenes and sesquiterpenes.

Black, white, long, and green peppers contain potassium, calcium, sodium, magnesium, and iron.

How Prepared and Consumed: peppers were used in ancient times to preserve meats. Today, pepper is the most important table spice throughout the world. It cuts across many ethnic cuisines. Peppers are used in marinades or spice mixtures, added during cooking or sprinkled on foods at the meal table to adjust flavor. Cracked black pepper is popular in marinades, salad dressings, and spice blends. Pepper pairs well with garlic, ginger, coconut milk, lemon, vinegar, pork, beef, basil, cilantro, seafood, eggs, creamy sauces, and vegetables, red wine, fermented soybeans, and soy sauce.

Black pepper is popular in North American, European, Southeast Asian, and South Indian cooking. In Kerala, it is an essential ingredient in chicken and meat dishes, soups, and fish curries, while white pepper is popular in European, Chinese, Japanese, and Thai cuisines. In Europe, pepper is added during and after cooking to foods including steaks, soups, soup stocks, vinaigrettes, cheeses (example, boursin), vinegars, pickles, or salads. With European cooking, black pepper is traditionally added to more hearty and stronger flavored foods such as steaks, pot roasts, hamburgers, soup stocks, meatballs, gravies, sausages, pepperoni, and cold cuts. Creole, Cajun, and Southern cooking in the United States uses black pepper abundantly.

Thais and other Southeast Asians use white and black pepper, coarse or finely ground, in many stir-fry dishes, grilled meats, soups, and curries. Jamaicans enjoy black pepper in goat curries and jerk dishes. In France, black pepper is added to steaks, charcuterie dishes, cold cuts, and pates and is the main ingredient in quatre epices. Black pepper can be cooked for a long time without losing its flavor, as seen in béarnaise sauce and black peppercorn sauce.

Arabs monopolized the pepper trade for years and carried it from the Far East to Europe. It is no surprise that it is a popular spice in their cooking and is found in almost all of their spice blends, including *zhoug/zhug*, *baharat*, *ras-el-hanout*, *berbere*, and *galat dagga*.

White pepper is popular in Europe and is usually used in delicate white sauces, clear salad dressings, cream soups, fish dishes, and wine, and is added more for visual reasons. In Thailand, it is added with fried garlic, chile pepper, and other spices to fried chicken and pork dishes.

Long peppers are used in curries and spicy pickles (or *achars*) of India and Indonesia. They are also popular in North African and East African cuisines. Ethiopians add it to meat and chicken stews and seasonings such as *berbere*. The leaves and roots are used to make locally brewed beers.

Cubebs were a popular substitute for black pepper in Europe in the seventeenth century, but they are not any more because of its strong bitterness. Today, they are an essential ingredient in the spice mixtures of Morocco and Tunisia, including *ras el hanout* and stews of Benin, in West Africa. Negro peppers are used in West African cooking especially added to sauces and stews in Senegal and Cameroon.

Tasmanian pepper is significant in Australian cooking. It is crushed and rubbed on meats which are then grilled, such as kangaroo steaks, emu hamburgers, and added with acacia, lemon myrtle, and tomato to soups, pasta sauces, pestos, and stews.

Spice Blends: mignonette pepper, berbere, ras-el hanout, baharat, galat dagga, steak au Poivre blend, jerk seasoning, curry blends, achar blend, pest blend, and Sarawak peppercorn blend.

Therapeutic Uses and Folklore: in ancient times, pepper was used for bartering, as a bride's dowry, a form of currency, and rent money. It was also used to preserve meats to prevent spoilage. Black pepper was traditionally used to treat headaches, constipation, and diarrhea. In India and China, pepper is widely used to improve circulation and to improve hypersensitivity to cold, coughs, asthma, kidney inflammations, and muscle and joint pains. Long pepper is used for those who suffer from sensitivity to cold temperatures.

Cubeb oil is an ingredient used in throat lozenges.

POPPY SEED

Poppy seeds are not fully formed until the plant matures, by which time the plant has lost all of its opium potential. Poppy seeds do not contain opium or narcotics, as other parts of the plant. The dried latex or exudates, from the immature poppy plant gives opium, which contains the alkaloid painkillers, morphine and codeine. The poppy plant has been valued since ancient times as a medicine and painkiller. In the Ebers Papyrus, the Egyptians described it as a sedative. As early as 1400 BC, the Greeks valued poppy seeds as a pain reliever. Later, the Egyptians produced edible oil from poppy seeds and mixed it with honey to make flavorful breads. Today, whole or ground poppy seeds are used in confectioneries, baked goods, and sauces. Muslim missionaries introduced it into India and the Far East. In India, the poppy plant was used as a narcotic beverage at the imperial courts during the Mogul era.

Scientific Name(s): Papaver somniferum. Family: Papaveraceae (poppy family). There are many varieties but P. somniferum is the commercially cultivated variety. India is also the largest producer of opium alkaloids.

Origin and Varieties: it is indigenous to Eastern Mediterranean and West Asia. It is now cultivated in India, China, Turkey, Holland, France, the United States, and Canada.

Common Names: opium seed, garden poppy, and maw seed. It is also called khash khash (Arabic), mekon (Armenian), affugoch (Assamese), posto (Bengali), bhainzi (Burmese), yeung shuk huk, ying shu ciao (Cantonese, Mandarin), slaapbol (Dutch), khash-khash (Farsi), pavot somnifere (French), mohn (German), paparouna (Greek), pereg (Hebrew), khas-khas (Hindi, Gujerati, Marathi, Urdu), mak (Hungarian), papavero (Italian), keshi (Japanese), za zang (Laotian), biji kas kas (Malaysian, Indonesian), kasha kasha (Malayalam), papoula (Portuguese), post (Punjabi), mak snotvornyj (Russian), ababa, adormidera, semilla de amapola (Spanish), vallmo (Swedish, Norwegian), kasa kasa (Tamil, Telegu), ton fin (Thai), hashas tahumu (Turkish), and cay thuoc phyen (Vietnamese).

Form: poppy seeds are tiny, kidney-shaped seeds with creamy, brown, red, bluish black or grey color, depending on their origins. When the seeds are ripe, they are used as flavorings and for thickening sauces. Poppy seeds are used whole, ground, as a paste, or as pale yellow nutty oil. The seeds have high protein content but lack the amino acids, methionine, and lysine.

Properties: they have a nutty and slightly sweet taste with a slightly smoky aroma.

Chemical Components: narcotic alkaloids are not present in the seeds that are used in foods. Poppy seeds contain no volatile oil but have 40% to 60% fixed oil that is high in unsaturated fatty acids, such as linoleic acid (60%), oleic acid (30%), and linolenic acid (3%). It is low in saturated fatty acids (about 10%). The oil has a pleasant aroma, quite similar to almond oil.

Poppy seeds contain protein and good amounts of lecithin, minerals (magnesium, zinc, iodine, copper, and manganese), and oxalic acid.

How Prepared and Consumed: the Egyptians mixed poppy seeds with honey as a dessert for their pharaohs, the Turks and Germans made a bread with poppy seeds and flour, and the ancient Indians mixed it with sugarcane juice for a confection. The paste is used in desserts and sauces in Turkey and in the Mogul cooking of northern India. Poppy seeds are roasted and ground to a paste that forms the basis of many Turkish dishes.

The creamy seeds are roasted to obtain a more intense flavor and are then ground with other spices to thicken and add flavor to meat and fish curries in India. The oil is used as a cooking oil and salad oil. The whole seeds are added to vegetables and curries to provide texture. In Japan, they are used as an ingredient in a popular seasoning called *shichimi togarashi.* Poppy seed oil is used as a cooking or salad oil in Europe and Asia.

In Europe and the United States, the seeds are sprinkled on breads, buns, bagels, cookies, and cakes. Eastern Europeans mix poppy seeds with honey or sugar to make cakes and fillings for strudel and croissants. In the United States, the seeds are used as toppings to provide nuttiness to rolls, breads, cookies, pastries, and cakes. Austrians grind poppy seeds and combine them with melted butter and powdered sugar and serve them with a yeast-based dumpling called *germknodel.* Mediterraneans and northern Europeans use poppy seeds in tuna fish and macaroni salad, sour cream and cheese dips, salad dressings, soups, cooked vegetables, boiled potatoes, and coleslaw. The Jewish holiday cake known as *hamentachen* is made with crushed poppy seeds, beaten egg, lemon juice, honey, salt, and water or milk.

Spice Blends: shichimi togarashi, hamentachen blend, fish curry blend, and germknodel filling blend.

Therapeutic Uses and Folklore: Egyptians, Greeks, Arabs, and Asians used the unripe seeds for their narcotic properties to relieve pain and act as a sedative. They are also used to treat constipation and bladder issues.

ROSEMARY

Rosemary is a flavoring that is deeply rooted in European cultures since 500 BC. Native to the Mediterranean region, it was grown around monasteries. It was noted for strengthening the memory, and ancient Greek students wore garlands of rosemary when taking exams so as to retain their knowledge. It was also a symbol of fidelity. During the Middle Ages, its name was changed from the Latin word *rosmarinus,* which means "dew of the sea," to *Rosa Maria* or rosemary, in honor of Mary, mother of Jesus. It was used in religious ceremonies, weddings, and funerals and as a symbol against evil. Today, it is still a popular spice in Europe, especially in Italy and France.

Scientific Name(s): Rosmarinus offinialis. Family: Labiatae or Lamiaceae (mint family).

Origin and Varieties: native to the Mediterranean, rosemary is cultivated in Algeria, Britain, France, Germany, Italy, Morocco, Portugal, Russia, Spain, Turkey, and the United States.

Common Names: it is also known as ikleel aljabal (Arabic), kngooni (Armenian), maid it heung, mi tieh hsiang (Cantonese, Mandarin), rosmarin (Danish), rozemarijn (Dutch), old man (English), eklil kuhi (Farsi), rosmarin (French), rosmarein (German), rozmari (Greek), rozmarin (Hebrew), rusmary (Hindi), rozmaring (Hungarian), ramerino (Italian), mannenro (Japanese), alecrim (Portuguese), rozmarin (Russian), romero, rosmario (Spanish), rosmarin (Swedish), romero (Tagalog), biberiye (Turkish), and la huong thao (Vietnamese).

Form: rosemary leaves are used fresh or dried, whole, chopped, crushed, or ground. They are small, narrow, and pine or needlelike. Fresh leaves are leathery, shiny, and dark green. Dried leaves are curved with dark green to brownish green colors.

Properties: when the leaves are crushed, they provide a strongly aromatic, piney, and tealike fragrance with a slightly sweet taste. Rosemary leaves have a slightly minty, sagelike, peppery, balsamic, and camphorlike taste with a bitter, woody aftertaste. The dried form retains its strong flavor notes. The stems and flowers are also aromatic.

Chemical Components: rosemary has 0.5% to 2.5% volatile oil, mainly 1,8-cineol (30%) (which gives rosemary its cool eucalyptus aroma), borneol (16% to 20%), camphor (15% to 25%), bornyl acetate (2% to 7%), and α-pinene (25%). Different varieties differ in flavor depending on their constituents. Its oleoresin is a greenish brown semisolid, and 5 lb. is equivalent to 100 lb. of ground, dried rosemary.

It contains vitamin C, vitamin A, potassium, sodium, calcium, magnesium, iron, and phosphorus.

How Prepared and Consumed: rosemary is a very popular spice in the Mediterranean region. It is used by Europeans in roast chicken, lamb, or veal, pot roasts, stews, potatoes, stuffings, marinades, vinegars, soup stocks, fruit-based desserts, breads, cold beverages, and sausages. Italians commonly use it with wine, honey, garlic, or chile peppers on grilled or roasted lamb, goat, veal, fish, shellfish, and rabbit dishes. French add it to quatre epices to flavor slow-cooked lamb and vegetable dishes. Rosemary pairs well with olive oil, lemon, cinnamon, clove, garlic, lamb, pork, chicken, tomato- and cream-based sauces, pizzas, and grilled vegetables.

In Tunisia, a strongly aromatic rosemary called kilil is used in a lamb tagine dish with paprika, cinnamon, onion, and black pepper.

Rosemary does not lose its flavor with long cooking. The whole leaves are crushed before use to release their aroma. Whole sprigs used in cooking are removed before serving. The stems (stripped of leaves) are used as aromatic skewers for broiling or barbecuing chicken, shrimp, or vegetables. The flowers are used in salads or in fruit fillings and purees.

Rosemary oleoresin is used as a natural antioxidant and helps preserve the color of processed meats. It is used sparingly because of its strong aroma.

Spice Blends: bouquet garni, herbs de Provence, chicken roast marinade, pizza sauce blend, and lamb tagine blend.

Therapeutic Uses and Folklore: rosemary has been used traditionally to treat whooping cough, fluid retention, poor circulation, jaundice, migraine, mental fatigue, panic attacks, irritability, and aching joints. The Romans used it for insect bites, the Greeks used it for curing jaundice, the Arabs used it to refresh the memory, and the French used it to sanitize the air in hospitals.

In ancient times, it was used to preserve meat and as a fumigant in hospitals to kill bacteria. Commercially, rosemary oleoresin is used to preserve the color of meat.

SAFFRON

Saffron derives its name from the Arab word *za'faran,* which means yellow. Called *kumkum* in Sanskrit, saffron's use as a dye, by the Sumerians, dates back to 2300 BC. The yellowish orange saffron color is auspicious to many cultures: the saffron paste or *kungumam* in Hindu temples, the Buddhist monks' saffron-colored robes, and the saffron cloaks of the Irish kings. Saffron originated around ancient Greece and Turkey, and was taken by Arab traders to Spain, Iran, and Kashmir (India), where it has become a popular coloring and flavoring. In the United States, the Pennsylvania Dutch brought saffron farming from Germany. They continue to grow saffron and use it in their cooking.

Saffron is one of the oldest and world's most expensive spices because it is harvested and processed by hand. The flowers are picked by hand, the stigmas are then separated from the rest of the flower, and finally dried and sold as saffron threads. About 80,000 crocus flowers (240,000 stigmas) produce 1 lb of dry saffron threads using about 10–12 days of labor. In the United States, one pound of high quality saffron sell retail for $3000 to $5000.

Scientific Name(s): Crocus sativus. Family: Iridaceae (iris family).

Origin and Varieties: saffron is indigenous to the eastern Mediterranean—Turkey, Iran, Greece, and the Middle East. Today many varieties are cultivated in Spain, Iran, Greece, Turkey, France, Kashmir, India, and the United States. Most U.S. saffron comes from Spain and is called saffron Mancha. The most prized saffron comes from Kashmir, India, Spain, and Iran, and the least expensive comes from Mexico. The U.S. saffron is emerging as equal in value to Iranian saffron. Because saffron is very expensive, whole threads are often adulterated with other parts of the flower, dyes, and other plants such as safflower (also called false or bastard saffron, or Mexican saffron), or the ground saffron is adulterated with turmeric, also known as Indian saffron.

Common Names: Spanish saffron, hay saffron, and crocus. It is also called za'faran (Arabic), saffron (Amharic), kerkum (Armenian), jafran (Bengali, Assamese), fan hong fah, fan huang hua (Cantonese, Mandarin), safran (Danish), saffraan (Dutch), zaafaran (Farsi), safran (French, German), safrani (Greek), za'afran (Hebrew), zaaffran/kesar (Hindi), safrany (Hungarian), kunyit kering (Indonesian, Malaysian), zaffarano (Italian), keshar (Marathi), safuran (Japanese), kesari (Nepalese), acafrao (Portuguese), shafran (Russian), azafran (Spanish), zafarani (Swahili), saffran (Swedish, Norwegian), kashuba (Tagalog), kungumapu (Tamil), kunkumapave (Telegu), ya faran (Thai), safran (Turkish), zafran (Urdu), and nghe tay (Vietnamese).

Form: saffron is dried stigmas of the flower, which are dark red to reddish brown wiry threads. Saffron is sold as whole threads or ground. Each saffron flower has three stigmas. For flavor release, whole saffron threads need to be soaked in warm to hot water or other liquids for at least ten minutes. Ground saffron has no flavor and is often adulterated with turmeric, marigold, and other ingredients.

Properties: saffron is deep burgundy, reddish orange, bright orange, or yellowish orange in color. It has a warm floral bouquet and a delicate, earthy, honeylike taste with bitter back notes. The flavors are stronger in the orange and red varieties from India, and milder in the yellow varieties from Spain. The deep red or orange types of saffron are of the best quality. The false types provide the yellow color but not the taste.

Chemical Components: saffron has 0.5% to 1.0% essential oil, mainly monoterpene aldehydes (such as safranal), terpenes (such as pinenes and cineol), and isophorone-related compounds. Its odor is due to 2,2,6-trimethyl-4,6-cyclohexidienal. The bitter taste is due to the colorless glycoside, picrocrocin. Its color is due to carotenoid pigments, especially the bright orange yellow, water soluble crocin. The other carotenoid pigments are alpha- and beta-carotene, lycopene, and zeaxanthin.

Saffron contains vitamins B1 and B2, vitamin A, calcium, sodium, manganese, potassium, and phosphorus.

How Prepared and Consumed: saffron is used sparingly for coloring and flavoring foods. The threads are crushed and then infused or steeped in hot water or milk for about twenty to thirty minutes for its color to be fully extracted. The colored infusion is then added to foods. Saffron can also be toasted dry and added directly to rice dishes. Saffron pairs well with savory and sweet dishes, such as curries, shellfish, garlic, chicken, cream sauces, soups, polenta, rices, seafood, puddings, flans, ice cream, and milkshakes. The English add it to scones, cakes, and breads.

Cumin, coriander, mints, rose essence, nutmeg, almonds, cashews, brown sugar, cardamom, milk, cinnamon, cloves, and raisins complement saffron. Acidic ingredients such as vinegar or lemon juice inhibit saffron's color development. If used at high levels, it can create bitterness in foods.

Saffron is an important ingredient in the fish-based dishes of the Mediterranean, such as *zarzuela de pescado* from Spain and bouillabaisse from France. It is an essential ingredient in many flavored rice dishes: paella in Spain, biryani and pillaos in northern India, *zafran pollous* in Iran, and risotto Milanese in Italy. Depending on the level of saffron used, the color of these foods varies from a deep orange to a turmeric-like yellow. Also, saffron is a common flavoring in India in desserts, including *kheer, kesari,* and *rasmalai,* ice-creams, and the yogurt drink, *lassi.* Saffron is an important flavoring and coloring in festive dishes of India, such as Mughlai style biryanis, pullaos, and meat dishes. In the United States, less expensive saffron is added to sauces for flavoring lobster and shrimp, to stewed potatoes and other vegetables, in mayonnaise, as a marinade for roasts or for desserts. The Arabs add saffron with cardamom to flavor coffees.

Spice Blends: paella blend, curry blend, biryani blend, bouillabaisse blend, rouille, kheer blend, kesari blend, and pollou blend.

Therapeutic Uses and Folklore: in India, saffron is regarded as a sacred color, and a wet paste is placed on foreheads when praying in temples. It is also used to symbolize the high castes in India. Other ancient cultures such as Greeks, Romans, Arabs, Chinese, and Persians have used it to perfume their baths. The yellow color was a symbol of royalty as well as piety, and it is still used as a natural dye to color robes for religious and royal occasions. It is also used as a hair coloring for royalty and as a symbol of hospitality for the wealthy.

It was used as a cure for kidney ailments and as an appetite stimulant in Greek medicine. In Spain, it is used as a folk remedy for lowering cholesterol and inhibiting tumor growth. In Ayurvedic medicine, saffron is used for urinary problems, as a digestive stimulant, and for relieving cramps, fevers, liver problems, and rheumatism. It is eaten with ghee to prevent diabetes, and to reduce joint inflammation and uterus pains.

SAGE

A member of the mint family, sage has been used since ancient times in Europe as a seasoning and as a medicine. Many Europeans called sage *herba sacra,* meaning "sacred herb" because of its powers with strengthening the memory and providing longevity. It was commonly taken as a herbal tea. The central Americans used sage for religious ceremonies and for medicinal use.

Scientific Name(s): Dalmation (English sage): *Salvia (S) officinalis;* clary sage: *S. sclarea;* Spanish sage: *S. lavandulaefolia;* Greek sage: *S. triloba.* Family: Lamiaceae (mint family).

Origin and Varieties: sage is native to the northeastern Mediterranean region. It is cultivated in Eastern Europe, China, Turkey, Greece, Italy, and the United States. There are many varieties—narrow leaf sage, broad leaf sage, garden sage, tricolored sage, purple leaf sage, golden sage, red sage, clary sage (France, Russia, and Morocco), Mexican sage, Spanish sage, and pineapple sage.

Common Names: Dalmatian sage, English sage, garden sage, common sage, and true sage. It is also known as maryameya (Arabic), yeghsbak (Armenian), bhui tulsi (Bengali, Hindi), lou mei chou, shao wei cao (Cantonese, Mandarin), salvie (Danish), salie (Dutch), mariam goli (Farsi), sauge (French), salbei (German), alisfakia (Greek), marva (Hebrew), zalya (Hungarian) salvia (Italian), seji (Japanese), kamarkas (Marathi), salvie (Norwegian), salva (Portuguese), sathi (Punjabi), shalfej (Russian), salvia (Spanish), salvia (Swedish), and shavliya (Turkish).

Form: common sage is slender, shiny, velvety, or furry green when fresh. It turns into a silvery gray when dried. The leaves are oblong or spear shaped and are covered with short, fine hairs. Sage is used fresh or dried, whole, minced, chopped, crushed, rubbed (ground coarsely), or finely ground. Greek sage has larger and thicker leaves.

Properties: sage's flavor varies from a mild, balsamic to a strong camphorous and herbaceous-like taste, depending on the variety. It provides a cooling sensation in the mouth. Common (or Dalmatian) sage is astringent, spicy, and bitter with a camphorlike aroma. Greek sage has stronger, camphorlike notes. Spanish sage is less bitter with a flavor in between the flavors of Dalmatian sage and Greek sage. Clary sage has muscatlike notes. Dried sage has a stronger flavor than fresh sage.

Chemical Components: sage has 1.5% to 3% essential oil, which is pale yellow to greenish yellow. Its composition differs in different varieties. Dalmatian sage has mainly thujone (28%), 1,8-cineol (12%), borneol (4%), camphor (23%), camphene (7.4%), α-humulene (5.31%), limonene (3.24%), β-caryophyllene (3.3%), and bornyl acetate (1.3%). Spanish sage lacks thujone but has more 1,8-cineol (27%) and camphor (20%). Greek sage has high levels of 1,8-cineol (39% to 67%) with smaller amounts of thujone (5%) and camphor (7%).

Sage oleoresin, which is dark green to brownish green and very viscous, is usually extracted from Dalmatian sage, and 7.5 lb. are equivalent to 100 lb. of freshly ground spice. It is used as a natural antioxidant because of its high phenol content.

Dried sage contains vitamin A, calcium, potassium, niacin, calcium, magnesium, phosphorus, and iron.

How Prepared and Consumed: sage is a popular spice in Italy, Greece, and other regions of Europe. It tends to dominate the flavor of a dish, so it is used sparingly in European cooking. Sage pairs well with onions, butter, tomatoes, garlic, fatty fish, game, pork, poultry, gravies, breakfast sausages, and turkey and goose stuffings. Italians use sage with veal, prosciutto, and red wine sauce in the the well-known *saltimboca alla Romana* dish. Sage is also used in pizzas or with meat, pork, and calf livers. Germans use sage in eel soups, English in stuffed duck or pork, while the French add it in charcuterie, sausages, and stuffing.

Middle Easterners use sage with fagioli in herb salads and in teas. In the United States, sage is used in chowders, pork, poultry stuffings, sausages, baked fish, and cheese.

Its oleoresin along with a small amount of ascorbic acid is used as a natural antioxidant to retain the color of processed meats.

Spice Blends: lamb tagine blend, charcuterie blend, saltimboca alla Romana blend, and goose stuffing blend.

Therapeutic Uses and Folklore: the Romans considered sage to be a sacred herb. The Greeks used it for snakebites. Persians and Europeans associated it with long life, and the British associated it with prolonged memory. Traditionally, sage is considered good for throat and respiratory infections, calming nerves, and reducing fevers. Sage helps soothe digestion, particularly after a rich, heavy meal. Sage also strengthens gums and whitens teeth.

SASSAFRAS

Sassafras is a native North American spice and is an important ingredient in Creole and Cajun cooking of Louisiana. French settlers in Louisiana introduced many local ingredients and flavorings from Native Americans, Spanish, and Africans and created spicier version of their dishes. One of the popular ingredients used is sassafras, which is added to their gumbos. The Choctaw Indians in Louisiana used ground or dried sassafras leaves (which are the major ingredient of file powder) to thicken stews and soups.

Scientific Name(s): formerly under *Sassafras (S) variifolium,* now called *S. albidium.* Family: Lauraceae (laurel family).

Origin and Varieties: sassafras is native to the eastern and southeastern United States (Louisianna).

Common Names: ague tree, cinnamon wood, saxifrax, file powder, gumbo file, and file (Creole). It is also called sasfras (Arabic), wong jeung, huang zhang (Cantonese, Mandarin), sassafras (Danish), sassafras (French), fenchelholzbaum (German), sasafras (Hebrew), szassza franz (Hungarian), sassfrasso (Italian), sassaforasu (Japanese), sassafras (Portuguese), sassafras (Russian), sasafras (Spanish), and cay de vang (Vietnamese).

Form: sassafras is light green in color and is available as fresh or dried leaves of varying sizes (1, 2, or 3 lobes) that are used whole or ground. Ground leaves are called file powder. The dried bark of root is also used as a flavoring.

Properties: the leaf has a pleasant odor that is astringent, with a bitter and anisiclike, lemony taste. The root has a woody and very bitter taste with a camphorlike odor.

Chemical Components: the leaf has about 2% to 3% essential oil, mainly -pinene, myrcene, phellandrene, citral, geraniol, linalool, and safrole. The bark has about 2% essential oil, mainly safrole (90%), which is not permitted as a food in several countries because it causes liver cancer. Safrole is removed from the leaf before it is used commercially in foods. The root has a high oil content of 6% to 9%, mainly α-pinene, phellandrene, eugenol, camphor, and thujone.

How Prepared and Consumed: Cajuns and Creoles use file powder (which may contain other spices such as coriander, allspice, and sage) to flavor and thicken meat, poultry, vegetable and seafood dishes, bisques, soups, and stews. It is used to thicken gumbo, a traditional Louisiana dish made with seafood, andouille sausage, okra, crawfish or chicken seasoned with thyme, black pepper, and paprika. File powder is added to gumbo right after it is cooked and before serving. It is blended in well. If any additional cooking is performed, file powder will give the gumbo a stringy texture.

Young sassafras leaves are chopped and used in salads.

The essential oil is obtained from the root, after safrole is removed, and this oil is used for flavoring the beverage, root beer, as well as meat products and confectionaries.

Spice Blends: file powder, gumbo blend, bisque blend, and root beer.

Therapeutic Uses and Folklore: the Choctaw Indians chewed sassafras roots to bring down fevers. Early European settlers in America used it in teas and cordials as a pain reliever, to remove kidney stones, to lower high blood pressure, relieve arthritis, gout, and eye inflammations.

SAVORY

Savory was introduced to England by the Romans, and it was named by the Anglo-Saxons for its strong spicy taste. Europeans used savory as a substitute for black pepper and so it came to be called "pepper herb." It is a popular ingredient for vegetable and legume dishes as it helps with their digestion. In ancient times, it was added with vinegar and used as a seasoning, as well as an aphrodisiac and cough medicine.

The German word for savory, *bohnenkraut*, means "bean herb" because they found that it reduces beans' flatulent properties as well as complements green beans, dried beans, and lentils. There are several varieties of savory including African and Iranian (sometimes referred to as *zatar*), but the two well-known types are summer savory and winter savory. Most commercial applications use summer savory.

Scientific Name(s): summer or garden savory: *Satureja (S) hortensis;* winter or mountain savory: *S. montana;* Spanish or Iranian type: *S. thymbra;* African: *S. biflora.* Family: Lamiaceae (mint family).

Origin and Varieties: summer savory is cultivated in France, Germany, Spain, Yugoslavia, Albania, Great Britain, Canada, North Africa, and the United States. Winter savory is cultivated in Spain, Eastern Europe, and Morocco.

Common Names: garden or summer savory is also called: tosinyi (Amharic), nadgh, zaatar (Arabic), kyngeni (Armenian), chubritsa (Bulgaria), fung leung choi, hsiang bao ho (Cantonese, Mandarin), bonenkrvia (Dutch), marzeh (Farsi), sariette (French), kondari (Georgia) bohnenkraut (German), throubi (Greek), satar (Hebrew), borsika (Hungarian), santoreggia (Italian), sabori (Japanese), sar (Norwegian), segurelha (Portuguese), chabjor (Russian), vitnisetraj (Slovenia), sabroso/tomillo salsero (Spanish), kyndel (Swedish), and dag rehani (Turkish).

Winter or mountain savory: also called sariette and savouree (French), santoreggia (Italian) and saborija (Spanish).

Form: savory leaves are small, narrow, and green when fresh and brown green when dried. Summer savory leaves are slightly larger than those of winter savory. Savory is used fresh or dried and whole or crushed.

Properties: summer savory has a fragrant herbaceous aroma with a sharp and slight peppery and bitter aftertaste. It is reminiscent of thyme, mint, and oregano. Winter savory has stronger, sharper, and spicier notes than summer savory. Spanish or Iranian savory has spicy thymelike notes while African savory has spicy lemony flavor.

Chemical Components: summer savory has 0.1% to 0.25% yellow to dark brown essential oil, while winter savory has 0.2% to 0.25% essential oil. The essential oil of summer savory mainly consists of carvacrol (3.4% to 50.4%), thymol (trace to 22.5%), γ-terpinene (2.1% to 60.3%), p-cymene (3.7% to 5.3%), limonene (0.2% to 5.3%), myrcene (0.5% to 2.8%), camphene, α-thujene (1.8% to 4.2%), borneol (trace to 34%). These chemical constituents vary widely, depending on its origins and types. European types have very different flavor profiles than the North African or Canadian types. African savory has mainly citral while Iranian has mostly thymol.

Savory contains niacin, vitamin A, calcium, iron, potassium, phosphorus, and magnesium.

How Prepared and Consumed: in the Mediterranean, savory is used mostly for vegetables, such as beans, cabbage, lentils, potatoes, and mushrooms. The Europeans use it as part of their herbes fines and bouquet garni blends to flavor sauces, game meats, and stuffing. Eastern Europeans including Bulgarians and Georgians add to grilled meats, potatoes, vegetables, and fish dishes. It pairs well with other sweet herbs, such as thyme, sage, and rosemary. Iranians and Arabs add to meat and vegetable dishes as well as for dips. Savory is used in scrambled eggs and other egg dishes, salads, vegetable juices, stuffing, charcuterie, and legume dishes. It enhances pork, chicken, hamburgers, fish chowders, soups, and sausages. Savory is normally added toward the end of cooking so its flavor is not destroyed. Commercially, savory is used in liqueurs, bitters, vermouths, condiments, gravy, soup mixes, and confections.

Winter savory goes well with stronger flavors such as game meats, liver pates, and mutton.

Spice Blends: fines herbes, herbs de Provence, charcuterie marinade, khmeli suneli, hamburger marinade, cabbage sauce blend, and kebab mariande.

Therapeutic Uses and Folklore: in ancient times, savory was used to reduce pain from bee and wasp stings. It was believed to stimulate the appetite, treat gastric upsets, and promote digestion, especially of difficult to digest foods. It was called "herb of amor" and was said to increase the sexual and love feelings of those who eat it.

Savory has antioxidant properties.

SCREW-PINE LEAF/PANDANUS/PANDAN LEAF

Daun pandan is the Malaysian and Indonesian name for this fragrant leaf. Screw-pine leaf was the name given by English traders who traveled to Asia. In Southeast Asia, pandan leaf is used to wrap chicken, meat, fish, and desserts before they are barbecued or steamed. They add distinct, sweet, floral-like notes to these products. Malaysians, Indonesians, and Thais add the bruised leaves or its extract to flavor rice dishes and glutinous and tapioca-based desserts and puddings. The whole leaf is used to wrap chicken and other meats before they are grilled or barbecued.

Scientific Name(s): Pandanus (P) odoratissimus (North India); *P. amaryllifolius* (Southeast Asia); *P. latifolius* (Sri Lanka); *P. veitchii.* Family: Pandanaceae (screw-pine family).

Origin and Varieties: screw pine is native to South Asia and Southeast Asia. There are a few varieties of screw pine that differ in flavor and appearance, depending on their origin. The most aromatic types are from Southeast Asia and Sri Lanka. Kewra is the flower of the South Asian variety of screw pine.

Common Names: pandan leaf. It is also known as kathey (Arabic), ketaky (Bengali), chan heung lahn, chan xiang lan (Cantonese, Mandarin) skrupalm (Danish), pandan (Dutch), pandanus (French), schraubenpalme (German), pandanus (Hebrew), rampe (Hindi), pandanuz (Hungarian), pandano (Italian), takonoki (Japanese), taey (Khmer), tay ban (Laotian), kaitha (Malayalam), daun pandan (Malaysian, Indonesian), skrupalme (Norwegian), pandano (Portuguese), rampe (Singhalese), pandano (Spanish), skruvpalm (Swedish), thazhai (Tamil), bai toey hom (Thai), and cay com nep (Vietnamese).

Pandan flower—*kewra* or *keora* (Hindi, Punjabi).

Form: Screw-pine leaf is long, thin, narrow, and green. It is sold fresh, frozen, or dried. The leaves and flowers also come as bright green extracts.

Properties: the dried leaves are less fragrant than the fresh leaves. The leaves have to be bruised or boiled in order to release their flavor. The leaves have a roselike, almondy, and milky sweet, vanilla-like flavor. The dried leaves have no flavor. The flowers are golden yellow and have a fragrant, strong, and sweet aroma.

Chemical Components: screw pine has low levels of essential oil, including 2-acetyl-1-pyrroline (which also gives the aroma in Thai and Basmati rice), styrene, linalool, and β-cayophyllene. It also contains piperidine-like alkaloids (pandamarine, pandamarilactones) that give screw pine its milky, floral-like taste.

How Prepared and Consumed: the leaf is used in curries of Sri Lanka and in Malaysian, Balinese, and Thai cooking. It is commonly used as a flavoring and coloring in Malaysian and Singaporean cooking, especially in Malay- and Nonya-style dishes. The screw-pine or pandan leaves are tied in a knot and placed in soups or stews that are being cooked. The leaf is also bruised or raked with the tines of a fork to release its aroma, pounded to release its aromatic juice, or even boiled to obtain its flavor. Pandan leaves are used as wrappers in Southeast Asian cooking to provide a distinct flavor to the foods. They are wrapped around chicken, pork, glutinous rice, fish, and desserts before grilling, roasting, barbecuing, or steaming.

Pandan leaves also enhance the flavor of seasoned rices, puddings, beverages, and curries. *Nasi lemak, nasi kuning,* and *nasi padang* are some of the fragrant pandan-flavored rices eaten in Malaysia and Indonesia. It pairs well with coconut milk, glutinous rice, lemongrass, milk, brown sugar, and turmeric. It also provides color to Indonesian, Thai, Malay- and Nonya-style glutinous rice-based desserts, candies, puddings, soups, and coconut drinks.

Screw-pine flower, which is more delicate and fragrant than the leaf, is used in North India to perfume biryanis. It goes well with rices, coconut, lemongrass, brown sugar, star anise, cumin, and nutmeg. Its extract, called *kewra,* is also commonly used to flavor Indian desserts such as *rasgulla* (cottage cheese in syrup), *gulab jamun* (fried cottage cheese in syrup), *rasmalai* (cottage cheese with condensed milk), cakes, and beverages.

The commercially available pandan leaf extract is much too bright green and does not totally capture its true flavor and color profiles.

Spice Blends: nasi lemak blend, nasi kuning blend, biryani blend, kueh lapis blend, rendang blend, rasmalai blend, and gulab jamun blend.

Therapeutic Uses and Folklore: in India, screw-pine leaves are sacred to the Hindu God, Shiva. In many Indian villages, the leaves are also tossed into open wells to scent the drinking water.

SESAME SEED

Sesame originated in India and is one of the oldest spices known since the Harappa civilization. Derived from the Greek word sesamon, it is a popular flavoring in the Middle East, Asia, Southern and Eastern Mediterranean, and northern Africa. The Assyrians believed their Gods drank sesame wine, while the Greeks and Romans used sesame on their long journeys to conquer new lands. There are three types of sesame, white, brown, and black, each being preferred by specific cultures.

Scientific Name(s): Sesamum indicum. Family: Pedaliaceae.

Origin and Varieties: sesame is indigenous to the Near East and northern India and today is cultivated in the Middle East, India, China, Korea, Egypt, East Africa,

Greece, and Mexico. There are many varieties, and, depending on their origins, they come with different colors and flavors.

Common Names: also called seli't (Amharic), benne (Wolof language of Sene-galese, West Africa), sem sem, juljulan (Arabic), sousma (Armenian), tisi (Assa-mese), til (Bengali, Marathi, Punjabi), wuh mah, zi mah (Cantonese, Mandarin), sesamzaad (Dutch), gingelly (English), khonjed (Farsi), till (French), sesam (Ger-man, Swedish), sesame (Greek), mitho tel (Gujerati), sumsum (Hebrew), gingili, til (Hindi), szezamfu (Hungarian), sesamo (Italian), goma (Japanese), khae (Korean), nga (Laotian), chitellu (Malayalam), bi-jan (Malaysian/Indonesian), gergelim (Por-tuguese), kunzhut (Russian), ajonjoli (Spanish), ufuta (Swahili), linga (Tagalog), ellu (Tamil), nuvulu (Telegu), nga (Thai), susam (Turkish), til (Urdu), and cay vung (Vietnamese).

Form: sesame seeds are small, flat, oval, unhulled seeds that are creamy white, brown, or black. Hulled seeds are pearly white in color. Sesame is used whole, ground, as a paste, or oil. The oil can be cold pressed or hot pressed or refined.

Properties: sesame has a nutty sweet aroma with a rich, milklike buttery taste. After roasting or toasting, the white seeds become golden colored and acquire a delicate almondlike flavor. Sesame oil is deep brown, aromatic, and nutty. Black sesame has a stronger, earthier, and nuttier taste than the other varieties.

Depending on regional preferences, sesame seeds are roasted or left unroasted before the oil is extracted, giving rise to oils with different flavor profiles.

Chemical Components: sesame has no perceptible volatile oil. It has 45% to 65% fixed oil, mainly olein, stearin, palmitin, myristin, sesamin, and sesaminol, with 30% protein. The oil contains mostly oleic acid (40%), linoleic acid (45%), and saturated acid (10%). When the seeds are roasted at high temperatures, furanes are formed—2-furyl-methanthiol, guaiacol, phenylethanthiol, and furaneol—which give roasted sesame oil its characteristic flavor. When sesame seed is mildly roasted, pyrazines are formed

Sesame has antioxidant properties due to sesamol.

How Prepared and Consumed: Romans, Arabs, Turks, Greeks, and Indians have eaten sesame with honey, jaggery, rice, dates, or licorice. African slaves brought it to the United States. White sesame seeds are used whole to add texture to bread and rolls (hamburger buns), biscuits, crackers, pastries, cooking oils, salad dressings, and margarine. Sesame pairs well with chicken, pork, breads, noodles, soy sauce, chickpeas, shallots, garlic, mint, cilantro, and mustard.

Middle Easterners use sesame to flavor a wide variety of foods. It is used in dips, sauces, breads, falafel, cakes, and sweets (*halva*). Unroasted white sesame seeds are ground to make a paste called tahini (sesame butter) that is used to flavor hummus or *baba ghanouj*. Tahini is also used as a spread for pita breads, a dip for vegetables, an accompaniment to kebabs and fruits, and as a sauce in cooking. *Zahtar*, a spice blend of toasted sesame seeds with thyme and sumac, is used as a dip and to add zest to vegetables, grilled meats, and breads. In Lebanon, sesame seeds are added to sugar syrup for a chewy sweet called *simsmiyeh*.

Asians prefer the more intense flavor of roasted sesame seeds. Indians dry roast the seeds to develop a more intense flavor and a crunchier texture before they are added to many Muslim-style savory dishes. Black sesame seeds are popular in

Japanese and Chinese cooking. Chinese and Japanese dry roast sesame seeds and sprinkle them as a garnish on confections, rice, meats, or steamed vegetables. They also grind them into a paste or extract the oil to flavor cold sesame noodles, sauces, and confections.

The hot-pressed unroasted oil, also called gingelly oil, is light and is different in flavor from the roasted sesame oil. It is used for Indian pickles and as a cooking medium in India, Burma, and Southeast Asia. Sesame oil from roasted sesame seeds adds flavor to many stir-fry dishes and baked goods in East Asia and Southeast Asia. This darker sesame oil is commonly used in Korean and Szechwan cooking with chile peppers, garlic, and ginger. For the best sesame oil flavor, the oil is added to a dish just before serving or is added toward the end of cooking. In Korea, dumplings called mandu contain fillings made of garlic, cloves, sesame seeds, roasted sesame oil, kim chee, and soy sauce.

In Mexico, sesame seeds are ground and added with chocolate, chilies, almonds, and spices for their famous moles.

Spice Blends: zahtar, tahini blend, shichimi togarashi, sesame noodle blend, mandu filling blend, stir-fry sauce blends, mole blends, and dumpling sauce blend.

Therapeutic Uses and Folklore: in Asia, sesame is thought to have magical powers and bring good luck. Sesame was traditionally used as a laxative, to relieve constipation and hemorrhoids. Gingelly oil (called nalanai in Tamil), commercially available in India and Southeast Asia, which possesses no sesame flavor, is used to massage hair and body before baths. It is believed to rejuvenate the body and mind.

Sesame has strong antioxidant properties.

SICHUAN OR SZECHWAN PEPPER/FAGARA

Fagara originated in China, where in ancient times it flavored foods and wines offered to the gods. The Chinese brewed it in teas and also gave it as a gift to friends. Today, fagara is commonly referred to as Szechwan pepper. It is a popular spice in South China, Japan, and North India, especially to flavor meat dishes.

Scientific Name(s): globally, there are a number of varieties including Sichuan, Japanese, Nepalese, Korean, Indian, or Indonesian, but the three most popular types are Chinese pepper: *Zanthoxylum (Z) piperitum or simulans*; Japanese pepper: *Z. sansho*; *Z. alatum*: Nepalese pepper; *Z. acanthopodium*: Indonesian pepper and Indian pepper: *Z. rhetsa*. Family: Rutacene (citrus family).

Origin and Varieties: szechwan pepper is indigenous to South China, Vietnam, Japan, and North and Western India.

Common Names: fagara, Chinese pepper, Japanese pepper, anise pepper, Nepalese pepper, Indonesian lemon pepper, Japanese prickly ash, and flower pepper. It is also called yan chiao, chi fa chiao (Cantonese), sechuan peber (Danish), sechuan peper (Dutch), poivre de setchuan (French), ksantosilum (Hebrew), anizs bors (Hungarian), Szechuan-pfeffer (German), tilfda (Hindi), andalimon (Indonesian), sansho (Japanese), tippal (Konkani), chopi (Korean), kok mak met (Laotian), kat-murikku (Malayalam), tirphal (Marathi), timur (Nepali), hua chiao, gan jiao (Mandarin), piment sechuan (Portuguese), sychuan skij perets (Russian), sezchuan peppar

(Swedish), chiit (Tagalog), ma lakh, mak kakh (Thai), yermah (Tibet), and dan cay (Vietnamese).

Sansho, sometimes called Japanese pepper, and *tilfda* (Indian pepper) are closely related to fagara. It is also called Japanischer pfeffer (German).

Form: fagara is a dark red to a reddish brown dried berry that is used whole, crushed, or coarsely ground. The seeds are round and hollow with a rough prickly exterior. They come as split-open berries, sometimes containing black bitter seeds that are removed before use. It is the berry that is used. The leaves (*kinome*) and young shoots are also used especially in Japanese cooking.

Properties: fagara is aromatic with woody, peppery, and bitter notes with undertones of citrus. The leaves have a minty lime flavor. Sansho is tangy and peppery and is characterized by biting taste and a numbing pungency. Korean variety has anisey floral scent and not as pungent; the Nepalese has citrusy notes, the Indian lemony pepper notes, and Indonesian variety has little pungency but with lime notes.

Chemical Components: the berry contains up to 4% essential oil, mainly terpenes such as geraniol, linalool, 1,8-cineol, citronellal, limonene, and dipentene.

Its pungency is due to the amides hydroxy α-sanshool and β-sanshool. Both of these amides degrade quickly after the pepper is ground. The Indian variety has mostly sabinene, limonene, α- and β-pinenes, ρ-cymene, and the terpinenes. The Nepalese has mostly linalool, limonene, methyl cinnamate, and cineol while Indonesian has mainly geranyl acetate, citronelol, and limonene.

Sansho leaves mainly have monoterpenes, such as citronellal, citronellol, and Z-3-hexenal, which give them their grasslike odor.

How Prepared and Consumed: Chinese dry roast fagara, add it to salt and grind this mixture coarsely for use as a dry condiment. The fagara is roasted to provide a more intense flavor bite to Szechwan dishes. It is an essential ingredient in Chinese five-spice blend, which commonly barbecues and roasts. In Southeast Asia, five-spice blend is used in Chinese-influenced meat, poultry, and duck dishes. Fagara pairs well with citrus, soy sauce, plum sauce, and black bean paste.

The Japanese grind sansho with nori and tangerine peel, black sesame seeds, and poppy seeds to make the popular seven spice blend called *shichimi togarashi*. This is sprinkled over grilled meats, soups, and noodles. A spicier and hotter version of the seven-spice blend is made with fagara or chile peppers. This spicy blend is a commonly used seasoning for noodle soups and as a table condiment. In Japan, sansho is sometimes preserved in soy sauce and is typically used with fatty foods. The Japanese garnish their soups with sansho leaves and add its shoots to misos. Koreans add it to meats, fish, vegetables, and kim chee.

In Goa, Kerala, and Gujerat, fish and vegetable dishes are seasoned with the Indian variety of fagara. It is usually not cooked but is added to the cooked dish just before serving. In Tibet, *momo* stuffing is made with yak meat, seasoned with the local variety of Sichuan pepper, garlic, ginger, and onion, then steamed and served with a condiment. The Nepalis frequently add it in their curries and pickles. In Indonesia, pork dishes, a favorite with the Bataks in North Sumatra, are flavored with Indonesian variety.

Spice Blends: Chinese five-spice, shichimi togarashi, Szechwan sauce blend, momo stuffing blend, pickle blend, and Chinese BBQ blend.

Therapeutic Uses and Folklore: Chinese used Sichuan pepper as a traditional cure for dysentery.

SORREL

The name sorrel is derived from the Germanic word *sur,* and the old French word *surele,* both meaning sour. It is an ancient herb used by Egyptians and Europeans to impart acidity to foods. Today, it is a popular flavoring for whitefish, soups, and salads in French cooking. Another sorrel called Jamaican sorrel is from the Hibiscus family and used in beverages and preserves of Mexico and the Caribbean.

Scientific Name(s): there are a few cultivated varieties of sorrel—garden, French, round leafed, and spinach dock. Garden sorrel: *Rumex acetosa* (6 inches, broad, arrow-shaped leaf); French sorrel: *R. scutatus* (buckler leaf); sheep sorrel: *R. acetosella* (smaller than garden sorrel but arrow shaped). Family: Polygonaceae.

Origin and Varieties: native to Europe and west Asia, sorrel is now cultivated in France, Egypt, and parts of Europe and the United States.

Common Names: garden sorrel, wild sorrel, French sorrel, sourgrass, and little vinegar. It is also called oseille (French), sauerampfer (German), hamtzitz (Hebrew), acetosa (Italian), and acedera (Spanish).

Form: sorrel has large, light to dark green, oblong-shaped, spinachlike leaves. It comes as fresh or frozen and chopped or whole.

Properties: its taste ranges from a refreshing, sharply acidic, or astringent spinachlike taste with bitter notes to a milder, lemony taste. The younger leaves are less acidic. The French variety has slight citruslike notes. The dried leaf loses its citrusy-like flavor.

The fresh leaf and flower are high in vitamin C, potassium, phosphorus, calcium, and magnesium.

Chemical Components: sorrel has a high level of binoxalate of potash, which gives it the acidic taste.

How Prepared and Consumed: Romans and Egyptians used sorrel in ancient times to offset rich, heavy foods. It is used typically in French and Egyptian foods, such as soups, sandwiches, salads, poached salmon, stewed or braised meats, and poached eggs. It goes well with fish, onions, pepper, potatoes, meats, pork, veal, eggs, salads, cream-based sauces, and goat cheese. Sorrel is pureed to flavor goose, fish, or soups or for use in condiments for meats. It is also used in teas. Tough meats can be wrapped in sorrel leaves to tenderize them before cooking.

Sorrel is cooked for a minimum time to preserve its fresh flavor. To prevent sorrel from blackening and developing a metallic taste, only stainless steel knives and noniron pots are used. Sorrel is a natural acidifier and can be a substitute for fresh lemon in salads, stews, and sauces.

Spice Blends: green sauce blend, potato soup blend, and meat marinade.

Therapeutic Uses and Folklore: the Romans and Greeks used sorrel to aid digestion and temper the effects of rich foods, to treat liver problems and for throat and mouth ulcers. It is traditionally used to treat scurvy and chronic skin conditions, and to lower fevers.

Jamaican sorrel (red sorrel, Florida cranberry, or roselle) is not a true sorrel. It was introduced to Jamaica from west Asia in the eighteenth century.

Scientific Name(s): Hibiscus sabdariffa. Family: Malvaceae.

Common Names: it is called roselle in English, rosella in Spanish, bissap rouge in Senegalese, karkadeh in Egypt, rozelle in French, karkadi in German, and carcade in Italian.

Form: Roselle is the fleshy young calyces surrounding the immature fruits. It also comes in dried form.

Properties: Jamaican sorrel is very acidic. It resembles cranberry in color and acidity.

How Prepared and Served: it is used to flavor drinks, jams, jellies, wine, and sauces in the Caribbean, Mexico, West Africa, and Egypt. Roselle is used fresh in salads, especially fruit salads, with cooked vegetables, and in sauces, stews, and pies or tarts. Roselle is also dried and used as natural coloring. The Caribbeans enjoy it as a traditional Christmas drink (also called sorrel) that is mixed with spices and rum. In Mexico, dried roselle is made into a refreshing drink called *aqua de Jamaica.* Africans add sugar to their roselle drink to tone down the sourness. Roselle is also used in curries and chutneys in India and Southeast Asia, and in foods and beverages of Cuba, Central America, Mexico, and Florida. In Egypt and other Middle Eastern regions, it is added to desserts, drinks, and sweets.

Spice Blends: sorrel drink blend, aqua de Jamaica blend.

Therapeutic Uses and Folklore: Indians, Mexicans, and Africans use it as a diuretic, to thin blood, and to lower blood pressure.

STAR ANISE

Star anise, a native spice of China, is a dominant ingredient in Chinese five-spice blend, and it is an essential flavoring for many Chinese, Malaysian, and Vietnamese foods. *Baht gokh* in Cantonese and *bao jiao* in Mandarin mean eight corners. *Jiao huei hsiang* in Mandarin means "eight horned fennel," since the Chinese perceived its taste to be similar to fennel. It is often adulterated with other "false" types of star anise from Japan.

Scientific Name(s): Illicium verum. Family: Illiaceae.

Origin and Varieties: star anise is native to southwestern China and northern Vietnam. It is now cultivated in China, Vietnam, Laos, Philippines, India, Japan, and Korea.

Common Names: Indian anise, Chinese anise, and Badian anise. It is also called albadyan (Arabic), baht gokh, bah jiao (Cantonese, Mandarin), stjerne anis (Danish), sternanijs (Dutch), badiyan (Farsi), anis etoile, anis de la Chine (French), sternanis (German), anison asteroeides (Greek), chakriphool (Hindi), kinai anizs (Hungarian), anice stellato (Italian), suta anis (Japanese), phkah cahn (Khmer), daehoihyang (Korean), bunga lawang (Malaysian, Indonesian), stjerneanis (Norwegian), anis estrelado (Portuguese), badyan (Russian), badian/anis estrella (Spanish), stjarnanis (Swedish), anasi poo, lavangai poo (Tamil, Malayalam), poy kak bua (Thai), cinanasonu (Turkish), and cay hoy (Vietnamese).

Form: star anise is an irregular star-shaped fruit with eight carpels joined around a central core, each carpel containing a seed. It is used in the following forms: dried whole, broken pieces, or ground. The fruits are picked when green, before they ripen and become sun dried.

Properties: the carpels are reddish brown and hard. Star anise has a licorice-like, sweet, and pungent flavor, similar to fennel and anise but stronger. It can leave a bitter aftertaste if used at high levels. Star anise gives a spicy, sweet flavor that becomes more intense as it is cooked. The seeds have less flavor than the pods, and the broken pieces of pods are less aromatic.

Chemical Components: the fresh fruit pod/pericarp has 5% to 8% essential oil, and the dried fruit pericarp/pod has 2.5% to 3.5%, mainly anethole (85% to 90%), α-pinene, phellandrene, ρ-cymene, 1,4-cineol, limonene, and *d*-terpineol. It has about 20% fixed oil.

How Prepared and Consumed: star anise characterizes Chinese, Vietnamese, and other Chinese-style cooking in Asia. It is used in marinades, barbecues, roasts, stews, and soups that require long simmering. The Chinese have introduced it to every region where they have settled. Star anise pairs well with roast poultry, roast pork, braised meats, and fish, and steamed or roast duck. It is an essential flavoring in the red cooking of the Shanghai region of China and is one of the five spices in the five-spice blend which is used as a marinade for meats, to flavor soups and sauces, and for batters.

In Southeast Asia, star anise is used in Chinese-style dishes, such as simmered beef, stir-fried vegetables, and steamed chicken. It is a popular spice in North Vietnamese beef noodle soups called *phos*. Malaysians and Singaporeans add it to their curries, soups, and sauces to give them unique tastes. Thais add to tea with milk, sugar, and other spices.

Brought to India by Chettiars of Tamil Nadu who traded in China and Southeast Asia, it is a popular spice in Kashmiri, Goan, and Chettinand cooking. It is also used in Caribbean masalas and spice blends.

The Europeans use star anise to flavor cordials, liqueurs, syrups, jams, and confectionaries.

Star anise pairs well with mint, cinnamon, ginger, soy sauce, orange, rose essence, curry leaves, chile peppers, curry powder, and black pepper. Typically, it is discarded before the dish is served.

Spice Blends: Chinese five-spice, hoisin, barbecue blend, Singapore pork curry blend, and Chettinand chicken curry blend.

Therapeutic Uses and Folklore: east Asians use star anise to relieve colic and stomach pains. They chew it after a meal to promote digestion and sweeten breath. Star anise is also used in teas to cure sore throats and coughs. Its oil is used as an ingredient in cough lozenges.

SUMAC

Native to Iran, sumac was used by Romans who enjoyed its sour taste and referred to it as Syrian sumac. Called "vinegar tree" by Germans and sour condiment by the Dutch, sumac is now a popular spice in Turkey, Iran, Egypt, and the Middle East.

It is found growing in the mountains around the Mediterranean. Native Americans prepare sour beverages with sumac berries.

Scientific Name(s): Rhus coriaria. Family: Anacardiaceae (cashew family).

Origin and Varieties: sumac is native to the Mediterranean region—Sicily, Middle East, and other regions of central Asia. There are many other varieties of sumac, Rh. glabra and Rh. Aromatica, and some ornamentals that are referred to as sumac in North America which are poisonous if eaten or are toxic to the skin. The bright, purplish red berries are safe, while the white or green white types are poisonous to touch.

Common Names: summak/summa (Arabic), kankrasringi (Bengali), sumac (Danish), zuurkruid (Dutch), shumac/Sicilian sumac (English), somagh (Farsi), sumac (French), sumach (German), roudi (Greek), sumac (Hebrew), kankrasing (Hindi), szomorce (Hungarian), sommacco (Italian), sumakku (Japanese), arkol (Punjabi), sumakh (Russian), zumaque (Spanish), karkhada garchingi (Tamil), karkararingi (Telegu), and sumak/somak (Turkish).

Form: sumac are small, red, dried berries that are sold whole but are usually ground and sometimes mixed with salt. The whole berries are soaked in water for about twenty minutes and are then squeezed to extract their juice before they are dried. The wild variety has green white berries that are poisonous.

Properties: sumac is reddish purple in color with a fruity, tart, and astringent taste.

Chemical Components: sumac has about 4% tannins (chrysanthemin, myrtillin, delphinidin) and high acid content, especially malic acid, others being citric, succinic, ascorbic, and fumaric, which gives its sour notes.

How Prepared and Consumed: the ground spice is a popular spice that is sprinkled over salads, meats, fish stews, and rice in Turkey, Iran, and other Middle Eastern regions. It is also commonly eaten with sliced onions, as an appetizer, and to season kebabs. Sumac is an important ingredient in spice blends of these regions, such as *zahtar* and *dukkah*, which are used as dips to perk up barbecued meats in the Middle East—Egypt, Syria, Israel, and Jordon. It is mixed with sesame, thyme, and other spices in Lebanon and Syria, boiled, and the thick extract is added as a lemon substitute for flavoring drinks, salads, vegetables, meat dishes, and fish stews.

Sumac goes well with pine nuts, cilantro, parsley, garlic, chile pepper, cumin, and coriander. It forms an important ingredient in spicy dips that are eaten with flat breads that are sold in the streets of Egypt and other parts of the Middle East. It is rubbed on roasted lamb, which is then grilled to give *m'choui* (Morocco) or *khouzi* (Egypt).

The Native Americans make a sour drink from ground sumac.

Spice Blends: zahtar, dukkah, and khouzi marinade.

Therapeutic Uses and Folklore: Middle Easterners use it to ease an upset stomach.

TAMARIND

Tamarind with origins from Africa or India is referred to as *tamr* Hindi in Arabic or *tamar* Hindi, in Farsi, meaning "date of India." Called *amlika* in Sanskrit, tamarind has been an important flavoring in Indian cuisine for a long time. Known as *asam koh, puli, or tamarindo,* it is now an important spice in many Asian and Latin American regions. Tamarind provides sweet sour notes to many South Indian foods such as sambars, fish curries, chutneys, and *rasams* and in Latin American beverages and candies. It is also one of the essential flavorings in Worcestershire sauce.

Scientific Name(s): Tamarindus indica. Family: Caesalpiniaceae (closely related to Leguminosae family).

Origin and Varieties: tamarind is indigenous to East Africa (Madagascar) and India, and is now cultivated in India, the Middle East, Africa, Southeast Asia, Central America, and the Caribbean.

Common Names: Indian date. It is also called tamr Hindi (Arabic), tetuli (Assamese), tentul, amli (Bengali), ma-gyi-thi (Burmese), loh tong jee, loh won ji (Cantonese, Mandarin), tamarind (Danish), tamarinde, asam koening (Dutch), tamar Hindi (Farsi), tamarinduz (Hungarian), tamarindi (Finnish), tamarin (French), tamarinde (German), tamarin (Greek), amli (Gujerati), tamrhindi (Hebrew), amli, imli (Hindi, Urdu), tamarindo (Italian), tamarindo (Japanese), ampil khui (Khmer), kok mak kham (Laotian), assam jawa (Malaysian, Indonesian), chinch, amli (Marathi), tamarindo (Portuguese), imli (Punjabi), tamarind (Russian), siyambala (Singhalese), tamarindo (Spanish), ukwaju (Swahili), tamarind (Swedish), sampalok (Tagalog), puli (Tamil, Malayalam), chinta (Telegu), makham (Thai), demir Hindi (Turkish), and me chua (Vietnamese).

Form: tamarind refers to the dark brown, dry, ripe fruits/pods or the fleshy pulp inside the pods. The pods are cinnamon colored with an inner dark brown, fleshy pulp that surrounds the black seeds. The pulp, which is fibrous and sticky, is the part used in cooking. The pods are dried, peeled, deseeded, and packed into blocks. Tamarind is sold as fresh whole pods, dried pulp slices, dried pulp concentrate or paste, or dried as a solid block. The concentrate has no seeds or fiber. It is also sold dried and ground. Tamarind juice is the strained liquid obtained when the paste is combined with warm or hot water.

Properties: the pulp has a sharp, sweet, sour, fruity taste and a sweetish, brown sugar-like aroma. It has acidic, molasses-like notes.

Chemical Components: the concentrate has 20% organic acids, mostly tartaric acid (12%), malic and succinic, and sugars, about 35%. The sugars are mainly glucose and fructose. It also has small amounts of terpenes (limonene, geraniol), methyl salicylate, safrol, cinnamic acid, and pyrazine. Tamarind is high in pectin.

Tamarind has calcium, phosphorus, potassium, and sodium.

How Prepared and Consumed: before use, tamarind pulp is soaked in hot water for about ten minutes and is then squeezed and strained to obtain its juice. The seeds and pulpy material are discarded. Tamarind enhances fish, meat, poultry, and vegetables, and goes well with chile peppers, legumes, shallots, vinegar, brown sugar, curry powder, and mushrooms. The young leaves and flowers are cooked as vegetables in India, and are used in soups and salads in Africa.

Tamarind juice to an Easterner is like lemon juice to the Westerner. It is used to provide sweet and sour notes to many Indian and Southeast Asian dishes. It tends to complement fiery hot dishes, so it is used in abundance in South Indian fish curries and pork vindaloos, *sambars* (stews), *rasams* (soups), chutneys, and Gujerati lentil, vegetables, and jams. Tamarind dip is made for samosas and *dahi wada* (dumpling of North India) that is smothered with yogurt and tamarind sauce.

Tamarind is also popular in Southeast Asian cooking especially Indonesian and Malaysian meat marinades and sauces to give sour sweet notes. It moderates the fiery notes of sambals, curries, and satay marinades. Many dips and condiments in Southeast Asia are prepared with tamarind juice. Tamarind is also added to Thai salads, *tom yom* soups and stir-fries, Chinese hot and sour soups, Malaysian and Indonesian satay marinades, sauces, and curries, and Vietnamese soups, stews, and curries.

Tamarind goes well with chilies, soy sauce, palm sugar, galangal, ginger, garlic, turmeric, clove, fermented shrimp paste, peanuts, and black pepper. The British love of Indian spices resulted in another popular condiment, Worcestershire sauce, which contains tamarind and other characteristic Indian flavorings.

Tamarind is popular in Mexican candies, Jamaican rices, and desserts. It is used in Puerto Rican juice drinks and as a cooling drink in India, Thailand, Latin America, and the Caribbean. The Muslims take a tamarind drink at Ramadan, their holy month. It is also used in Angostura bitters and sherbets. In Senegal, Africa, it is taken with sugar as a thirst-quenching beverage called *dakhar.*

Because tamarind is high in pectin, it can be used as a natural stabilizer in foods and beverages.

Assam gelugor is used in Southeast Asian and South Indian cuisines. It is not tamarind even though it has a similar flavor.

Scientific Name: Garcinia atroviridis.

How Prepared and Consumed: assam gelugor is dried and thinly sliced, and during cooking, it swells up and is discarded before the food is served. It is commonly used in Indonesian, Malaysian, and Singaporian cooking. *Assam gelugor* should also not be confused with *asam kandis* fruit (scientific name: *Garcinia globulosa*), popular souring fruit used in Indonesia and Malaysia, or *kokum* (scientific name: *G. indica*) that is commonly used in Gujerati and Kerala cooking.

Spice Blends: sambal blends, south Indian curry blends, rasam blend, vindaloo blend, rendang blend, Worcestershire sauce, tom yom blend, and Nonya laksa blend.

Therapeutic Uses and Folklore: in Thailand, traditionally, tamarind is used to prevent dysentery and reduce blood pressure. In India, tamarind is used to relieve asthma and fevers and is also used as a mild laxative. According to the Ayurvedics, it heals wounds and joint pains, sore throats, and bronchial disorders.

In India, tamarind is also mixed with salt and becomes an excellent brass polisher.

TARRAGON

Also called *estragon,* tarragon derives its name from the Greek word *drakoon,* the Arabic word *tarkhun,* and the Spanish word *taragoncia,* all of which mean little dragon and which describe its coiled, serpentine root. It is native to Southern Russia and Western Asia, and the Mongols introduced it to the West. Today, it is a popular spicing in southern European cooking, especially in French cuisine.

Scientific Name(s): French, German tarragon: *Artemisia dracunculus;* Mexican or Spanish tarragon: *Tagetes lucida.* Family: Compositae or Asteraceae (sunflower family).

Origin and Varieties: tarragon is native to eastern and central Europe, southern Russia, and western Asia. It is now cultivated in southern Europe, Russia, and the United States. There are two varieties: French tarragon (also called German tarragon) and Russian tarragon (withstands cooler climate), also known as false tarragon. In the United States, French tarragon is commonly used. Mexican tarragon, also referred to as Mexican marigold, Spanish tarragon, *tagete, yerba anis,* or *yauhtli,* has been used since the Aztecs, for religious and medicinal purposes. It is grown in Mexico and southern United States.

Common Names: French or German tarragon: estragole and little dragon. It is also called tarkhun (Arabic), ngao hou, ai hao (Cantonese, Mandarin), esdragon (Danish), dragon (Dutch), tarkhun (Farsi), rakuna (Finnish), estragon (French, German,), taragon (Hebrew), tarkony (Hungarian), estragon (Italian), esutoragon (Japanese), estragon (Norwegian), estragao (Portuguese), estragon (Russian), estrago'n (Spanish), dragon (Swedish), and tarhun (Turkish).

Form: tarragon has long, narrow, grayish green leaves. Fresh, it is used whole, chopped, or minced. Dried, it is used whole, crushed, or ground. Sometimes the stems are included with the leaves.

Properties: Fresh tarragon has stronger notes than the dried form. French tarragon has a cool, sweet, licorice-like aroma with slight bitter tones. Its taste is green and herbaceous, with anise- and basil-like notes. It is more delicate in flavor than the Russian variety. The Russian variety has larger leaves, lacks the anisic taste, and is slightly bitter and harsh in flavor. Dried tarragon has haylike tones and has less flavor than the fresh form. Mexican tarragon has pleasant aniselike notes, quite similar to French tarragon.

Chemical Components: fresh French tarragon has 0.5% to 2.5% essential oil, which is pale yellow to amber in color and consists mainly of 60% to 75% estragol (also called methyl chavicol). It also has 10% anethole (which mainly contributes to tarragon's aroma), α- and β-pinenes, camphene, phellandrene, limonene, and myrcene. There is less volatile oil (about 0.3% to 0.8%) in the dried herb. Russian tarragon has little essential oil (about 0.1%), mainly sabinene, methyl eugenol, and chemicin. Its harsh notes are due to quercetins and petuletin. It does not have estragol (so lacks the sweet notes) and thus, is considered inferior to French Tarragon.

The oleoresin is dark green and viscous, and 2 lb. are equivalent to 100 lb. of ground, dried leaves.

Tarragon contains niacin, vitamin A, manganese, potassium, calcium, iron, magnesium, phosphorus, and sodium.

How Prepared and Consumed: in Europe, tarragon is popularly used to flavor many sauces. It is a favorite spice in France and characterizes French Dijon mustard and sauces based on sour cream, eggs, and mayonnaise, such as tartar, béarnaise, and hollandaise. It is also used in cream soups, salads, omelettes, and gravies. These sauces are added to broiled, baked, or fried fish, meat, and chicken. Tarragon is an important component of fines herbes and is one of the optional ingredients of herbs de Provence of France.

Armenians use tarragon on vegetables, fish, and meat dishes. In the United States, it is used in vinegar, tartar sauce, eggs, chicken, and seafoods.

Tarragon enhances roast chicken or turkey and can also be used in stuffings. It is combined with vinegar and capers in salad dressings. Tarragon pairs well with salmon, chicken, veal, lobster, scallops, lentils, vinegar, lemon juice, and mild leafy spices such as parsley and chives. Cooking intensifies and changes its flavor, so it is usually added to a dish toward the end of cooking to retain its characteristic aroma and taste.

Spice Blends: fines herbes, béarnaise sauce blend, Dijon mustard blend, tartar sauce blend, and herbs de Provence.

Therapeutic Uses and Folklore: the Arabs used tarragon in ancient times to cure insomnia and to dull the taste of medicines. In traditional medicine, tarragon is used to increase appetite, to flush toxins from the body and as a digestive stimulant, especially by heavy-meat-eating cultures. Tarragon has also been used as an anesthetic for aching teeth, sores, or cuts.

THYME

Thyme comes from the Greek word *thymon,* which refers to its strong smoky odor. The Egyptian word for thyme, *tham,* means strong smelling, and it was mainly used in embalming. During the Middle Ages, thyme was used as a symbol of courage and love. In the Middle East regions, the word *zahtar* for thyme is also the name for majoram, savory as well as a mixture of spices. In the United States, it is an important leafy spice for New England clam chowder and Cajun blackening spice mixture for meats and seafood.

Scientific Name(s): garden (common thyme): *Thymus (T) vulgaris;* white (or Spanish thyme): *T. zygis;* wild (or creeping thyme/serpolet): *T. serpyllum;* Spanish origanum: *T. capitatus;* lemon thyme; *T. citriodorus;* Moroccon thyme: *T. satureioides;* broad leaf thyme: *T. pulegioides;* herba barona, caraway thyme. Family: Labiatae or Lamiaceae (mint family).

Origin and Varieties: thyme is native to the Mediterranean regions and is now cultivated in France, Spain, Italy, Portugal, Germany, Morocco, Algeria, Egypt, England, the Caribbean, and the United States. There are numerous varieties of thyme, each different in flavor based on the climatic conditions: garden thyme, wild thyme, lemon thyme, orange thyme, anise thyme, caraway thyme, and Moroccon thyme. The United States mainly uses the Spanish and Moroccon varieties.

Common Names: garden thyme, red thyme, and French thyme. It is also called tosinyi (Amharic), za'tar (Arabic), tsotor (Armenian), bhak liew heung, bai li hsiang (Cantonese, Mandarin), timian (Danish), tijm (Dutch), zatar (Farsi), thym (French), thymian (German), thimari (Greek), timin (Hebrew), banajwain (Hindi), timian (Hungarian), timi (Indonesian), timo (Italian), taimu (Japanese), timian (Norwegian), tomilho (Portuguese), timyan (Russian), tomillo (Spanish), timjan (Swedish), dag kekigi (Turkish), hasha (Urdu), tymyan (Ukranian), and hung tay (Vietnamese).

Wild thyme: creeping thyme. It is also called wilde tijm (Dutch), awishan shirazi (Farsi), serpolet (French), quendal (German), serpillo (Italian), kryptimian (Norwegian), and serpoleto (Spanish).

Form: The fresh leaves are green, pointed, oval, slightly rolled, and covered in fine hairs. They can be broad leafed or narrow leafed or large or small based on the variety. They are used fresh or dried. The dried form is sold whole, minced, or ground.

Properties: Thyme has a piney, smoky taste with a herbaceous and slightly floral aroma. Some have bitter, slightly lemony, and minty notes. It turns black in an acidic medium (tomato sauce) and loses its aroma quickly with heat. Its flavor is retained better when dried. Moroccon thyme is broader and shorter leafed with higher volatile oil content than Spanish thyme.

Chemical Components: The essential oil content of thyme ranges from 1.5% to 5% and has mainly phenols. Common thyme yields about 1.5% to 2.0% essential oil, which is colorless to pale yellowish red and contains predominantly thymol (45%), and carvacrol (18%). Spanish thyme (which provides 90% of the world's thyme oil) has 12% to 61% thymol and 0.4% to 20.6% carvacrol, 1,8-cineole (0.2% to 14.2%), ρ-cymene (9.1% to 22.2%), linalool (2.2% to 4.8%), borneol (0.6% to 7.5%), α-pinene (0.9% to 6.6%), and camphor (0% to 7.3%). Most other varieties, such as Moroccan, lemon, or orange thymes, have lesser thymol. Lemon thyme has mainly geraniol and citral with little thymol.

Wild thymes yield less volatile oil, about 0.5%.

Its oleoresin is green to brownish green and is viscous, and 4 lb. of oleoresin are equivalent to 100 lb. of freshly ground thyme.

Thyme contains vitamin A, niacin, calcium, sodium, potassium, magnesium, phosphorus, and sodium.

How Prepared and Consumed: thyme is added during cooking or toward the end of cooking or is even sprinkled over the dish at the serving table. It pairs well with fatty foods, such as mutton, goose, duck, pork, tomato sauces, stuffings, roasts, fish, and wine.

Thyme is a very important spicing in European cooking, especially in the southern areas of Europe and England, for stuffings, vegetable soups, mutton stews, beef bourguignonne, long-simmered poultry, fish, meat, and game. It is used to flavor pickled olives and the liqueur benedictine, and it is combined with marjoram in many dishes. It is an important ingredient in herbs de Provence and bouquet garni. The French tie up thyme with other herbs in bundles and add this to soups, sauces, and stews. The bundles are removed before serving. In Central Europe, it flavors soups, meats, seafood, eggs, sausages, and cheese.

In the Middle East, thyme is found in spice mixtures such as *zahtar* from Jordan and *dukkah* from Egypt. It is used with roasted sesame seeds, black pepper, and other spices to flavor meat or is eaten with bread and olive oil.

In the United States, thyme is a prominent flavoring in Manhattan and New England clam chowders, other seafood dishes, and in poultry stuffing and sausages. Cajun and Creole cooking use thyme as part of a coating mix for fish or meat before they are blackened. The blackening spices become dark brown but not charred, so their flavor is retained.

It is also a popular spice in the Caribbean, where the broad-leaved variety of thyme is combined with allspice and other spices to make jerk seasoning, stews, and curries.

Spice Blends: bouquet garni, dukkah, zahtar, Cajun spice, herbs de Provence, curry blend, and jerk seasoning.

Therapeutic Uses and Folklore: thyme was used by the ancient Greeks to purify temples and by Romans to infuse bath waters. In traditional European medicine, thyme was used to treat infections, digestive complaints, and respiratory ailments, such as bronchitis, laryngitis, and whooping cough. It has been used to treat circulation problems, relieve muscle pains, and as a tranquilizer, as an expectorant, and to help hangovers.

Thyme is also used as a mild insect repellent. It has good antibacterial properties.

TURMERIC/TUMERIC

Called yellow root in many European languages and referred to as yellow ginger in Chinese, the name turmeric comes from the Latin *terra merita,* meaning deserving earth. The use of turmeric dates back to 1500 BC when turmeric was mentioned as *haridra* in the Vedas, the sacred scriptures of the Hindus. Known to Westerners as "Indian saffron," turmeric is an important spice in India where it is used to flavor and color foods, as well as added to cosmetics. It is a must in curries, stews, and sauces of India and Southeast Asia.

It is used in religious ceremonies and is a traditional Ayurvedis medicine for treating digestive complaints, colds, and wounds. Its leaves are popular in Indonesian cooking, for wrapping meats and fish, before they are cooked.

Scientific Name(s): Curcuma (C) longa or C. domestica. Family: Zingibearaceae (ginger family).

Origin and Varieties: indigenous to South and Southeast Asia, turmeric is cultivated in India (Alleppey and Madras), Sri Lanka, Java, Malaysia, China, Peru, and Jamaica. There are numerous varieties in India alone, and the most valued turmeric is Allepy, which has the best color value and flavor.

Other types of turmeric, including C. zanthorrhiza called *temu lawak* in Java and C. zoedoaria, known as zedoary or white turmeric, are used abundantly in Southeast Asian cooking. Zeodary is discussed separately.

Common Names: Indian saffron, yellow ginger. It is also called ird (Amharic), kurkum (Arabic), toomerik (Armenian), haladi (Assamese), halud (Bengali), sa nwin (Burmese), wong yeung, yu-chin, (Cantonese, Mandarin), gurkemeje (Danish), geelwortel (Dutch), zardchubeh (Farsi), curcuma, safran des Indes (French), gelbwurz, Indescher safran (German), kourkoumi (Greek), halad (Gujerati), kurkum (Hebrew), haldi (Hindi), kurkuma (Hungarian), curcuma (Italian), ukon (Japanese), romiet (Khmer), khimin khun (Laotian), kunyit (dried), kunyit basah (fresh) (Malaysian, Indonesian), shuva, manjal (Malayalam), curcuma, acafrao da India (Portuguese), haldi (Punjabi), zholtyj imbir (Russian), kaha (Singhalese), cu'rcuma (Spanish), manjana (Swahili), gurkmeja (Swedish), dilaw (Tagalog), manjal (Tamil), pasupu (Telegu), haladi (Urdu), khamin (Thai), gaser (Tibetian), hint safrani (Turkish), and cu nghe (dried), bot nghe (fresh) (Vietnamese).

Form: turmeric is a rhizome with a brownish yellow skin and a bright yellowish orange interior. It is cured (boiled or steamed) to intensify its aroma and color, then dried and sold whole or ground into powder. It is used fresh (chopped, grated, cut) and dried (whole or powder). Its color varies depending on its origin. The fresh form looks like ginger except it is thinner and its inner flesh is yellow. In Indonesia, the fresh leaves are used whole as wrappers or are chopped and used in local dishes.

Properties: fresh turmeric has a bright orange flesh. It has a fragrant musky and earthy aroma with gingery, slightly bitter, peppery notes. It loses its aromatic flavor during storage and when exposed to light. The dried rhizome is lemon yellow to orange yellow in color. Madras turmeric is lighter and brighter yellow, while Alleppey tumeric is a brownish yellow color on the outside with a deep orange color to yellowish orange on the inside.

Turmeric leaves are large and aromatic.

Chemical Components: turmeric contains three components: essential oil (1.5% to 6%), with coloring compounds called curcuminoids (3% to 8%), and starch. The essential oil contains mainly turmerone (30%), dihydrotumerone (25%) and zingiberene (25%), along with sabinene (6.0%), 1,8-cineole (1.0%), borneol, sesquiphellandrene, curcumene, bisabolene, and α-phellandrene. The essential oil is highly aromatic. The Alleppey turmeric has 3.5% to 6.9% essential oil, with 5.3% to 6.5% curcumin, while the Madras turmeric has 2.2% essential oil, with 2.1% curcumin. Javanese variety has 6% to 10% essential oil. The chief coloring component is curcumin.

The oleoresin is a deep red to orange red viscous oil and contains 15% to 25% volatile oil and coloring matter. About 8 lb. are equivalent to 100 lb. of freshly ground turmeric.

Turmeric contains vitamin C, potassium, iron, sodium, phosphorus, magnesium, and calcium.

How Prepared and Consumed: turmeric pairs well with cilantro, kari leaf, galangal, ginger, mustard seeds, nutmeg, coconut, pandan leaf, bay leaf, lemongrass, cumin, clove, coriander, mustard seeds, dill, and black pepper. It is an essential ingredient in all curry powders and is used in most Indian dishes to help digest the complex carbohydrates.

In Asian Indian and Southeast Asian cooking, turmeric is a popular spice for providing color and flavor to curries (especially vegetarian curries), pickles, soups, lentils, vegetables (especially potatoes and cauliflower), fried fish, pullaos, and desserts.

In Thai, Malaysian, and Indonesian cooking, turmeric is commonly used freshly grated in yellow and red curries, yellow rice (*nasi kuning*), stews, *laksas*, and vegetables. It is combined with chile peppers, lemongrass, lesser galangal, cinnamon, clove, ginger, and coriander to create *bumbu*, a spice blend used for *rendangs*, *gulais*, and other fiery sauces and soups.

In Western countries, turmeric imparts a bright yellow coloring to mustard condiments and sauces. It is also used in deviled eggs, Worcestershire sauce, chicken stock, potatoes, cheeses, yogurt, sausages, egg salads, pickles, relishes, spreads, and beverages.

Today, in the United States, turmeric oleoresin, extracts, or curcumin (mixed with solvent and emulsifiers) are used commercially as natural colorants.

Spice Blends: curry blends, laksa blend, pulao blend, nasi kuning blend, vegetable spice blend, and bumbu blend.

Therapeutic Uses and Folklore: turmeric is often used as a dye in the Middle East, in China to color Buddhist robes, and by Indonesians as body paints and to ward off evil spirits. Chinese herbalists and Indian Ayurvedics use turmeric to treat liver problems, high cholesterol, and digestive problems. Turmeric is also used for healing bruises and sores and to prevent hair growth. Turmeric helps inhibit blood clotting, strengthens the gall bladder, and treats skin diseases.

Traditionally in Europe, turmeric is not recommended for pregnant women because it may induce uterine stimulation, but in the Indian culture, it is given to heal the uterus after childbirth and to help restore loss of blood. In Ayurvedic medicine, it is used as an anticancer agent and to lower cholesterol. Curcumin is found to have strong anti-inflammatory properties.

With ongoing research, turmeric is found to have good antioxidant for many cooking oils and antimicrobial properties against Salmonella typhi, Staphyloccus, Vibrio cholera, and Micrococcus pyogenes.

WASABI

Wasabi or Japanese horseradish is a popular flavoring in Japan. Served with raw fish or mixed with sauce for tempura, it is also an important flavoring ingredient for Japanese snacks. In Japan, it is eaten fresh but sold as a pale green paste or powder around the world.

Scientific Name(s): Wasabia japonica. Family: Brassicaceae (cabbage family).

Origin and Varieties: wasabi is indigenous to Japan and is now also grown in the United States.

Common Names: Japanese horseradish. It is also called saan kwai, shan kui (Cantonese, Mandarin), Japansk peberrod (Danish), Japanse mierikswortel (Dutch), raifort du Japon (French), Japanischer kren (German), vasabi (Hebrew), Japan torma (Hungarian), and Japansk pepparrot (Swedish).

Form: wasabi is a thick rhizome that is creamy white when young and brownish green when mature. It comes freshly grated, as a pale green dried powder or as a green paste.

Properties: dried wasabi root does not have a pungent taste but develops it when it is mixed with water. The flavor is sharp, clean, burning, pungent, and nutty with a biting aroma like horseradish, but it comes through with a sweet note. The heat perceived is instantaneous and stronger than horseradish. Dried wasabi is less aromatic but more pungent than the fresh form.

Chemical Components: the essential oil contains sinigrin and glucocochlearin. When the tissues are cut or bruised, or when water reacts with wasabi powder, enzymatic reactions occur that break down the sinigrin to allyl isothiocyanates which give the penetrating odor.

How Prepared and Consumed: wasabi is used coarsely grated or as a smooth paste, often alongside soy sauce, with Japanese foods such as sashimi, sushi, soba, and tofu. At the meal table, sashimi or raw fish is first dipped in soy sauce and then into wasabi paste before it is eaten. Sushi, a glutinous rice seasoned with vinegar and sugar and wrapped in dried toasted seaweed (*nori*) or stuffed with raw fish, other seafood dishes, fresh or pickled vegetables, and omelettes, are served with wasabi paste. Sometimes, wasabi paste is placed between the raw fish or vegetables and glutinous rice in sushi.

Wasabi pairs well with soy sauce, brown sugar, glutinous rice, chicken, ginger, and fish. It does not retain its flavor in cooked foods, so fresh wasabi is used toward the end of cooking or is added to cold foods. The fresh paste also loses its flavor quickly. Wasabi powder is mixed with warm water and let to stand for ten minutes to develop its flavor. It can be used in marinades, barbecue sauces, creamy salad dressings, soups, and steamed fish.

Spice Blends: wasabi snack blend, Japanese sauce blend, and pickle blend.

Therapeutic Uses and Folklore: the Japanese believe it can kill parasites.

WATERCRESS

A favorite with the English, this leafy green spice can be traced back to the Greeks, Romans, and Persians who used it for medical purposes, as a treatment (with vinegar) for insanity, as a stimulant, and as a breath freshener. It belongs to the cabbage family, along with mustard, horseradish, and wasabi. Soldiers and sailors ate it to treat scurvy. Referred to as the poor man's food in England, in the early 1800s, watercress sandwiches were a staple breakfast for the working class. Brought by European immigrants to the United States in mid-1800s, it has become a popular garnish and vegetable for North Americans. It is also a popular seasoning with Chinese and Vietnamese who add it to soups and stir-fries.

Scientific Name(s): Watercress: Nasturtium officinale; garden cress: Lepidium sativum. Family: Brassicaceae (cabbage family).

Origin and Varieties: Watercress is native to western Asia but cultivated in Europe, United States, and Asia. In many languages watercress means water, as it is grown in water. Other botanically related plants are garden cress (or pepper cress), winter cress, bitter cress, and nasturtium or Indian cress (more of an ornamental plant originating in South America).

Common names: barbeen (Arabic), jhri godem (Armenian), sai yeung choi (Cantonese), waterkers (Dutch), tazer alaf shesmeh (Farsi), isovesi krasi (Finnish), cresson de fontaine (French), brunnenkresse (German), nerokardamo (Greek), garga hanazir (Hebrew), selada air/ayer (Indonesian, Malaysian), votakuresu (Japanese), vaistino rezuiko (Lithuanian), shui jee cai (Mandarin), bronnkarse (Norwegian), agriao (Portuguese), kress vodianoj (Russian), vodna kresa (Slovenian), berro di agua/crenchas (Spanish), kallfrane (Swedish), lampaka (Tagalog), phakat nam (Thai), su teresi (Turkish), and cai soong (Vietnamese).

Form: the watercress leaflets or clusters of leaves are enjoyed fresh. Dried leaves do not have the flavor as fresh leaves do. With garden cress, the flowers and unripe fruits, too, are eaten. Watercress has a crunchy texture which is appealing for salads.

Properties: The fresh leaves have a refreshing, sharp, and savory aroma with a peppery, pungent taste.

Chemical Components: Its aroma is due to isothiocyanates which are formed from its precursors, glucosinolates from chopping or chewing. Gluconasturtin in watercress is converted to 2-phenylethyl isothiocyanate (PEITC), a volatile that is sensitive to heat and moisture.

It is high in potassium, vitamin A, vitamin C, vitamin E, iron, folic acid, and calcium.

How Prepared and Consumed: Europeans and North Americans enjoy watercress in sandwiches, in potato salads, in omelets, as cottage cheese spreads, or as garnishes in soup and scrambled eggs. It is pureed and made into watercress soup, a favorite with the English who claimed it to cleanse the blood. The French add it to fines herbes, many white sauces, and flavored vinegars. It adds crunchiness to salads, soups, and sandwiches. Westerners enjoy it fresh while Asians cook it. It is a popular vegetable in Asia, where it is added to stir-fries and soups. As a simple stir-fry, rice wine, sugar, and salt are added. Or it is blanched, chopped, and flavored with sesame oil, garlic, and miso.

Spice Blends: herbes fines, watercress soup blend, stir-fry blend, and blend for omelet.

Therapeutic Uses and Folklore: It was believed to be an aphrodisiac and a stimulant by the Arabs, Greeks, English, and many other ancient cultures. Hippocrates used it as a blood purifier and for bronchial disorders, and to increase stamina. The Persian soldiers ate it to prevent and treat scurvy. Made into tea, it was taken to ease aches, pains, and migraines. Today, in South America, it is believed to be an antitumor agent, and North Americans are researching its PEITC's effect in preventing lung diseases such as cancer and emphysema through tobacco smoking.

ZEDOARY

Called white turmeric, zedoary is a close relative of turmeric, and originated in India or Southeast Asia. Zedoary has its origins from the Arabic word *zadwar* and Farsi word *zedwaar*. It flavors curries in Southeast Asia and South India, while in China and Japan it is used as a medicine. In Thailand, it is prized as an aromatic vegetable. The Arabs introduced this spice to the West. It is called *amb halad* in many Indian languages because of its mango-like aroma.

Scientific Name(s): Curcuma (C) zedoaria. Family: Zingiberaceae (ginger family).

Origin and Varieties: zedoary is native to Northeast India and Southeast Asia but today cultivated in India, China, and Southeast Asia. There are many varieties which vary in size and color but two types are commonly used, the long, slender turmeric-like C. zerumbet and the small, round, gingerlike C. rotundae

Common Names: white turmeric, wild turmeric. It is also called zadwaar (Arabic), ngo seut, i zhu (Cantonese, Mandarin), zadoarwortel (Dutch), zedwaar (Farsi), zedoaire (French), zitwer (German), amb halad, kachur (Hindi), feher kurkuma (Hungarian), kencur zadwar, kunir putih, kentjur (Indonesian), zedoaria (Italian), gajutsu (Japanese), khi min khay (Laotian), zedoari (Russian), kunchor (Singhalese), cedoaria (Spanish), zitherrot (Swedish), khamin khao (Thai), cedwar (Turkish), and nga truat (Vietnamese).

Form: zedoary is used fresh and dried, and sliced or ground. It is sold as whole fresh or dried, sliced, and ground.

Properties: it is a pale yellow, large and fleshy tuberous rhizome with many branches and large, long, fragrant leaves. The inner slices of the fresh rhizome are bright orange with a hard texture, while the exterior is dark brown. Its flavor is gingerlike, bitter, and camphoraceous, with a musky mango, turmeric aroma. The dried zedoary has a yellowish to grayish white interior.

Chemical Components: the essential oil is viscous and dark green with about 79% sesquiterpenes (mainly germacrone-4, 5-epoxide, furanodienone, curzerenone, curcumenol, isocurcumenol, and zederone), 1,8-cineol, camphor, and α-pinene.

How Prepared and Consumed: fresh zedoary is added to pickles and poultry, lamb, and fish dishes of South India. In Indonesia, the dried rhizome flavors seafood dishes and meat curries while the fresh aromatic leaves and shoots are added to fish dishes and to wrap foods for grilling or baking. Zedoary is used at low levels because of its bitter flavor.

In Thailand, the young rhizome is eaten as a vegetable or used in curries, condiments, and spice mixtures. In China and Japan, it is used in liquors and as a medicine.

Spice Blends: pickle blend, fish marinade, and chicken curry blends.

Therapeutic Uses and Folklore: zedoary is used to treat colds in India and is used as a digestive aid in Indonesia and Thailand. It is valued as a blood purifier and heals cuts and wounds.

FIGURE 1 Spices, chile peppers and flavorings.

FIGURE 2 Spices in a mortar/molcajete/batu lesung.

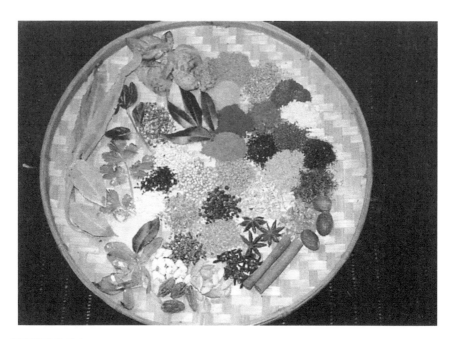

FIGURE 3 Spices on a tray.

FIGURE 4 Top right, clockwise: red cayenne peppers, bunga kantan/torch ginger bud used in Malay curries and sauces in Malaysia, small lime (limau kasturi), eggplant.

FIGURE 5 Fresh chile peppers, from top left, first row: New Mexican habaneros; second row: jalapenos, serranos, cayennes; third row: cubanelle, Thai peppers, bird peppers/cili padi.

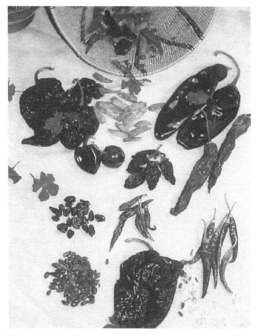

FIGURE 6 Dried chile peppers, from top left, first row: mulatto, Tabasco, New Mexican; second row: cascabel, chipotle, pasilla; third row: chiltepin, cili kering/fried cayenne; fourth row: pequin, ancho, chile de arbol.

FIGURE 7 From left: fenugreek seed/methi; right: fenugreek leaf/sag methi.

FIGURE 8 From left: coriander seed; right: coriander leaf/cilantro.

FIGURE 9 Spices in masala dani/spice box, from left to right, clockwise: green cardamon, fenugreek, cumin, fennel, black peppercorn, coriander; center: ground tumeric.

FIGURE 10 Kari/curry leaves/karivepillai.

FIGURE 11 Clockwise from left: brown cardamon, green cardamon, white cardamon.

FIGURE 12 Clockwise from top left: nigella/kalonji, black cumin, ajowan, anise; center: fennel.

FIGURE 13 Clockwise left to right: broken mace blades/aril, whole nutmeg, ground nutmeg.

FIGURE 14 Tumeric: whole, ground.

FIGURE 15 "Wet" market in northeast Malaysia.

6 Emerging Flavor Contributors

INTRODUCTION

Other significant flavor, color, and texture contributors in addition to spices are gaining in importance. This chapter describes the roots and rhizomes, nuts, seafoods, flowers, wrappers made from edible and inedible leaves, and husks and processed "skins" from beans and grains that can be used to contribute taste and texture. Flowers impart visual appeal and flavor to applications. Roots and rhizomes are versatile ingredients that can add intense, pungent notes or simply provide texture and consistency to a product. Wrappers are natural cooking vessels that enhance the taste of the foods they contain.

Spices can also be combined with many ingredients, such as fish sauce, shrimp pastes, fruits, vegetables, legumes, and nuts to provide characterizing tastes or textures to a product. In the cuisines of many cultures, these flavorings are paired with chilies, coriander, lemongrass, onions, and spices to create distinct flavors. These flavorings can also be salted, fermented, dried, or pickled and added with spices to create unique seasoning blends (some of which are discussed in Chapter 7, "Emerging Spice Blends and Seasonings").

Other flavorings such as vanilla, chocolate, tea, and coffee contribute sweetness and bitterness to foods and beverages. They have been used to balance savory and spicy profiles in some traditional ethnic cuisines. Finally, preparation and cooking techniques help create new flavor profiles and contribute to a product's final taste.

Table 21 gives examples of these other flavorings from around the world.

These "other flavorings" also provide nutrition and valuable phytochemicals that are being tapped as therapeutic aids and natural cures. Fruits, vegetables, nuts and seeds, seafood, and other flavorings have phenols, isoflavones, isothiocyanates, antioxidants, glucosinolates, omega 3 fatty acids, and flavanoids.

ROOT/TUBER/RHIZOME FLAVORINGS

Roots, tubers, and rhizomes provide consistency, flavor, and color to foods. Some roots provide intense, pungent notes, such as licorice, ginger, horseradish, wasabi, coriander, and turmeric. Others serve merely as a bland starchy base, such as cassava (manioc/yuca), *arracachia* (Peruvian carrot), *yautia*, or yam.

Roots are a good source of complex carbohydrates and easily digestible starches. They can be used as natural thickening agents or natural starch modifiers. Many roots such as jicama, yam, or lotus root are boiled, roasted, stir-fried, or stewed.

TABLE 21
Other Flavorings Used around the World

Flavorings	Product Name/Region
Roots, tubers, rhizomes	Lotus root (Asia), yautia (Latin America, Caribbean), parsley root (Mediterranean), daikon (Japan, Korea), taro (Southeast Asia, Latin America, Hawaii, Caribbean)
Flowers	Rose petal (India, Southeast Asia, Mediterranean), orange blossoms (Mediterranean), jasmine (Southeast Asia), lotus (China, Southeast Asia, Sri Lanka), zucchini blossoms (Mexico, Italy, France)
Wrappers	Corn husk, leaf (Latin America, Caribbean), banana leaf (India, Southeast Asia, Caribbean, Latin America), lotus leaf (China), salam leaf (Indonesia), hoba leaf (Japan)
Soybean	Fermented pastes (jiangs); salted beans (China, Southeast Asia); misos (Japan); taucheo (Malaysia, Singapore); toenjang (Korea); oyster sauce (China, Southeast Asia, Korea)
Soy sauce	Tamari, Shoyu (Japan); jiangjou (China); kecap soy masin, kecap soy pekat, kecap manis and kecap asin (Indonesia, Malaysia, Singapore, Philippines); kanjang (Korea)
Fish/shrimp flavorings	Nam pla (Thailand); nuoc mam (Vietnam); trassi (Indonesia); blacan, petis (Indonesia, Malaysia, Singapore); bagoong, patis, (Philippines); mam cho (Laos); pra hoc (Cambodia); dashi, shotturn (Japan); saeujot (Korea); yu lu, yu jiang (China); ngapi ye (Burma); Bombay duck (India); maldive (Sri Lanka); bacalao (Mediterranean, Caribbean, Brazil); liquamen (France)
Fruits	Ponzu (Japan); orange peel (China); plum, lichee (Southeast Asia); olives, pomegranate (India, Mediterranean); mango, kokum, coconut (India, Southeast Asia, Caribbean)
Vegetables	Pickled/salted: radish, cabbage, burdock, mustard greens (China, Japan); kimchee (Korea, Japan); mushrooms (Japan, United States, China, Southeast Asia); tomato (Mediterranean, Asia, United States), collard greens (United States, Africa, Asia)
Nuts, seeds	Pumpkin seeds/pepitas (Mexico); sesame seed (Asia, Mexico); candlenuts (Malaysia, Singapore, Indonesia); almonds, pinenuts/pinones (Italy, United States); walnuts (China); almonds (India, United States)
Dairy products	Ghee, yogurt, buttermilk (India, Southeast Asia, Middle East, Africa), cheeses (global)
Sweet and bitter flavorings	Vanilla, cocoa, chocolate, coffee, tea (United States, Europe, Africa, Mexico, Asia, Middle East)
Cooking techniques	Broil, grill, steam, stir-fry, braise, pickle, ferment, claypot, Tandoor, tagine, paellera

Others are pureed and act as nondescript bases for sauces and as thickeners for soups and curries.

Traditionally, many cultures, especially Native Americans and Asians, used roots as cures. Native Americans used American ginseng to ease childbirth, treat nosebleeds, and as a tonic for mental problems. East Asians take panax ginseng to provide energy and lotus root to relieve stress. Roots and tubers are a good source of phytochemicals. Sweet potato (also called batata and boniata) contains carotenoids; yam has plant sterols; licorice contains triterpenoids and glycyrrhizin; horseradish and wasabi contain isothiocyanates; burdock contains high levels of iron, vitamin A, and inulin, an easily digestible starch that provides a medium for the gut bacteria; and jicama has high levels of potassium, iron, vitamin A, and calcium.

Roots, Tubers, and Rhizomes in Ethnic Cuisines

Roots, tubers, and rhizomes were traditionally used to add flavor, color, and texture to foods by many indigenous groups such as Mayans, Aztecs, Asian Indians, and Africans who could not afford or procure meat or fish. They use potatoes, yams, sweet potatoes, and malanga to provide texture and to add bulk to soups, stews, and sauces. Yuca is mashed and made into flour for cakes and breads.

The Amazonian Indians from South America used roots and tubers as a staple flavoring in their foods. The Arawaks and Caribs in the Caribbean extracted juice from grated cassava root, boiled it with chile peppers, and flavored many of their dishes. This was called "coui" by the Caribs, and today, cassareep by Brazilians and Caribbeans. It is mixed with chile peppers, cinnamon, clove, or brown sugar to flavor sauces, rice, beans, and stews (such as pepperpot) in Brazil, Jamaica, Trinidad, and Barbados.

In Puerto Rico and Thailand, the coriander root is used to create pungent recados, sauces, and curries.

Many roots are popular in China and Southeast Asia as foods, especially in soups and beverages, such as ginseng, ginger, or galangal. The Balinese grind the root of salam leaf plant to create intense spice pastes called *jangkap*. Chinese use galangal for their five-spice sauces, and Thais use coriander root for their spicy red curries.

Table 22 is a chart of the more popular roots and rhizomes, their flavoring properties, and their ethnic preferences.

Specific Roots, Tubers, and Rhizomes as Ingredients

Several of the more important roots and rhizomes are explained in detail below.

Liquorice/Licorice/Chinese Sweetroot

Licorice comes from the root of a legume that is indigenous to China and the Middle East and is also cultivated in Spain, Sicily, Turkey, and South Russia. There are two varieties, Spanish licorice and Russian licorice. It has a medicinal, bittersweet taste, and its aroma is similar to anise or fennel, but stronger.

The scientific name for licorice is *Glycyrrhiza glabra*. Glycyrrhiza is derived from the Greek word *glukos,* meaning sweet. The Romans changed the name to *liquiritia,* which subsequently evolved into licorice.

Licorice is also called gancao (Chinese), reglisse (French), jethimadh (Hindi), liqueriziz (Italian), arpsous (Arabic), and orozuz (Spanish).

Licorice is an essential ingredient of the Chinese five-spice blend, which is used in sauces and barbecues. It is also used in soft drinks, ice cream, candy, smoothies,

TABLE 22
Global Roots and Tubers and Their Flavoring Properties

Roots	Flavoring Property	Ethnic Preference
Licorice	Bittersweet flavor	China, Southeast Asia, Korea
Burdock	Chewy texture	Japan, Mediterranean
Celeraic/celery root	Chewy, crunchy texture	Mediterranean, United States
Wasabi	Pungent flavor	Japan, United States
Lotus root	Crunchy texture, sweet flavor	Asia, United States
Parsley root	Aroma	Middle East, Mediterranean
Daikon	Crispy texture	Japan, Korea, United States
Sweet potato/boniato	Sweet flavor, mealy texture	Southeast Asia, Americas
Taro/eddo/coco-yam	crispy, creamy texture	Southeast Asia, Hawaii, Latin America, Caribbean
Yam	Mealy texture	Africa, Asia, United States
Yautia/malanga	Creamy texture, color	Latin America, Caribbean
Cassava/manioc/yuca/tapioca	Mealy texture	Caribbean, Southeast Asia, Latin America

drinks, and beer. In the United States and Europe, licorice is used in cough syrups, confectionaries, and lozenges. It also masks bitterness in medicines.

Licorice can be a noncaloric sweetener but must be used at very low levels in foods so it does not impart a licorice taste.

The root has about 20% to 30% of water-soluble extractive and 4% of a glycoside called glycyrrhizin, which is fifty times sweeter than cane sugar. It has the sweet and slightly astringent-like flavor of anise. Licorice is sold as peeled or unpeeled dried whole root, in pieces or powdered. "Block juice" is the black brittle concentrated extract from the root that is free of insolubles. It consists of sugars, starches, and gums in addition to 12% to 20% glycyrrhizin.

Licorice contains vitamin C, niacin, sodium, potassium, magnesium, and calcium. It has high phenols and triterpenoids. The Chinese have used licorice since ancient times to treat many ailments, such as ulcers, sore throats, or circulation problems.

Lotus Root

Lotus root is used by Chinese, Indians, and Southeast Asians for soups, curries, stews, and as a stir-fried vegetable. It has a potato-like skin and is cream to salmon colored. When it is cut crosswise, it shows a starlike pattern with symmetrical round openings. It is slightly sweet and has a texture like water chestnuts. Lotus root is available fresh or dried. It can be seasoned and eaten as an appetizer or ground and made into flour or sauces for Japanese and Chinese dishes.

Lotus root is also eaten pickled or as a garnish for salads. It is grated or sliced for stews and soups or candied and eaten as a sweet.

Ginseng Root

The ancient Chinese believed that ginseng root had the spirit and power of God to strengthen the body and cure illnesses. Ginseng is from the Chinese word *jen shen,* which means "essence of the earth from the form of man."

East Asians take ginseng as a "cure." The Chinese and Koreans add ginseng to chicken and vegetable dishes, soups, stews, and teas to relieve stress, reduce serum cholesterol, and strengthen the nervous system. *Tong shui*, which are sweet snacklike tonics containing ginseng, are taken to increase energy.

There are three major varieties of ginseng—Asian, American, and Siberian. Asian ginseng (panax ginseng) from Japan and Korea is aromatic and has a sweet licorice-like taste. It is stir-fried, boiled, or roasted. Traditionally, the root is chewed or brewed into a tea or tonic. In the United States and Europe, Asian ginseng is used in beverages, pasta sauces, and candies. It is eaten to boost energy, stimulate the immune system, and increase stamina. The other two ginsengs are Siberian and American; the latter is used by Native Americans to increase mental strength, to induce childbirth, and to treat nosebleeds. In Chinese traditional medicine, Asian ginseng is classified as "yang" and is used to increase energy and strength and to promote appetite. American ginseng is classified as "yin" and is used to reduce fatigue and increase immunity. Siberian ginseng is used to increase energy and to treat lower back pains.

FLOWER FLAVORINGS

Many cultures around the world, including Mediterranean, Latin American, and Asian, use flowers whole, dried, or in essence form to give unique flavors and colors to foods and beverages. Flowers with vivid colors and fancy shapes, such as violets, pansies, carnations, primroses, and orchids, create visual effects for entrees and desserts. Other flowers, such as jasmines, roses, chamomiles, elderflowers, geraniums, nasturtiums, and marigolds have unique aromas.

Flowers obtained from spices, fruits, and vegetables are fried, stuffed, crystallized, or chopped to provide floral, sweet, bitter, and other unique tastes to many dishes. Lime, plum, and orange flowers (or blossoms) with strong tastes or delicate aromas are commonly used in desserts, teas, and liqueurs. Spices such as rosemary, chive, dill, thyme, ginger, and garlic have blossoms that add a floral scent to foods and beverages.

Flower petals are served fresh in salads, lightly fried as appetizers, crystallized for cakes, infused for teas, or dried to blend with spices. Flowers tend to blend well with many spices, each flower complementing certain spices. For example, rose complements star anise, mint, turmeric, caraway, coriander, and chile pepper, while orange blossom complements clove, bay leaf, ginger, cinnamon, and nigella.

Since ancient times, blossoms have been popular not only for flavoring but also for healing. Jasmines, marigolds, violets, and roses have been traditionally used for their medicinal properties. Marigold has been used as a digestive stimulant and to heal tumors and wounds; rose hips, which are high in vitamins A and C and bioflavanoids, have been used to treat cold symptoms and relieve rheumatic pains;

and violets, which are high in vitamins A and C, have been used as expectorants. Flower blossoms and oils have also been popular in massage and aromatherapy.

FLOWERS IN ETHNIC CUISINES

In Europe, since medieval times, flowers have been distilled for their aromas and candied or boiled into jams and syrups. The English and Persians have favored their extracts. Today, in Asia, Latin America, the Middle East, and Europe, flowers serve important culinary purposes. They provide color, visual appeal, or fragrances to a wide number of applications such as cordials, butter, ice creams, salads, soups, terrines, syrups, jams, jellies, sorbets, cakes, confectionaries, and sauces.

Flowers such as rose petals, marigolds, zucchini blossoms, pansies, banana blossoms, and jasmines perk up salad dressings, vegetables, curries, stews, chutneys, puddings, sweets, liqueurs, and custards of India, Thailand, Mexico, the Middle East, and Europe. They are used fresh, battered, or cooked.

In China, jasmines and chrysanthemums flavor teas and wine; lotus petals are added to soups, noodles, and salads; and dried lily buds add texture and mild aroma to stir-fries and soups. In northern Africa and India, rose petals are added to teas, desserts, and spice blends.

SPECIFIC FLOWERS AS INGREDIENTS

Nasturtiums are orange or yellow in color and have a slight pepperlike flavor. They add visual appeal to salads and sandwiches and add flavor to vinegars and marinades of the Mediterranean. They enhance the color of spreads and margarine.

Europeans and Mexicans color rice, salads, and sweet dishes with marigold and violets. Marigold is crushed to give a saffronlike color and a slightly bitter herbal note to salads, soups, and entrees of Europe. Violets are highly fragrant, and in Europe, they are used in sweets, chocolates, and chilled soups. In Mexico and many other regions, marigold oil is added to chicken feed to provide a yellow color to chicken and eggs.

The fragrant bunga kantan (or wild ginger bud) is a characteristic flavoring in Nonya and Malay sauces and curries of Malaysia and Singapore. In Japan, it is called *mioga* bud, and it garnishes meals to provide visual enhancement.

Dried lily buds (*dok mai jeen*) are commonly used in Cantonese and Southeast Asian soups, stir-fried noodles, and sauces.

Elder flowers have a muscatel aroma and are used in England to flavor wines, cordials, stewed fruits, jams, jellies, and syrups.

The scientific name for Rose is *Rosa damascena,* and it belongs to the Rosaceae family. Rose is also called gul (Arabic), gulab (Hindi), rosa (Italian), bunga ros (Malaysian), ros (Swedish) and irosa (Tamil). Its petals emit a perfumed, sweet, and floral aroma. Rose is a popular flavoring among many ethnic groups. It is sold as dried petals or as a distillate called rose water.

Rose has about 1% or less essential oil consisting of geraniol (75%), citronellol (20%), neroli, and damascenone. Its odor is due to 2-phenyl ethanol.

Rose water (or rose essence) is greatly enjoyed in the Middle East and India. In Turkey, it is used in candies, a drink called *loukoum*, liqueurs, and jams. Called *ma'el-ward*, Arabs commonly add it to meats, sweets, and sauces. It is a common flavoring in desserts and preserves around the world, such as kheer, gulab jamun, and rasgullah in India, jams and rice puddings in Iran and Turkey, and in many Malaysian, Singaporean, and Thai rice desserts.

Rose petal is used in spice mixtures, sweets, coffees, wine, vinegars, and yogurt drinks of the Middle East, North Africa, Iran, North India, and Southeast Asia. It is a common ingredient in Indian sherbets and sweet desserts, in curries and biryanis of North India, and in some pork dishes of northern China.

Orange blossoms, called *zhaar* in Moroccon and *ma'el zahr* in Arabic, are added to fruit salads, stews, sorbets, and candies. They have a slightly bitter taste so are added sparingly. The essential oil of orange blossom, called orange oil (or neroli oil), is used in blancmange, puddings, and pastries of the Mediterranean. In the Middle East, it is given to revive persons from fainting spells.

Lavender's scientific name is *Lavendula angustifolia,* and it is a member of the Lamiaceae family. It is also called lavand (French), lavendel (German), la-vanda (Russian) and lavendel (Swedish). It has a strong, perfumed aroma and is popular as a garnish. Lavender is indigenous to western Mediterranean regions and is commonly found in French spice blends.

Lavender has 1% to 3% essential oil, mainly linalyl acetate (30% to 50%), linalool (20% to 35%), cineol, β-ocimene, camphor, and caryophyllene.

Jasmine and chrysanthemum are popular flavorings of teas and beverages in Japan, China, and Southeast Asia. In Thailand, jasmine essence is mixed with coconut milk to flavor many cakes and desserts.

Roselle or Jamaican sorrel (hibiscus family) is very acidic and flavors jams, cold beverages, and wine in the Caribbean and Mexico. It is also used in Central America, India, and Southeast Asia for sauces, curries, chutneys, and soups.

Lotus flower is used to scent Chinese and Vietnamese soups or Sri Lankan curries. To Buddhists, it is a symbol of beauty and purity.

Squash or zucchini blossoms are found in Mexican salads and appetizers and are becoming trendy in U.S. restaurants. They go well with spices such as cardamom, clove, nutmeg, cinnamon, mint, basil, and sorrel. They also tend to lose aroma after a few days, especially with prolonged heat, and so are typically added to foods toward the end of cooking or after cooking. Zucchini blossoms are also commonly used in Italian and French cooking. They are braised, stuffed, lightly fried, or battered and deep-fried. In Mexico, Oaxacans chop the flowers and add them to soups and sauces or saute them with poblano peppers, cilantro, and onions and add them to quesadillas, tacos, and enchiladas. In Italy, these blossoms are added whole or chopped to risottos, pastas, and salads and are an essential ingredient in *frittata alle erbe.*

The beautiful shape, thin papery texture, and subtle flavor of zucchini blossoms lends them to "stuffed" concepts. In France, they are stuffed with cheese, spices, and herbs, dipped in milk and flour, and sauteed or stuffed with cooked seafood and herbs and steamed or deep-fried. They are also used as wrappers, especially in the

United States, to provide texture to other foods. They can be battered lightly and deep-fried to provide crunchiness in a meal.

WRAPPER FLAVORINGS

Many cultures from Asia, Latin America, and the Mediterranean use wrappers as cooking and serving utensils. Wrappers can be edible or nonedible and provide distinct flavors, textures, and colors to the foods they contain. Some wrappers make the food succulent and tender. Others give crunchiness and crispiness. These wrappers include a variety of leaves, flowers, corn husks, cactus fiber, seaweeds, hollow bamboo stems, dried rice paper, and soy milk skin. They are used with meat, poultry, fish, vegetables, rice, cheeses, or spice pastes before foods are barbecued, deep-fried, grilled, baked, roasted, or steamed. This is the Asian, Latin American, or Native American equivalent of cooking en papillote.

Edible wrappers are used to wrap fresh ingredients and are eaten whole as salads or appetizers. Examples of such wrappers are lettuce, grape leaves, nori, and shiso. Others, such as corn husks, lotus leaves, or banana leaves are inedible.

Depending on the type of wrapper, varying amounts of color, flavor, or texture are contributed to the finished product. Today, these wrappers are becoming a trendy restaurant item to provide fresh-cooked aromas and textures at a meal table. They are being used to create authenticity or an enhanced visual appeal to a meal.

There are several basic categories of wrappers: Corn husks: they are sold fresh or dried. Before use, they are brought to a boil in water and left submerged for about an hour to make them more pliable. They are then pat dried. Leaves: leaves used as wrappers include pandan leaf, lotus leaf, corn leaf, banana leaf, hoba leaf, turmeric leaf, grape leaf, lettuce leaf, spinach leaf, cabbage leaf, *paan* leaf, *hoja santa*, *salam* leaf, and papaya leaf. Seaweeds: in Japan, nori (seaweed) is lightly toasted and used to wrap glutinous rice, which is called sushi. In China, seaweed is used to wrap fried foods such as tofu, fish, or vegetables. They provide texture, color, and a seaweed-type flavor. Maguey/cactus fiber provides succulence and moistness to steamed pork and other meats in Mexico. Hollow bamboo segments are commonly used by indigenous groups in Southeast Asia to cook rice and meats. Flowers: zucchini blossoms, rose petals, ginger flower, and banana blossoms provide fragrant aromas and colors to meals in many parts of the world.

WRAPPERS IN ETHNIC CUISINES

Different types of wrappers are used in ethnic cuisines. Mexicans use corn husks and a variety of leaves, such as maguey, banana, corn, or *hoja santa* to prepare foods. The Japanese use *hoba* leaf and nori to wrap beef, chicken, sticky rice, seafood, and vegetables. Southeast Asians and Indians commonly use banana, pandan, or *salam* leaves to wrap fish, rice, meats, or vegetables before steaming, grilling, or barbecueing. The following is a list of geographic regions that use different wrappers.

Greece

Tender grape leaves are blanched and used to stuff rice, lentils, or beans with nuts, spices, and raisins, with the resulting food called *dolmades*. They give a lemony tang to the food.

France

Zucchini blossoms are used to wrap cheese or fish with herbs and spices.

Middle East

Meloukhia leaf is a deep green, spinachlike leaf, popularly eaten in Egypt, Palestine, and Tunisia (in North Africa). It has a glutinous texture that gives the finished product a gelatinous, viscous texture. It is mainly used with meat, chicken, or duck soups, as a main vegetable entree with boiled rice, or chopped finely and cooked as a side dish, accompanied by onions, hot chilies, meat, and rice. It is sold frozen or dried and can be used as a wrapper.

India

Banana leaf is a large green leaf that provides moistness and a delicate flavor to foods. It is cut up in pieces and used to wrap whole fish or fillets to be baked, chicken to be grilled, or rice before it is steamed. In Asian Indian cultures, especially with the vegetarian communities, it is used as a plate for serving a whole meal. Hindus serve meals at temples on banana leaves because they signify cleanliness and purity.

Paan leaf is used to wrap a dry filling of chopped betel nuts, pieces of coconut kernel, spices, and sugar. It is like an "after dinner mint," to freshen breath as well as aid digestion. The leaf is wrapped around the filling and tied with a clove.

China

Lotus leaf is a large and round inedible leaf. It is commonly used in Cantonese-style cooking for stuffing rice, dumplings, and fish that are then steamed or grilled. It provides a sweet tealike flavor to foods. It is sold dried and is soaked before use.

Bamboo leaf is a greenish ribbed leaf with a magenta tinge that is used to wrap sliced fish, meat, or rice before steaming, boiling, or grilling. It gives a smoky aromatic flavor to foods. Before use, it is soaked in warm water until it softens.

Soymilk skin, called *yuba*, *fuchok*, or *fu zhu*, is dried bean curd that is creamy yellow in color. It is used in Chinese cuisine to wrap seasoned fish, rice, vegetables, or meats before baking or grilling. It provides a firm yet chewy consistency to foods. Dry yuba needs to be soaked until fully hydrated before use. Similarly, rice paper is commonly used to wrap freshly cooked rice, vegetables, and meats before serving. It provides a tender, smooth, slightly chewy texture to the finished product.

Japan

Hoba leaf is a large leaf used in Japanese cuisine to wrap beef and chicken for grilling.

Southeast Asia

Thais, Vietnamese, Malays, and Nonyas in Malaysia and Indonesians commonly use banana leaf for wrapping fish, chicken, sausages, and meats before grilling and for wrapping glutinous rice and desserts before steaming. In Thailand and Malaysia, banana leaves also serve as natural and elegant plate liners that provide visual appeal to meals. It is an indispensable ingredient in Balinese cooking for wrapping food before roasting or steaming.

In Malaysia, Singapore, and Thailand, fish, chicken pieces, and rice are wrapped in pandan leaf (or screw-pine leaf) before grilling, barbecuing, or steaming. The fragrant Nonya's and Malay's fish and rice dishes commonly use pandan leaf. Local condiments and desserts are also served in pandan leaf.

Salam leaf is a favorite wrapper of the Balinese in Indonesia for fish and ground meat cooked with spicy condiments. Indonesians also use the long, narrow, fragrant turmeric leaf to wrap spiced fish or meats. Cabbage or local salad leaves are also used to wrap beef before grilling. In Thailand, glutinous rice with dried shrimp is steamed in lotus leaf which adds a unique flavor to this dish.

In Sarawak, Sabah, Sulawesi, and West Sumatra, indigenous populations stuff hollow bamboo segments with glutinous rice and finely chopped meats or poultry seasoned with spices and place them directly in open fires to cook. These bamboo segments provide a slightly smoky aroma and a distinct texture to the finished foods.

Caribbean

In many Caribbean regions, whole fish, corn dough, rice and beans, and vegetables are wrapped in corn husks before cooking. In Cuba and Puerto Rico, tamales (masa or corn dough filled with seasoned chicken, beef, fish, or cheese), fish, chilies, or vegetables are steamed or grilled in corn husks. Jamaicans, Puerto Ricans, and Dominicans create their *picadillos* by stuffing spicy ground meat into corn husks that are then steamed.

Latin America

Fresh corn leaves and husks or dried corn husks are used in Mexico, Central America, and South America for steaming, grilling, baking, simmering, or roasting tamales, vegetables, moles, meat, fish, recados, and other seasoning pastes. Before use, the dried corn husks are soaked in hot water until they soften. In Michoacán, Mexico, masa with fillings are wrapped in fresh corn leaves (or *hojas de milpa*) before they are steamed to make tamales.

Maguey leaf from the Agave plant (not a cactus) is a common wrapper in the Yucatan and other regions of Mexico. It has a thin, transparent skin called *mixiote* that is like a pliable parchment paper, which wraps seasoned goat (barbacoa), pork, and chicken before steaming or barbequing.

Banana leaf is used to wrap and slowly roast pork, venison, and other meats. In the Yucatan, banana leaf is used to wrap marinated whole pig that is cooked in a pit to make the local speciality called *cochinita pibil*. Banana leaf is sold fresh or frozen. Before use, it is steamed for about twenty minutes to make it pliable and soft.

Hoja santa (also called *acuyo*/holy leaf/*hierba santa*) is a large green leaf in which corn dough with seasoned fish, moles, and meats are wrapped before they are steamed or roasted. The leaf gives an aniselike flavor to the cooked foods.

SEAFOOD FLAVORINGS

In days before refrigeration, and even today in many Asian, European, and Caribbean cultures, fish and other seafoods are dried, salted, or brined. This is an economical way to preserve them as well as to add flavors and textures to foods. Many indigenous cultures combined them with spices to "spike up" a meatless meal. The ancient Greeks used a fish seasoning called *garon*, and the Romans used fish and its entrails with salt, called *garum*.

Seafood flavorings are essential to the cuisines of many countries, particularly in Asia, to enhance main entrees and side dishes. Seafood flavorings come in whole, ground, sauce, or paste forms. They are obtained from dried, smoked, salted, or fermented fish, oysters, shellfish, scallops, and squid. Their preparation varies in different regions of the world. The paste is the solid portion of the dried and fermented product, and the sauce is the liquid portion.

The fish used as flavorings are usually small fish, such as anchovies or whitebait (also called ikan bilis in Malaysia and Indonesia or karuvaddu in Sri Lanka, South India, and Malaysia). In the Caribbean and Mediterranean, the larger codfish is also used. Brazilians use whole dried shrimp or codfish (*bacalhau*) to flavor many dishes. In Europe, anchovies and pickled herring are used as flavorings and appetizers.

SEAFOOD FLAVORINGS IN ETHNIC CUISINES

European Seafood Flavorings

Small bony fish or tiny shrimp, along with the bones and heads, are used to make fish stocks that are the basis of many fish soups, stews, and sauces. The French and Italians use cured anchovies with salt, capers, olive oil, and spices to flavor many of their soups, stews, salads, and marinades.

In southern France, a popular spread called *anchoiade* is a salted paste made with anchovy fillets, black pepper, olive oil, garlic, and vinegar. In Piedmont, Italy, anchovies with garlic, butter, and olive oil are made into a dip for raw vegetables called *bagna cauda*. Anchovies are also added to tomato sauces that are then used as spreads. *Liquamen*, a fish sauce that is a by-product from anchovy manufacture, is used for intensifying the flavor of seafood dishes in southern Europe. In Holland and Belgium, anchovies are marinated in wine vinegar with onions and lemon rind and used as a condiment. The Swedish use pickled herring with sugar, vinegar, allspice, onions, bay leaf, and dill to flavor many of their cream sauces.

Salt-dried cod, bacalhau or bacalao, is a popular flavoring for sauces, paellas, and vegetables in the Basque region of France, Spain, Portugal, and Italy. The Portuguese enjoy cod in a variety of ways—mixed with potatoes, eggs, parsley, and onions, flaked and made into fillings for savory pastries, or as cod balls. In Spain, bacalao is prepared with garlic and pil pil, as codfish balls with sweet spices and parsley, stuffing in pastas, and added to croquette. In Liguria, Italy, salt cod or baccala is softened in water and cooked with parsley, capers, dried mushrooms, pine nuts, and vegetables for sauces or is combined with flour to make fritters. It is also soaked in milk and cooked slowly in olive oil, spices, and vegetables to make a codfish stew called *stoccafissa* (or *stocchefisso*).

African Seafood Flavorings

In West Africa, dried and smoked fish or shrimp are used to flavor many dishes. Some examples include dried, smoked shrimp used in the Gulf of Benin, dried, smoked mollusk and guedge, (a dried, smoked fish), both used in Senegalese stews and sauces. In Egypt and Sudan, a seasoning sauce called *faseekh* is made by fermenting small fish with salt.

Asian Seafood Flavorings

East Asian Seafood Flavorings

Fish sauce and pastes or dried whole seafood, called *yu-jiang* and *yu-lu* in China, or *shotturn*, *dashi*, or *gyomiso* in Japan, are used abundantly to season cooking sauces or soups. Oyster sauce, a popular cooking sauce, contains oyster juice (from cooked, dried oysters), fermented soybeans, sugar, starch, and other ingredients.

Southeast Asian Seafood Flavorings

Fermented and dried shrimp are a staple flavoring and food in Southeast Asia. Southeast Asians use fermented or salted, dried seafood to add zest and enhance rice, noodles, vegetables, and seafood dishes and condiments. *Ikan bilis* is used as a seasoning as well as a crunchy side dish and appetizer in Malaysia and Singapore. In Vietnam, Indonesia, Philippines, and Malaysia, a meal does not taste right without salted or fermented seafood condiments. *Kecap ikan*, a fish sauce, is commonly used in many Indonesian and Malaysian dishes. The Vietnamese *nuoc mam*, Thai *nam pla*, Malaysian *belacan*, or Filipino *patis* are essential ingredients for enhancing many sauces, curries, salads, marinades, and condiments. These fish or shrimp flavorings are unacceptable by themselves, but when cooked and added to other dishes, they round off other flavors or add zest to sauces, soups, or salads. They come with different flavor intensities and consistencies as viscous pastes or as thin liquids. They can be light colored with mild flavors or dark colored with heavy, pungent flavors.

Seafood flavorings come as dried whole pieces or ground, as pastes, and as sauces. Southeast Asians use the pastes and sauces as condiments or seasonings, much like the French use roux or wine, and the Chinese and Japanese use soy sauce or rice wine. They flavor broths, soups, stir-fried noodles, and sauces or are used as marinades. They become part of Japanese dashis and Chinese stir-fries, as well as pad *thai, nuoc cham,* or *sambal belacan.* Other dishes using these flavorings include: Vietnamese *pho bo* (beef noodle) soup, Nonya laksa (fragrant curry noodles), Laotian *laab* (minced beef salad), Malay *sambal udang* (fiery shrimp dish), Burmese *balachaung* (condiment), Thai *nam prik gaeng ped* (red curry paste), and Filipino chicken *binakol.* The Cristangs (of mixed Portuguese and local ancestry) from the Malacca region of Malaysia use a popular condiment called *chinchaluk,* made from salted and fermented krill (a type of shrimp) with chile peppers, shallots, and lime juice.

Commercial fish sauce is most commonly made from fish but can also be made from shrimp. It adds a subtle flavor, enhances other ingredients, or mellows stronger flavors. It is added as a freshly prepared table condiment, as a dipping sauce for spring rolls, as a marinade, or toward the end of cooking. Though the sauce has a fishy aroma and salty taste, the fishy aroma disappears during cooking and adds a delicate overall flavor to the product. It pairs well with chile peppers, lemongrass, lime juice, sugar, garlic, coconut milk, cilantro, spearmint, basil, and galangal.

Fish sauce (*nam pla*) is the main flavoring in Thai cooking. It is a light brown extract made from salted and fermented small fish. The fish are dried in the sun for a few hours then layered in barrels with salt and water and kept submerged. These barrels are exposed to the sun, whereby the mixture ferments for about three months. Then, the liqueur is drawn off slowly, put in a ceramic urn and aged another month before bottling. The first extraction from the fermented fish is used in a fresh batch of dried fish and fermented for another three months. This method produces the saltier, darker, and stronger fishy note that is so desired in local cooking.

Dried whole shrimp or prawns (boiled and dried), called *kung haeng* in Thai, *udang kering* in Malaysian, or *tom kho* in Vietnamese, are commonly used for vegetables, fried rice, noodles, stir-fries, condiments, soups, and sauces. They are soaked in warm water for about five minutes and then used unpeeled or peeled. These dried shrimp are used whole, chopped, sliced, or pounded. They can be deep-fried and served as an appetizer with a dip or added to vegetables, sambals, rice, and noodles. They are pinkish brown in color and perk up blander ingredients such as noodles, rices, tofu, cassava, potatoes, and cabbage. They can be ground to a coarse powder and sprinkled over salads or stir-fried for sauces and soups.

Shrimp paste is a moist greyish to a hard, crumbly brownish black block. It is a shrimpy, salty, and pungent seasoning and is generally sold as a purplish pink to beige paste. The aroma is intense and overpowering, but when cooked, its raw taste and aroma disappear and it develops a more fragrant flavor. Shrimp paste is generally cooked with other ingredients, but if it is used in uncooked foods, it is toasted before being added to the food. This toasting creates a more acceptable flavor in the finished product. It does not dominate the flavor of the dish but enhances the other flavors.

Shrimp paste adds pungency and a fragrant aroma to condiments, sambals, soups, salads, and stews. The heavier dark to black shrimp paste, called *petis* or *hay koh,* is a molasses-like seasoning. It is a favorite of the Nonya and Muslim cooks in Malaysia, Singapore, and Indonesia, who use it in noodle soups, stewed noodles, and spicy salads called *rojak.*

Some of the more popular fish and shrimp flavorings include the following:

(1) Small fish/anchovy sauces and pastes: patis (Filipino), sauejot (Korean), nam pla (Thai), nuoc mam (Vietnamese), and mamcho (Laotian). Also, heavier fish sauces include bagoong Balayan (Filipino), ngapi (Myanmar), and liquamen (southern European).

(2) Shrimp paste: bagoong alamang (Filipino), trasi (Indonesian), belacan (Malaysian), kapi (Thai), and mam ruoc (Vietnamese). Heavy, darker pastes include petis (Indonesian) and hay koh (Singaporean).

(3) Whole dried fish or shrimp: ebi (Indonesian), dried bonito flakes/kat-suobushi (Japanese); whole salted fish/ikan bilis (Malaysian, Indonesian), salted fish/pla haeng (Thai), dried shrimp/udang kering (Malaysian), salted shrimp/kung haeng (Thai); salted shrimp/tom kho (Vietnamese), maldive fish, and nettali karuvaddu (Sri Lanka).

These dried and or fermented sauces and pastes are rich in protein, B vitamins, calcium, and omega 3 fatty acids.

South Asian Seafood Flavorings

Bombay duck, a Goan condiment from southwest India, is a dried, gelatinous, spicy fish paste used to perk up sauces or condiments. It is also eaten as a snack. In Sri Lanka, pounded and dried maldive fish are popularly used in a variety of *sambols* (local condiments) combined with green ginger, chile peppers, garlic, lemongrass, cinnamon, onions, coconut, and lime juice. *Sambols* are prepared in different ways and are commonly eaten with local breads such as *thosai*, *puttu*, or even bread and butter, especially for breakfast. Dried small fish called *nethali karuvadu* are combined with green chile peppers, shallots, sugar, and vegetables for sauces, curries, and as condiments.

Caribbean Seafood Flavorings

In the Caribbean, salt cod fish (saltfish or *bacalao*) is a common base flavoring for stews, soups, or made into spicy fritters called *bacalaitos* or *accra*. It is shredded and added to vegetables, such as ackee, made into souffles and pies, and flavors rice and beans, eggs, and many root vegetables. It is mixed with habanero peppers, onions, tomato, and olive oil to create buljol, which is served for breakfast or brunch, in Trinidad. In Puerto Rico, bacalao is combined with sweet chile peppers, tomatoes, pimientos, and garlic to flavor mashed potatoes, yautia, and cabbage.

Brazilian Seafood Flavoring

Salted codfish (*bacalhau*) and dried whole shrimp are common flavorings in many stews and sauces. A *refogado* is made with salted codfish or dried shrimp, onions, tomatoes, chives, parsley, chile peppers, and sweet bell peppers and is used to flavor potatoes, spinach, coconut milk, rice, omelets, squash, and fish.

FRUIT FLAVORINGS

Fruit flavorings come from the edible fleshy pericarp, sometimes the receptacle, and the skin of the fruit. Fruits are used fresh, dried, pickled, pureed, ground, or chopped. Whole fruits are chopped, sliced, or grated in salads, stews, and curries. With some fruits, only the rind is used. Pureed, they provide flavor, consistency, and color to many dishes and condiments. The flavor profile varies between ripe and unripe fruit and also changes when the fruit is dried or cooked.

Nuts are fruits that will be discussed separately. Tomato, avocado, breadfruit, and cucumber are fruits but will be placed with vegetables because they function as vegetables.

Traditionally, North Americans and Europeans have used ripe or dried subtropical or temperate fruits to provide characterizing fruity flavors or sweetness to beverages, desserts, and confectionaries. Fruits that are typical of North American and European cuisines include berries (cranberry, blueberry, strawberry), citrus, grapes, stoned fruits (apricot, cherry), plums, melons, and pomes (apples, quinces, pears).

Fruit flavorings can create new flavors, balance flavors, or tone down pungent flavors. They provide natural colors and visual appeal and can also become natural acidulants. They can be used to flavor or color vinegars, cooking oils, and sauces. Fruits complement many spices, nuts, fish flavorings, and chile peppers.

Fruits provide tartness, sweetness, fruitiness, consistency, and color to dishes. Sweetness in fruits is due to sugars, mainly fructose and glucose, and varies with their ripeness. Carambola (or star fruit) can be mildly sweet when ripe and sour when unripe. Because of their tartness, fruits can be natural acidulants. The acidity is due to organic acids such as malic, citric, or tartaric. Lemon, lime, sour orange, tamarind, carambola, kalamansi, and pomegranates are used to provide acidity and astringency to foods and can be tapped as "natural" souring agents for foods and beverages to replace citric or malic acids. At the same time, they will add other enhancing notes to the food—sweet, fruity, astringent, or floral. Fruits can be further explored as natural coloring agents, antioxidants, and antimicrobials in beverage and food applications.

Mediterranean, Caribbean, and Asian cuisines combine fresh, dried, or pureed fruits to complement their savory and hot flavors and to create visual appeal. Tropical fruits also provide the variety that consumers seek. Fruits are added fresh or cooked to main dishes, side dishes, and condiments or serve as attractive serving vessels for condiments, salads, and all other foods.

Mango, banana, fig, lime, dates, papaya, pineapple, guava, rambutan, star fruit, guarana, and lychee (or litchee), are many examples of tropical fruits. Many unique fruits such as kokum, pomegranate (*anardhana*), green mango, *belimbing wuluh*, green papaya, yuzu orange, sour jackfruit, olive, and tamarind are commonly used in Asian, Caribbean, and Latin American cuisines and are also becoming known in the United States. These tropical fruits, such as mango, acerola, or melon, differ in their flavors, colors, and textures, depending upon their regional origins. For example, mango can have a pulpy or stringy texture and can be sweet and fragrant or sweet and tart.

Today, North Americans are adding fruits to savory dishes and to create unique salsas, condiments, soups, or salads. Fruits complement many savory ingredients. Apples and apricots go well with pork; cherries and plums with duck and star anise; cranberries with turkey and nutmeg; gooseberries with mutton and coriander; orange peel with chicken, duck and basil; pineapple with pork, ham, and lemongrass; dried or pickled plums with beef and shiso; pomegranate with legumes and bay leaf;

prunes with pork and ginger; mango with lamb curry and chile peppers; coconut with pungent laksas; kokum with fish and chutneys; olive with garlic and tomato; or kalamansi with steak and pork.

Consumers are also seeking "natural" ways of staying healthy. Fruits are great sources of vitamins A, C, K, and E, minerals, and other important nutrients.

Fruits in Ethnic Cuisines

Fruits, when ripe or unripe (green), give flavor and color to many ethnic savory dishes. Mango is used fresh or dried (unripe) in Indian cooking and provides tartness and flavor to curries and chutneys. Mandarin orange peels give sweet and citrusy notes to Chinese stir-fries and sauces. Papaya, carambola, and guava add unique fruity notes to hot Caribbean sauces and beverages. Grapefruit, sapote, and tomatilla spike up Mexican condiments and salsas, while passion fruit and jackfruit provide consistency to Indonesian *sambals* and *gulais*.

Southeast Asians, Indians, Caribbeans, or North Americans are fond of the flavor of unripe (or green fruits) and sour flavors and use them to flavor curries, soups, and salads, including green papaya (Thailand, Vietnam), green mango (India, Caribbean, Southeast Asia), green jackfruit (Vietnam, Indonesia), plantain (Caribbean, Latin American), plum (Japan, China), ackee (Caribbean), belimbing wuluh (Malaysia, Indonesia), green carambola (Caribbean, Southeast Asia, Florida), and green apple (United States, Europe).

Unripe carambola is used as a flavoring in many Southeast Asian and Caribbean foods. *Belimbing* or *bilimbi*, smaller varieties of carambola, are added to soups, pickles, and marinades in Malaysia, Indonesia, and Singapore and curries of South India and Sri Lanka. *Kamia*, a relative of star fruit, is popularly used as a souring agent for marinades, stews, salads, and soups in the Philippines.

Dried fruits such as amchur, prunes, figs, dates, and raisins and dried persimmons and pomegranate add flavor to many Mexican, Asian, and Mediterranean dishes. Prunes, dates, and figs are popular flavorings in Middle Eastern rices, meats, and desserts. In Mexico, prunes and raisins are blended to add flavor and texture to desserts, moles, meat picadillos, flans, and rice puddings. Fruits are also candied and used in many dishes of the Mediterranean.

Table 23 describes various fruits, their flavors, and other qualities they add to foods and the ethnic cuisines in which they are typically found.

Specific Fruits as Ingredients

Below is a detailed description of some of the more significant fruits for food applications.

Mango

Mango is indigenous to India, and the word "mango" is derived from the Tamil word "mangai." There are many types of mangoes that vary in size, shape, flavor, texture, and color. They are grown in India, Southeast Asia, the Caribbean, the Pacific Islands, Africa, and Latin America.

TABLE 23
Global Fruits and Their Flavoring Properties

Fruit Flavorings	Flavoring Property	Ethnic Preference
Star fruit/carambola	Sour, fruity	Caribbean, Mexico, Southeast Asia
Belimbang, Bilimbi	Sour, fruity	South India, Southeast Asia, Sri Lanka
Kalamansi/ponzu	Sour, fruity	Philippines, Japan, Indonesia
Mango, Guava	Sour, floral, sweet	India, Caribbean, East Asia, Southeast Asia
Rambutan	Fruity sweet	Southeast Asia
Guarana	Caffeine, astringent	South America
Persimmon	Astringent, sweet	East Asia, Southeast Asia
Pomegranate	Fruity, sour, sl. sweet	Mediterranean, South Mexico, India
Plum	Sour, fruity	Japan, China
Atemoya	Creamy	Mexico, Caribbean
Cherimoya	Creamy	South America, Mexico
Cranberry	Sour, color, acidic	United States
Langsat/duku	Sweet, sour	Southeast Asia
Fig	Sweet, fruity	Mediterranean, United States
Jujube	Sweet, astringent, acidic	China
Acerola/West Indian cherry	Acidic, fruity	South America, Caribbean
Papaya	Sweet, floral	Southeast Asia, Caribbean
Pineapple	Sweet, fruity	Southeast Asia, Caribbean
Orange (Seville, Mandarin)	Sweet, sour, fruity, bitter	Latin American, Asian, Mediterranean

Mangoes' colors range from a yellowish red to yellow to a yellowish green. The inner flesh is generally yellow. Fresh mango has a sweet to slightly tart taste, depending on the variety. Indian varieties are much sweeter and aromatic and less fibrous than other varieties.

Mango has a high level of beta-carotene (an average-sized mango has 8000 IU), vitamin C, dietary fiber, and potassium. Traditionally, it was used to help the body fight infections.

In the United States and Europe, mango is generally made into jams, sherbets, and ice cream. In Asia, fresh mango is sliced and eaten after a meal, added to salads, or made into desserts or juice beverages. In the Caribbean, it is made into beverages or added to sauces.

Mango pairs well with coriander, chile pepper, mint, ginger, tamarind, and kaffir lime leaf and is generally added to poultry, lamb, and vegetable dishes.

The dried green mango, called *amchur* in Hindi, is sliced and ground into a powder. The powder has very little aroma but gives tartness to foods. It has a sour, astringent taste and is especially popular with vegetables and pickles. It is also used in chutneys and in marinades for the barbecued meats (tandooris) of North India. Amchur tenderizes the meats and chicken cooked in the tandoors, thus making them succulent and juicy. Its flavor is due to citric acid and terpenes such as ocimene, myrcene, limonene, with aldehydes and esters. The dried pieces are pickled and become a popular condiment in every Indian home. It is also an essential component of a green spice blend called *chat masala* that is commonly used in North India.

Green mango is also eaten with fermented fish paste or is mixed with salads in the Philippines and Bali. In Malaysia, Indonesia, and Singapore, it is mixed with chilies, brown sugar, and other fruits and vegetables for a spicy salad called rojak. In the Caribbean, it is added to curries and chutneys, especially in Trinidad and other islands where there is a substantial Indian population.

Pomegranate

Pomegranate is native to Persia. They are large beige to red skinned fruits with numerous pulpy seeds that have a sweet-sour taste. It is these seeds that are used in culinary applications. Pomegranate is called *anardana* in Hindi, *anar* in Farsi, *rumman* in Arabic, *melogranate* in Italian, *mattalam* in Malayalam, or granada in Spanish. Pomegranate juice, its seeds, and paste made from the seeds (*pekmez*) are essential flavorings in Turkish, Iranian, and other Middle Eastern regions. They are used as marinades and in sauces and desserts. The dried seeds are sprinkled over salads, hummus, and desserts. They give a sweet, sour, fruity taste to meats and vegetables.

Dried pomegranate is darkish red in color and adds sourness to cakes, desserts, chutneys, and curries. In North India, pomegranate seed juice is used to marinate meats and is added to desserts. In Punjab, pomegranate seeds are dried, crushed, or ground and used as a garnish and souring agent in chutneys, *pakora* fillings, *paratha* stuffing, and curries. They are commonly used in the vegetarian cooking of Gujerat, which is noted for its spicy sweet flavors. The vegetarian Jain community of Gujerat uses pomegranate seeds as toppings on fiery legume curries.

In Mexico, pomegranate seeds are used as garnishes for a white sauce topping on stuffed poblanos. This is a renowned dish of the Puebla region called *chile en nogada*, and the red color of pomegranate seeds gives visual appeal and provides a bittersweet flavor to the product.

Papaya

Papaya or pawpaw is a juicy, deep yellow to orange fleshy fruit with centrally located black seeds. The ripe fruit has a sweet taste and enhances many spices, such as chile peppers, onions, ground mustard, cardamom, nutmeg, mint, or cinnamon. The juice of the unripe fruit is used as a meat tenderizer in many Southeast Asian regions. In the Caribbean, papaya is added with habaneros, mustard, and vinegar to create a hot fruity sauce. A variety of green papaya, called *du du xanh*, is peeled and shredded to give a crunchy texture to spicy salads of Thailand and Vietnam.

Papaya is high in vitamin C, beta-carotene (about five times that of orange), and fiber. Traditionally, it was eaten to help resist infection, aid digestion, and help prevent hardening of the arteries.

Orange

Oranges are native to India but spread to China and were brought to the Mediterranean by the Arabs. There are many types of oranges commonly used to flavor savory foods—*naranga* China (sweet), *naranga agria* (Seville, sour/bitter orange), Mandarin orange, bergamot orange, and tangerine. Sweet oranges, such as Spanish, Mediterranean, blood or navel types, have a firm, tight skin. They are common eating oranges and are used in desserts, beverages, and baked products. The sour and bitter types of oranges are commonly used to flavor meats, seafood, and salads. Seville oranges, or *narangia agria*, are used for marmalades, cakes, stir-fries, and condiments of Asia and the Mediterranean, and is a popular flavoring in Latin American marinades, ceviches, and stews. Mojos in Cuba, adobos in the Dominican Republic, ceviches in Ecuador, and recados in Yucatan contain this sour, bitter orange juice.

Mandarins and tangerines are loose-skinned oranges that have less acidity than the sweet orange. Bergamot orange is a variety of bitter orange.

The orange peel or rind is strongly aromatic with a bittersweet taste. It has about 2.5% essential oils, mainly limonene, citral, bitter glycosides, and rutin. The highly scented flowers of oranges are extracted for their essential oil, called neroli oil.

Orange peel, fresh or dried, whether sliced, grated, or ground, is an important flavoring in many European and Asian cuisines. In France, it is added to bouquet garni, fish, and meat dishes. It adds a fruity, bitter flavor to the many hot, sweet, and sour Szechwan dishes of China. It pairs well with star anise, ginger, soy sauce, fennel, mustard, brown sugar, and rosemary.

Dried orange peel is a common ingredient in many Chinese-based sauces for chicken and beef that require long simmering or stir-frying.

Yuzu orange, called *som na* (in Thai), *cidra* (in Spanish), and *turung* (in Arabic) is commonly used in Japanese, Southeast Asian, and Mediterranean cuisines. The rind of yuzu orange is an essential ingredient in the Japanese spice mixture shichimi togarashi. Kumquats are tiny orangelike fruits but are not related to oranges. The rind and peels are used in Chinese, Southeast Asian, and Mediterranean cooking and also as a flavoring in many European liqueurs.

Oranges are high in vitamin C, beta-carotene, potassium, and flavanoids. They are used in traditional medicine to protect against cancer, lower LDL, and prevent colds.

Lemon and Lime

Lemon is derived from the Persian word limun, or Arabic word lemun. Lime and lemon are fundamental ingredients in Southeast Asian cooking. In many regions, lime is often confused with lemon, and they have similar names. Lime juice and lemon juice have long been used to flavor many cuisines. Lemon is called *limon* (Spanish), lemun (Arabic), *limun* (Farsi), zitrone (German), *ning meng* (Chinese), citron (French), *limau* (Malaysian), *elimicham* (Tamil), and *manao leung* (Thai).

Lemons are rich in essential oils (6%) that are mainly composed of limonene (90%) and citral (5%), with some glycosides and coumarin. Lemon is the principal souring agent in the United States and in Europe. Its peel is aromatic and bitter and is also used as a flavoring for fish soups, stews, couscous, beverages, and desserts. In India and Iran, lemon peel is ground and used to flavor rice.

Lime is called *limon agria* (Spanish), *lai meng* (Chinese), *limette* (German), *limau nipis* (Malaysian), and *manao* (Thai). Limes are green to greenish yellow with thin, tight skins and are generally smaller than lemons. Typically, lime has a sour and slightly bitter juice and, depending on their origins, can have a lemony flavor.

Kalamansi, also called *limau kesturi, jeruk Cina*, or *ponzu*, is a less acidic and more fragrant small lime that is used in noodles, soups, sambals, and desserts and as marinades in Southeast Asian and Japanese cooking.

Kaffir lime does not have juice, but its peel and leaves add zest to the curries and sauces of Malaysia, Indonesia, and Thailand.

Other Tropical Fruits

Salty pickled sour plum (*umeboshi*) gives a fruity aroma to plain glutinous rice which is taken for breakfast in Japan. It aids digestion of a meal. Plum sauce, made from salted plums, chilies, and vinegar, is a popular dip that is added to many Cantonese poultry and pork dishes.

The small, pale green astringent *belimbing wuluh* (relative of star fruit) is an acidic fruit that provides tang to the fish dishes, *laksas*, and *sambals* of Malaysia, Indonesia, and Singapore. A popular flavoring in the Philippines, it is pickled, or added to soups, stews, and condiments. The juice of the sour kalamansi (called *jeruk Cina* in Indonesia) is used to marinate fish and meats of these regions. It also adds aromatic citrus notes to noodles, soups, and desserts.

Jackfruit or *nangka* (Indonesian and Malaysian) is used green or unripe to flavor stews, sambals, or curries. It has a banana- and cantaloupe-like flavor.

The soft seeds of the Mahlab cherry (a black cherry), either ground or whole, provide a delicate aroma and bitterness to Turkish and Middle Eastern sweets, biscuits, breads, and stews.

Olives

Called *oliva* (Italian), *zaytun* (Arabic), *aceituna* (Spanish), *zeytin* (Turkish), or *saidun* (Tamil), olives are indigenous to Greece, Italy, Spain, northern Africa, and other regions of southern Europe and western Asia. One of the oldest known fruits, they were valued by the Egyptians, Romans, and Greeks, who decorated Olympic winners with olive branches.

Olive is a fleshy fruit (a drupe), and there are many types that vary in size, color, oil content, and flavor. Examples include *sevillano* from California, *manzanillos* from Spain, kalamata from Greece, *liguaria* from Italy, and nicoise from France.

Olives are picked at various stages of maturity. They are tiny, hard, and green when they begin to ripen, become purple before they are fully ripened, and are plump and black when fully ripened. Each type goes through different curing methods after a lye or water wash. Brine is used for the green (unripe) olive, while salt is used for the fully ripe olive.

Olives are sold whole and pickled (in brine), crushed as a paste, fermented, or extracted as olive oil. Raw olives are very bitter but lose their bitterness and develop an aroma after several months of curing. Green olives are picked and treated with lye solution before pickling, which removes the bitterness. In Greece, the fully ripened or black olives develop a strong flavor when they are treated with salt, which also removes the bitterness. Pickled olives are black or green depending on whether they were picked ripe or unripe. Leafy spices are added to pickled olives to enhance their flavor and create new flavors.

Olives contain a glycoside, called oleuropein, a phenol that causes the bitterness in the unripe fruit, and that decreases during ripening. The oil (25% to 55%, depending on the variety) is from the mesocarp. The oil mainly contains oleic acid, linoleic acid, and palmitic acid, with some eicosenoic acid and palmitoleic acid. The green coloring of the olive is due to chlorophyll.

Olives are an essential flavoring in Mediterranean cuisines, especially green olives that are used in cold (or uncooked) dishes such as salads and in sauces, breads, and spreads. They are even eaten as a snack. Olives are commonly served as tapas or mezzes accompanied by cheese, breads, nuts, or tomatoes. Black olives are used in many cooked dishes of Italy as pizza toppings and in tomato sauces. In Provence, France, black olives are an essential ingredient of tapenade, a spread with garlic, anchovies, and capers.

In Morocco and Tunisia, olives are cooked with lemons, chicken, and spices in tagines. Olives add slightly sharp, tart, and bitter notes to salt cod, potatoes, chicken, and eggplant. In Spain, the pits are removed, and the green olives are stuffed with sweet red pepper, capers, anchovies, or almonds. In Mexico, olives are chopped and combined with ground meat and spices for sweet, savory *picadillos*. In the Veracruz region, they give a sour note to the popular red snapper dish.

Olive oil comes as heavy and fruity or light with a delicate perfumed note, depending upon its stages of extraction or pressing of the oil. These different types of oils are used for specific applications and are often added at the end of cooking to add flavor.

Olive oil is used for frying, cooking, and as a flavoring. It adds flavor to hummus, zhoug, aioli, and pasta or is used as a dip for breads and salads. It goes well with garlic, tomato, sesame paste, beans, many leafy spices (basil, thyme, oregano, rosemary, tarragon and parsley), and green chile peppers.

Kokum/Kodampoli

Kokum fruit, scientific name *Garcinia indica,* is a popular flavoring in India, especially in Kerala, Gujerat, and Maharashtra regions. In Kerala, it is known as *kodampoli* or referred to as "fish tamarind" and is used for many fish dishes instead of the tamarind. It has a pinkish purple or deep purple skin that is sticky to the touch when ripe. It is available as a whole fruit, dried rind or skin, or as a paste. It has a fruity, sour, slightly smoky, and slightly salty taste. Its sourness is due to malic and tartaric acids.

Kerala is noted for its hot and sour flavors, and *kokum* is used to provide the sourness to fish, vegetables, coconut-based curries, and many chutneys. It is generally discarded before the dish is served.

Garcinia globulosa (or called *asam kandis* in Indonesia) is a small, thin-skinned fruit that is used as a substitute for tamarind in Indonesia, especially in Sumatra. The skin of the fruit, which is bitter, is dried into a hard black piece that is used in many of the local dishes. It is believed to relieve skin allergies and sunstroke.

Coconut

Coconut's scientific name is *Cocos nucifera.* Called "shripal" in Sanskrit or "fruit of the Gods," it is a holy fruit for the Hindus and is used in Hindu religious ceremonies, weddings, and prayers. It is broken as an offering to the Gods, and its creamy firm kernel (meat) is eaten as a blessing. Hindus believe that the three eyes on the coconut represent the Hindu trinity of Lord Shiva, Vishnu, and Brahma. It is also called kelapa (Malay), kokos (German), coco (French), polgaha (Singhalese), thenga (Tamil), nalikeram (Malayalam), nariel (Hindi), coco fruto (Spanish), or maprow (Thai).

Coconut provides many flavorings—milk, water, vinegar, and a fermented beverage called "toddy" or "*arrak*." The inner fleshy pulp (endosperm) is scooped and eaten fresh when the coconut is green or unripe. When the coconut becomes ripe, the endosperm or meat is grated and used fresh or dried (called *kopra* or copra) to thicken sauces and to provide texture or flavor foods. The dried coconut is sold as desiccated and comes in three grades—fine, medium, and coarse.

Coconut juice is extracted with water from the grated coconut or endosperm and is called coconut milk or *santan* in Indonesia and Malaysia. It is sweet, milky white, and nutty and is a popular flavoring in South India, Southeast Asia, and the Caribbean.

Coconut milk is available canned, dried, or as a paste concentrate that can be reconstituted. The copra can be sold desiccated, toasted, creamed (canned sweetened or unsweetened), or as an extract (concentrated form) and contains 60% to 70% oil, mainly lauric (40% to 55%), myristic (15% to 20%), caprylic, capric and palmitic acids, with small amounts of oleic acid. Coconut oil is the most popular frying medium in South India. The coconutty flavor is due to delta-lactones, especially gamma-deca lactone, while the toasted coconut flavor is due to maltole and cyclotene.

Coconut flavoring as coconut milk is indigenous to Southeast Asian, Sri Lankan, Caribbean, and South Indian cooking. It is an essential ingredient in the local curries, marinades, soups, stews, sauces, breads, and sweets. The sweet flavor of coconut milk tones down intense heat, as well as rounds off flavors.

Coconut cream is a must in many dishes of Kerala, especially in the local fish curries. The dessicated coconut gives texture and visual appeal to many dishes and steamed breads. *Avial*, a popular local vegetable dish, contains toasted, desiccated coconut and dried chilies, kari leaf, asafetida, turmeric, toasted lentils, and other spices.

In Sri Lanka, coconut milk is used with toasted spices to mellow and give body to hot curries. Coconut milk is also used in the steamed, fermented breads called "hoppers," which are made from rice flour, coconut milk, and yeast and are served for breakfast.

In Indonesia and Malaysia, coconut milk and grated coconut are essential flavorings in *rendang* (which is beef simmered in coconut milk), curries, and vegetables. They are also popular in desserts and cakes such as *wajik*, *dodol*, black rice pudding, and *dadar* (pancakes with a sweet coconut filling).

In Thailand, coconut milk is an essential ingredient in curries, including red, green, yellow, Mussaman or Panang. Coconut pairs well with fish sauce, belacan, turmeric, lemongrass, tomato, galangal, basil, eggplant, potatoes, chocolate, fruits, vanilla, pandanus, glutinous rice, and *gula melaka*.

Palm Sugar

Palm sugar is made from the sugary sap of the sugar palms of Southeast Asia and India and the flowers of the coconut palm. Palm sugar (called *gula melaka*, *gula merah*, *nam taan* peep, or *gula jawa*) flavors the sweet and savory dishes of Thailand, Malaysia, Singapore, and Indonesia. It is sold as golden to light brown compressed cylindrical packs or thick pastes. It has a caramel, maple-sugarlike flavor. The darker types are stronger in flavor. It is used commonly in desserts and puddings in Thailand (with water chestnuts, pumpkin, and bananas), in Nonya cooking (with adjuki beans, coconut milk, and pandan leaf essence) and in Malay desserts (with jackfruit, banana, and glutinous rice). Thai sweet dumplings called *kanom tom daeng* are served with a sauce of palm sugar, coconut milk, and caramelized sugar. *Gula melaka* is also the name of a Malaysian dessert that contains palm sugar, coconut milk, pearl sago, and pandan leaf.

Jaggery, used commonly in Indian desserts, comes from another palm called *palmyrah*. Jaggery is light brown to dark brown, has less sweetness than cane sugar, and has an aromatic flavor and caramel-like taste. The Mexican *piloncillo* or *panela* are sold as large, flat, compressed cakes and small or large cones. To use these hard sugars, they must be grated, chopped with a knife, or softened in water, the latter method being the best.

Since palm sugars are unrefined sugars, they contain many minerals and vitamins.

VEGETABLE FLAVORINGS

In the West, vegetables traditionally have been boiled or served raw and have been eaten mostly because of their nutritive value. In many regions around the world, such as India or Southeast Asia, certain cultures survived only on vegetarian meals, either for economic or religious reasons. These cultures learned to season vegetables to give them great flavors. Asian preparation techniques, such as sauteing, grilling, steaming, baking, braising, or currying, have created many unique flavors and textures in vegetable dishes. Many vegetables contain volatile oils that give sauces and beverages pungent aromas and savory sweet or bitter tastes.

Today, with consumer's interest in a healthy lifestyle, vegetable extracts and purees can be substituted for meat and chicken stocks or pastes for flavoring soups, sauces, or stews. They are a good source of fiber, vitamins, especially C and A, minerals, and antioxidants.

The cruciferous vegetables such as broccoli, cabbages (green savoy, red, Chinese cabbage or bok choy), collard greens, mustard greens, kale, kohlrabi, and cauliflower are high in vitamins, monoterpenes, thiols and indoles. Broccoli contains high levels of calcium, pantothenic acid, vitamin A, thiols, and indoles and sulfur compounds that have antimicrobial activity. Cauliflower has high levels of iron, boron, thiols, indoles, and sulfur compounds. Carrots contain a high level of carotenoids and antioxidants.

Spinach contains iron, carotenoids, and chlorophyll and is an especially good source of lutein. Bitter melon has polypeptide P (an insulin-like compound that can replace insulin); eggplant is high in bioflavanoids; plantains have high folic acid, vitamin C, and potassium; and pumpkin has folic acid, vitamin A, and vitamin C. Mushrooms, especially Asian types, are rich in zinc, pantothenic acid, and folic acid.

VEGETABLES IN ETHNIC CUISINES

Table 24 lists various vegetables, their sensory qualities, and their ethnic preferences.

Specific Vegetables as Ingredients

Some vegetables that add significant flavorings are described in detail below.

Tomato

Tomatoes are a global flavoring. Tomatoes are eaten raw or cooked and are used chopped, pureed, as paste, juiced, or whole. They provide sweetness, acidity, and color to many ethnic foods, such as salsas, sambals, chutneys, curries, pasta sauces, pizzas, or sweet and sour sauces.

The volatile oils in raw tomato include 3-hexenal, 3-methylbutanal, betaionone, 3-hexenol, and eugenol. Canned tomatoes primarily have dimethyl sulfide. Tomatoes are high in lycopene.

Southeast Asians add tomatoes to most dishes to add a sweet acidic flavor or to balance the pungent fiery notes of sambals whose main ingredients are chile peppers and fermented shrimp. The Chinese add them to characterize their sweet and sour sauces. Indians add tomatoes to chutneys and curries to balance the spiciness, and Mexicans add them to their salsas to give sweet notes. Sometimes, Southeast Asians add ketchup as a source of tomato flavoring in their sweet-sour sauces, sambals, and curries.

Tomatilla or Mexican green tomato is not a true tomato but comes from the cherry family. It has a sweet-sour taste, with a fresh haylike aroma. A calyx surrounds the tomatilla and, as it grows, the calyx also grows and becomes a brownish and

TABLE 24
Global Vegetables and Their Flavoring Properties

Vegetable	Flavoring Quality	Ethnic Preference
Beet	Color, sweet	Mediterranean, United States
Bitter melon/karela	Bitter	India, Southeast Asia
Broccoli	Color	United States, Mediterranean
Cabbage	Texture, color	Mediterranean, Asia, United States
Carrot	Color	United States, Asia, South America, Mediterranean
Collard greens	Color	Southern United States, Africa
Eggplant	Color, texture	Asia, Mediterranean
Plantain	Mealy texture	Caribbean, Latin America
Pumpkin	Texture, color	Caribbean, Asia, Latin America
Spinach	Color	United States, Europe, Asia
Tomato	Acid, color	Asian, Global
Shiitake mushroom	Meaty, texture	East and Southeast Asia, United States
Chayote/cho cho/christophene	Texture	Caribbean, Latin America, Southeast Asia

papery form that encloses the tomatilla. Tomatilla is green to pale yellow in color and is eaten when unripe. It was a staple flavoring in Mayan and Aztec cooking, and today, is used in Mexican, Central American, and South American green salsas (or salsa verdes). In the United States, it is popular in Southwestern dishes.

Mushrooms

Mushrooms are great flavorings for many foods, especially to provide meaty, earthy profiles to vegetarian cooking. Mushrooms are fungi and come fresh or dried in many varieties, each with its own unique flavor. In ancient times, the Egyptians only served mushrooms to the pharaohs, and the Romans and Greeks saw them as food fit for their Gods. In Asia, they were eaten as a tonic for longevity. The French began cultivating mushrooms in the seventeenth century and called them *mousseron*, which means "growing on moist moss."

Mushrooms come wild or cultivated. Each mushroom has its own type of flavor. The common cultivated field mushrooms are the button mushrooms, which have very little flavor when raw, but which develop a rich nutty and earthy flavor when cooked. They have a white, tan, or brown color. They provide texture to salads, soups, or curries. Cup mushrooms, which have stronger flavors, are good for stuffing, while the more intense flavored flat mushrooms are used in sauces, soups, or stews. Cremini mushrooms, which are tan to dark brown, are a variety of button mushrooms that has a richer flavor and a firmer texture.

Portobellos or brown mushrooms are bigger than the creminis and have more intense and steaklike flavor. They have thick, flat, deep brown caps. They are popular grilled or roasted.

The cultivated wild mushrooms include shiitake, which has a meaty flavor and is used in East Asian dishes, and the oyster mushroom, which has a chewy texture and subtle flavor that it loses when overcooked. Other types include the boletus (called porcini in Italian, due to their resemblance to pigs, and cepe in French), which are usually dried. These dried types are dark brown to reddish brown in color and have a rich, earthy, and meaty flavor.

Chanterelles (from Europe), also called *girolle*, are a "poor man's truffle." They are orange, bright yellow, or red, have an apricot-like aroma, and a nutty taste that mellows with long cooking times. Morels, or sponge mushrooms, are conical shaped and spongy looking. The dried types have a stronger, earthier, and smokier flavor. They are popular in Europe, especially in French cuisine.

Truffles, called the black diamonds, are the king of wild mushrooms. There are many varieties, such as the black truffle (in Perigord, France) or the white truffle (in North Italy). Truffles are said to have the best flavor, and they are eaten raw or slightly sauteed to intensify their flavor.

There are many Asian mushrooms that provide great textures to soups and stir-fries of East Asia. Shiitake, which is also called the Chinese or Oriental black mushroom, is dark brown to almost black, with an umbrella shape. It has a slightly smoky and meaty flavor, and when cooked, gives a firm chewy texture.

Cloud ears, also called wood fungus, wood ears, black mushrooms, or tree ears, are shriveled and greyish brown with an earthy flavor. They are popular in Chinese and Korean cuisine. Enokis, also called golden mushrooms, are creamy white and have a mild flavor with a crisp texture (sold as *enokitaki* which are clusters of cream-colored stalks with tiny caps). They are popular in Japanese soups, sauces, and sushis.

Oyster mushrooms, also called maiitake, are fan shaped and cream colored, with a delicate flavor and chewy texture. Other Asian mushrooms include *nameko* (with reddish brown cap and a slippery texture), *reishi* (in Japanese soups), paddy straw (used in Chinese stir-fries), *shimeji* (aromatic and meaty), and white *matsutake* (also called Japanese oysters mushrooms) that are marinated, or used as toppings for seafoods and steaks.

Huitlacoche, or corn smut that grows on corn kernels, is eaten as a delicacy in Mexico. It is bluish gray and has a smoky, meaty flavor. It pairs well with chile peppers, chocolate, caramelized onions, vanilla, and *epazote*. It enhances meats and corn-based foods.

Mushrooms can be used whole, stuffed, chopped, sliced, julienned, or pureed. They are available fresh or dried. Mushroom extract is from the dried mushroom and is a dark brown viscous paste with a rich meaty flavor. The earthy and meatlike flavor of mushrooms enhances the flavor of soups, stews, and sauces. Mushroom pairs well with meats, poultry, potatoes, ginger, olives, garlic, cilantro, cumin, coriander, turmeric, and black pepper. Cooking releases the true flavors of mushrooms. Most dried mushrooms have more intense flavors than fresh mushrooms.

Fresh mushrooms contain about 7% sugar, 3% to 5% protein, 0.4% fats, 88% water, and some volatile compounds. Dried mushrooms have no sugars but are high in folic acid, potassium, and phosphorus.

The flavor of mushrooms is due to glutamic acid (which is the major component of MSG). Mushrooms have two distinct flavors, when fresh (or uncooked) and when they are cooked or dried. Certain mushrooms can take on the flavor of the food with which they are cooked. Unsaturated alcohols are found in fresh mushrooms. Fresh shiitake has not much of an aroma, but the dried shiitake has a smoky, sulfury aroma due to 1-octane-3-ol, 3-octanone, 2-octenol, lenthionine, and octanol.

Chinese dried mushrooms, such as cloud ears, have a mild flavor but when reconstituted add a glutinous, chewy texture for soups and stir-fries. Dried morels enhance cream sauces, rice, polenta, eggs, or butter. Dried shiitake, a popular ingredient in Chinese and Japanese dishes, has a tough and fibrous texture when reconstituted. Dried boletus goes well with pasta sauce or risottos. Porcinis taste great when grilled, marinated, or slowly braised or sauteed with leafy spices, garlic, and shallots and added to risottos, vegetables, and salads.

Dried mushrooms are soaked in boiling water for twenty minutes, strained, and then used. The strained water is sometimes added to soups and stews or in marinades. The mushrooms can be coarsely blended or ground to powder and added with other ingredients to vegetable dishes to give meatier, more in-depth notes. Sauteing and long cooking bring out the flavors of most mushrooms, and are popular cooking methods used in Chinese and Japanese cuisines.

Mushroom extracts are added to soy sauce, chile pastes, soups, salad dressings, and stews to provide brothy beeflike notes. They can enhance fermented soy blends and are used to coat fish and vegetables to give a meaty brown color. They can enhance mayonnaise, rice products, noodles, couscous, eggplant, and tomato. They

TABLE 25
Global Mushrooms and Their Culinary Uses

Mushroom	Ethnic Preference	Culinary Uses
Chanterelles	Europe	Omelet fillings, with scrambled eggs
Button	United States and Europe	Soups, sauces, stir-fries
Shiitake	China and Japan	Stir-fries, soups, sauces
Enoki	Japan	Soups and salads
Cloud ears	China, Korea, Southeast Asia	Stir-fries, soups, noodles
Truffles	France	Grilled or smoked meats, toppings for pasta, as spreads
Huitlacoche	Mexico	Spreads, quesadilla fillings, mousse, soups
Morels	France	Stuffing, soups, sauteed in butter

can be blended with olive oil, nuts, and leafy spices to create intense spreads or pestos (Table 25).

Asian mushrooms have been used traditionally to thin blood, lower cholesterol, and stimulate the immune system.

LEGUME FLAVORINGS

Legumes, also called pulses, are the edible seeds and pods of certain plants, including beans, soybeans, lentils, peas, and peanuts. Legumes are popular in vegetarian cooking as a source of protein. They are combined with rice, meat, or vegetables as main dishes, or used whole, pureed, or ground into flour for snacks, soups, stews, desserts, noodles, or crepes.

Certain legumes are very flavorful and can add variety to a meal, such as black beans, masoor dal, or chickpeas. Legumes can provide different textures and mouth-feel properties depending on cooking techniques. They can give background notes to a sauce and, therefore, replace chicken- or meat-based roux. Other types are economical because they are nonassertive in flavor and can "stretch" a meal. Soybeans are made into fermented pastes and sauces that are commonly used as seasonings in Chinese, Korean, Japanese, and Southeast Asian sauces, condiments, soups, and stir-fries.

Beans originated in the New World, and today they are a staple item in Asian, Mediterranean, Caribbean, Latin American, and North American cooking. There are many types of beans or lentils that vary in flavor, color, texture, size, and shape.

Legumes in Ethnic Cuisines

Beans are known by different names, depending on the culture or region, and include *frijoles, habichuelas, granos, pois, kacang,* or *dals.* White beans (lima, navy, white kidney, cannellini), pintos, black, red or pink beans, red kidney beans, and black-eyed peas are commonly consumed legumes in the United States.

Black-eyed peas, called cowpeas in the United States, pigeon peas, and gandules in the Caribbean, were brought here by the West Africans. They are now a staple in

Creole, Southern regional, and Caribbean cooking. Chickpeas or garbanzos are widely used in Mediterranean and Indian cuisines. *Channa dal*, the split version of *kabuli channa*, and *kala channa*, the smaller and darker variety of chickpeas, are favorites of North India. They have meatier and sweeter tastes than the yellow split pea and are used as braised vegetables or in dal soups.

Lentils, *lentejas*, or dals, are the most commonly eaten legumes in India. They come in white, pink, yellow, or brown colors. The word dal is used for beans, lentils, or peas. Lentils are aromatic and delicate in taste when cooked. Texture and consistency differ among lentils. Dals are generally available hulled and split and come with different names and colors. Sometimes the same split dal has a different name from the whole dal. The unhulled lentils look attractive, but take longer to cook and are harder to digest. Once they are hulled (skin removed) and split, they become easier to cook and digest. These forms are popular in India. The split *chana, masoor, urad*, mung, and *toovar* dals are commonly used in Indian vegetarian curries, seasonings, and sauces. They are also pureed or processed into flour for snacks, soups, and desserts. The urad and toovar dals are toasted (dry roasted) and combined with spices to create spice mixtures that enhance many sauces, soups, and chutneys (Table 26).

Specific Legumes as Ingredients

Soybean

The soybean, referred to as one of the "grains" in China, is ubiquitously used in Asian cooking. Fermented soybeans paste (misos, *tan* or *tow jiew, toenjang, taucheong*, and *dou bain jiang*) and soy sauce (called *jiangyou, shoyu, namsiew, kecap*, or *taoyu*) made from soybeans are essential flavorings for condiments and many dishes in East Asia and Southeast Asia. East Asians have developed their own unique processes for fermenting soybeans to achieve varying flavoring end products. The yellow or brown, sweeter and less bitter tian mian jiangs used in stir-fried vegetables of northern China, and the more pungent black, bitter la ba jiangs for seafood, pork, and beef dishes in Shanghai style as well as Southeast Asian dishes, are the more commonly used pastes. These fermented pastes are smooth or chunky, hot, salty, sharp, or sweet depending on the added ingredients and how they are prepared.

Misos are used abundantly in Japan to flavor soups, stir-fries, or salads. They have varying flavor profiles and colors depending on other added ingredients and the method of preparation. Miso made from only soybeans is dark red with a salty taste; miso made from soybeans and barley is dark brown and semisweet, while miso made from soybeans and rice is light yellow and sweet. Different misos are used for different applications. Many foods and appetizers use misos to enhance their tastes, such as shumai (or steamed Japanese dumplings) that are dipped into a miso-based sauce. In Korea, fermented soybean paste or toenjang is flavored with garlic to perk-up many dishes including wrapped rice and as dipping sauce for meats.

Soy sauce originated in China over 2500 years ago. The fermented or brewed soy sauce that is commonly used in Asian regions has a characteristic umami (brothy or MSG-like) taste. Its unique flavor is due to over 300 chemicals that it contains—amino acids, alcohols, organic acids, esters, sugars, and salt and are thin, thick, salty,

TABLE 26
Legumes and Their Ethnic Preferences

Legumes	Ethnic Preference
Beans	
Soybeans	China, North India, Southeast Asia, Korea, Japan
Adjuki beans	Japan, China, Korea, Southeast Asia
Black beans	Cuba, South Mexico, Venezuela, Brazil South, Northeast and Southwest United States
Red beans	South, Northeast and Western United States, Caribbean
Pinto beans	Northern Mexico, Puerto Rico, Southwest United States
Red kidney beans/rajma dal	Caribbean, North India, South and Southwest United States, Mediterranean
Lima beans	Southern United States, Peru, Caribbean
Flageolets	Mediterranean, Italy, France, Northeast United States
Lentils	
Chana/Bengal gram/yellow split pea	India, Southeast Asia, Mediterranean
Masoor/red/pink (whole or split)	India, Egypt, Spain, Mediterranean
Urad/black gram (whole or split)	South India, Malaysia, Singapore
Toor/toovar/translucent yellow dal/arhar	India, Southeast Asia
Mung bean/green gram/Pale yellow when split	India, Southeast Asia, Mediterranean
Peas	
Chickpeas/garbanzo/kabuli channa	Middle East, North Africa, India, SE Asia, United States, Europe
Pigeon peas/gandules	Creole, Southern United States, Caribbean
Black-eyed peas/chowla dal (split)	United States, North India, West Africa, Brazil
Yellow, green, brown split peas	Mediterranean, U.S., Asia
Kala channa (whole dark brown)	North India, Southeast Asia

sweet, garlicky or hot, depending on local preferences and applications. It is used to flavor vegetarian foods that do not use garlic or shallots.

Adjuki bean, which has a sweet taste, is ground for desserts and fillings for buns and pastries. In China and Southeast Asia, it is seasoned with star anise, rice wine, ginger, and shiitake mushrooms and is added to pork, vegetables, or glutinous rice.

Nutritionally, soybeans are a good source of protein and fiber and have little or no fat. They are also a good source of folic acid and minerals such as iron, calcium, and phosphorus. They contain isoflavones that help prevent breast and colon cancer, slow calcium loss from bones, lower serum cholesterol, and help alleviate menopausal symptoms.

NUT FLAVORINGS

Nuts are seeds or fruits that consist of kernels and are surrounded by shells. Examples include almonds, candlenuts, cashew nuts, pistachios, hazelnuts, walnuts, macadamia nuts, chestnuts, Brazil nuts (also called cream nuts, butternuts, or Para nuts), kola nuts (or cola nuts), ginkgo nuts (ya-chio), jackfruit nuts, peanuts (that are legumes), coconuts (a drupe), pecans, and *pili* nuts (Java almonds). Peanuts and Brazil nuts are not true nuts, but they are included here because they function as nuts in foods. (Nuts and seeds are included together here because they have similar applications and functions. Sometimes, seeds are classified with nuts and vice versa.)

Nuts have been used since ancient times as food and as an oil. The Romans used almonds, the Aztecs used peanuts and pecans, Asian Indians used cashews, and South American Indians favored Brazil nuts. The Arabs, Persians, and Turks used nuts to thicken sauces and to garnish many of their dishes.

Nuts provide consistency, flavor, and visual appeal. Some can be eaten raw, like pine nuts, sunflower, lotus, or pumpkin seeds. Some possess sweet tastes, such as chestnuts or cashews, and can be used in desserts or as a crunchy garnish. Nuts can be roasted, boiled, or steamed to provide toasted flavors to soups, stews, curries, ice cream, or for cakes, in which nuts such as pistachios, lotus, hazelnuts, jackfruit nuts, and almonds are used. Roasting brings out their full flavor. They can be added during cooking, toward the end of cooking, or as a topping.

Nuts can be used whole, chopped, sliced, slivered, flaked, or ground. Candlenuts, macadamia nuts, peanuts, lotus, and Brazil nuts are ground coarse or fine to provide a creamy texture and a rich flavor to sauces, stews, curries, and many vegetarian dishes. They go well with chocolate, cinnamon, chilies, saffron, raisins, cardamom, coconut, rice, caramelized onions, and citrus.

Nuts can be boiled, steamed, roasted, stir-fried, tumised, or curried. In European regions, they are mostly used in baked goods, snacks, and confectionaries for textural effects. In many other cultures, nuts are used as flavor and texture contributors for sauces and curries. Nuts can even be used in beverages. The kola nut is indigenous to West Africa and is used to provide the "cola" type of flavoring to many carbonated soft drinks.

Seeds are also used as flavorings in foods. Popular seeds include pine nuts (*pignolias/pinons*), pumpkin seeds, *chia* seeds, flaxseeds, lotus seeds, sunflower seeds, poppy seeds, watermelon seeds, and sesame seeds.

Nuts and seeds have many health benefits. They are good sources of protein, vitamin E, calcium, magnesium, phosphorus, folic acid, niacin, and pyridoxine (vitamin B6). Nuts are an excellent source of arginine. They are high in fat, generally 50% to 70%. The fat is nearly all monounsaturated (with walnuts having polyunsaturated fats). Almond is a good source of omega-3 fatty acid and linolenic acid (60%). Ginkgo nuts have a high amount of flavonoids and traditionally have been used to help prevent memory loss and improve blood circulation.

Pumpkin seed has high magnesium, zinc, and omega-3 fatty acid. Sunflower is high in vitamin D and B complex vitamins, and pectin.

Nuts in Ethnic Cuisines

Nuts, such as almonds, walnuts and pistachios, are an essential ingredient in Middle Eastern cuisines. They are used in rice dishes, chicken stuffings, and pastries. Walnuts are commonly used in Turkish tarator and baklava. Pistachios are indigenous to the Middle East. They have a bright green color and are used in a dessert called *locoum* (known as Turkish Delight), cakes, sausages, and grilled meats. In North India, almonds, cashews, and pistachios are commonly used in biryanis, vegetarian dishes, desserts, puddings, and ice creams. They are also used to thicken curries.

Roasted peanuts (a legume by classification) provide the dominant flavor in Southeast Asian sauces, such as satay sauce, stir-fried noodles, and laksas. Called groundnuts in Africa, they are also a staple ingredient in West African cuisines. They are used to provide flavor and consistency to sauces, stews, soups, and curries.

Cashews and peanuts are used in Szechwan and Southeast Asian stir-fries and sauces with chile peppers, sugar, and spices. Ginkgo nuts have a mild, grassy, cheeselike flavor and are grilled or boiled by the Chinese and Japanese. Chestnuts are typically used in sweet dishes, as garnishes, and in the soups of southern Europe. Hazelnuts are roasted and used in romesco sauce of Spain and the meat dishes and confectionaries of the Middle East. Nuts are also ground into flour and are sometimes added to make polenta or crepes, as thickeners for soups and stews, and in desserts.

Pine nuts, also called *pignoli* in Italian, or also called Indian nuts, are indigenous to the Mediterranean and the United States. They are used in a variety of Italian dishes, pestos, desserts, and in Middle Eastern minced meat stuffings and fruit dishes. In Lebanese cuisine, they are toasted and sprinkled over lamb patties. Walnuts are ground and used in pestos of Italy and Provence.

In Mexican cooking, pumpkin seeds, sesame seeds, and pine nuts are toasted or fried and then finely ground to thicken and provide flavor to sauces and moles. *Pipian*, containing ground pumpkin seeds, is a popular sauce in Oaxaca and other parts of Mexico. *Pipian* is also mixed with smoked pasilla, chipotles, cinnamon, and aniseed and added to corn dough and then steamed in *hoja santa* leaves to create tamales. Walnuts go well with fruits, salad dressings, stuffings, and fillings and are an important ingredient in *chile en nogada* of Puebla, Mexico. Almonds are also used in the famous moles of Oaxaca and in chocolate beverages.

Kola nuts are ground and used as flavorings in beverages of West Africa (from which the cola-flavored beverages sprouted globally). They were the original ingredient (along with coca leaf extract from Peru) in the Coca-Cola® beverage and are still used as a flavoring in many cola-type beverages. Their taste is slightly bitter and sweet. They have a high level of caffeine and tannins, and, thus, are used as a stimulant in West Africa. They were traditionally chewed by West Africans to relieve fatigue.

In Europe, hazelnuts, almonds, and chestnuts are used in desserts, confectionaries, and pastries. In the United States, nuts and seeds are added to give texture to burgers, pastries, rolls, and grain-type breads. Pecans, indigenous to the southern United States, are used in stuffings, cakes, pies, and ice cream. Sweet almonds are used in butter, pastries, cakes, liqueurs, and pralines.

Table 27 lists some popular nuts and seeds, their properties, and the region in which they are preferred.

Specific Nuts as Ingredients

Some of the more important nuts and seeds are examined in detail below.

TABLE 27
Global Nuts and Seeds and Their Flavoring Properties

Nuts/Seeds	Flavoring Quality	Ethnic Preference
Almond	Texture, sweet	India, Middle East
Walnut	Texture, flavor	Mediterranean
Macadamia	Creamy, sweet	South America, Asia, Middle East
Lotus	Texture	China, Vietnam
Peanut	Creamy, toasted	Southeast Asia, Africa, China
Pecan	Sweet, pulpy	United States
Pinon	Texture, buttery	Italy, United States
Pumpkin	Flavor, oil, texture	United States, Europe, Mexico
Sunflower	Nutty, texture, oil	Asia, Europe, United States, Asia
Watermelon	Nutty, flavor	West Africa, China
Candlenut	Nutty, consistency	Southeast Asia
Kola nut	Bitter, caffeine	West Africa, Brazil, Java
Coconut	Creamy sweet	India, Southeast Asia, Caribbean
Pistachio	Texture	Middle East, India, Iran
Ginkgo biloba	Nutty	China, Japan, Korea

Almonds

The scientific name for almond is *Prunus amygdalus* (*P. amygdalus*), which is the soft interior of a fruit from the Rosaceae family. There are two varieties, a sweet type (*P. a.* var. *dulcis*) and a bitter type (*P. a.* var. *amara*). Almonds are also called badam (Hindi), almendra (Spanish), mandorla (Italian), lawz (Arabic), and amande (French). Indigenous to North Africa, central and western Asia, almonds are now cultivated in Spain, Italy, and the United States (California).

Almonds have a bland fixed oil of 54% and 18% protein. The deoiled press cake of bitter almonds has 0.5% to 0.7% volatile oil. Bitter almond contains amygdalin, which breaks down to benzaldehyde and hydrocyanic acid (prussic acid). These are poisonous compounds that are removed through heat before it is used as an oil. The oil of almond is colorless to pale yellow and consists mostly of benzaldehyde. Bitter almond has about 3% to 5% amygdalin, while sweet almond has only small amounts of amygdalin. Bitter almond is distilled for essence that is used to flavor cookies, cakes, and marzipan. Almonds have about 50% fat, mainly oleic acid (80%), linoleic (15%), and palmitic (5%).

Almonds are blanched and toasted before use. They are popular ingredients in Middle Eastern and North Indian cooking. They thicken curries, are fried in ghee for biryanis and pilafs, and are mixed with saffron for *kheer*. Sweet almond is mostly used for desserts and confectionaries. It is toasted until golden brown and sprinkled over chicken in Tunisia, and over sauces and condiments in North India. In the West, almonds are mixed with rose water and sugar in a confectionary called *marzipan*.

Almonds are a good source of tocopherol E, folic acid, niacin, phosphorus, calcium, omega 3 fatty acids, and arginine.

Pine Nuts/Pignoli Nuts

Pine nuts are seeds from the pine tree species, of which there are many varieties. The Greeks, Romans, Chinese, and Navaho Indians used them for flavoring, as offerings to their Gods, or in magical potions. In the United States, they are also called Indian nuts. Pine nuts are small, thin, and white and have a delicate buttery taste and a soft texture. They can be eaten raw or lightly roasted, and they are used in pestos, sauces, curries, desserts, vegetables, and as garnishes in Mediterranean, Mexican, and Middle Eastern foods. They are high in protein and potassium.

Candlenuts

Candlenuts are also called *buah keras* (or "hard fruits" in Malaysian or Indonesian) or Indian walnuts. They are finely ground and used to thicken and enrich curries, laksas, and sambals of Southeast Asia. They are oily fruits that are also used abundantly in Hawaii and other Pacific Islands. They contain a poisonous residue, which, after extraction, is inactivated at high heats and long cooking times.

Candlenuts have a walnutlike flavor and are roasted, cracked open, and ground with shallots, garlic, or chile peppers for sauces. They are named candlenuts because at one time they were threaded into the rib of a palm leaf and then burned as candles.

Pumpkin Seeds/Pepian/Pepito

Pumpkin seeds can be eaten raw or toasted and are excellent as snacks, on salads, in breads as purees for sauces and cakes, or used as a cooking oil. They are pureed and commonly used in Mexican moles, sauces, salads, and stews.

Pumpkin oil (dark green) is used in India, Europe, and America for sauces, pasta, and vegetables. It contains omega 3 and omega 6 fatty acids.

SWEET AND BITTER FLAVORINGS

There are many other flavorings, such as ghee (clarified butter), yogurt, vanilla, cacao, chocolate, coffee, and tea, that are used around the world to flavor sweet and savory products and are used as beverages and cooking oils. While some of these ingredients are familiar to North Americans, their use has been substantially limited to beverages and sweet products. As North American exposure to emerging ethnic cuisines grows, we can anticipate finding these ingredients as enhancing flavors of savory applications with much greater frequency.

SWEET AND BITTER FLAVORINGS IN ETHNIC CUISINES

Next, vanilla, cacao and chocolate, and coffee will be discussed. These popular ingredients provide characterizing creamy, sweet, bitter, or aromatic flavor profiles to many ethnic dishes. For example, Indian, Middle Eastern, and African cuisines often utilize the sweet, creamy, buttery flavors of vanilla and ghee. The bitter notes offered by chocolate and coffee provide a balance or background roasted notes to the spicy profiles of Mexican moles, the sweet, pungent notes of Indonesian sambals, and the sour tastes of Caribbean *colombos*.

While not technically flavorings, barbecuing, roasting, and other cooking and preparation techniques also add flavors to prepared foods. In addition to the flavoring ingredients, the next section discusses how various cultures use these preparation techniques to add flavor to foods.

Specific Other Flavoring as Ingredients

Vanilla

Vanilla is the second most expensive flavor after saffron. It was called tlilxochitl by the Aztecs, which means "black flower." The Aztecs flavored their chocolatl (cocoa beans-corn-chile pepper-honey-annatto drink) with ground vanilla pods. Their ruler Montezuma offered this drink to the Spanish invader Hernan Cortez who then introduced it to Europe.

Vanilla is derived from the Spanish word vainilla, meaning "little pod." It is derived from the fruit of the orchid vine. Synthetic vanilla, made from wood wastes or other beans, is much cheaper but lacks the flavor of true vanilla.

Scientific Name(s): Vanilla planifolia (or *V. fragrans*): French or Bourbon vanilla; *V. pompono:* West Indian variety, also called Antilles vanilla or Guadeloupe vanilla; and *V. tahitiensis:* Tahiti vanilla. Family: Orchidaceae.

Origin and Varieties: vanilla is indigenous to southeast Mexico and is now cultivated in Madagascar, Reunion Island, Indonesia, Tahiti, Comoro Islands, Caribbean islands, and Central America. French vanilla beans (from Reunion, Comoro islands, and the northwest tip of Madagascar,) supply about 75% of the world's demand for vanilla. These beans are called Bourbon vanilla because they were first grown on Isle de Bourbon, which is now known as Reunion Island.

Common Names: vanila (Arabic), vanille (Dutch), vanille (French, German), vanil (Hebrew), vanilla (Hindi, Urdu, Malayalam), paneli (Indonesian), vaniglia (Italian), banira (Japanese), waneela (Malaysian), hsiang ts'ao (Mandarin), vanilje (Norwegian), baunilha (Portuguese), vanil' (Russian), vainilla (Spanish), vanilj (Swedish), vanikkodo (Tamil), and waneela (Thai).

Form: vanilla is the cured, dried fruit or seedpod of the creeping orchid. Vanilla bean is a dark brown, narrow, long, and waxy wrinkled fruit. Vanilla comes as whole beans, ground, and as extracts. Fresh vanilla pods do not have any flavor because vanillin, which is responsible for its flavor, is bound as a glycoside and needs to be cleaved by enzymatic action. Vanilla is expensive because it is pollinated and harvested by hand.

Powdered vanilla is obtained by mixing the ground beans with sugar or by extracting the flavor from the beans and then mixing with sugar. Vanilla flavor is extracted with alcohol.

Vanilla is sometimes adulterated with other ingredients, such as tonka beans. Synthetic vanilla is made from wood or pulp waste and lacks true vanilla flavor.

Properties: the flesh of the cured vanilla bean has a perfumed, tobacco-like aroma. Pure vanilla has a mellow, rich flavor. Vanilla extract, prepared by steeping the macerated beans in alcohol, has a delicate sweet aroma and a mellow aftertaste. Various grades of vanilla beans are blended to obtain vanilla for ice cream and other products. Synthetic vanilla (from pulp waste) or imitation vanilla (from clove oils) has a heavy aroma and a grassy and bitter aftertaste.

Generally, vanilla has a sweet, delicate, rich and creamy flavor. It enhances the flavor of other ingredients such as fruits, chocolate, butter, chile peppers, or spices. Different vanillas exhibit varying flavor profiles: Bourbon vanilla, now grown on Reunion Island and Madagascar, has smooth, rich, and creamy flavors. The Reunion type is sweeter and more delicate in aroma than Madagascar vanilla. Indonesian (Java) vanilla is smokier, woody, and has stronger notes. The Tahiti type is more floral, fruity, and perfumed. Mexican vanilla has a creamy, delicate, and mellow aroma, similar to Madagascan vanilla.

Chemical Components: fresh vanilla has no taste. When picked, the green pods have no vanillin because the vanillin is bound to a sugar molecule. Vanilla is released through enzymatic reaction during the curing process to give aroma and taste. The curing process differs, depending on the region, blanching for Bourbon vanilla or steaming for Mexican vanilla, thus resulting in different quality beans. Enzymatic action of beta-glucosidase on the precursor, glucovanillin, gives vanillin and sugar. Vanillin appears as a white crystalline covering on the surface of the beans. It is developed through curing (sweating, fermenting, and drying) the beans. The fermented bean contains about 2% vanillin.

Other components are glucovanillin, vanillic acid, anisic acid, aldehyde, *p*-hydroxy-benzaldehyde, *p*-hydroxy methyl ether, phenols, alcohols, lactones, acids, and esters. Vanilla has about 25% sugars, 15% fat, 6% minerals, and 30% cellulose. Natural vanilla extract includes hydrocarbons, aldehydes, acids, carbonyls, lactones, furans, alcohols, esters, and phenols.

Oleoresin vanilla is derived by evaporation under vacuum of the vanilla extract. The oleoresin is diluted with solvents to give one-, two- or ten-fold strengths.

Vanilla contains calcium and phosphorus.

How Prepared and Consumed: Europeans introduced the world to vanilla through sweet dishes such as cakes, cookies, puddings, custards, sweet sauces, sweets, yogurts, and ice cream. Vanilla is also used in soft drinks, liqueurs, and chocolates. It pairs well with sugar, milk, chilies, almonds, chocolate, fruits, tea, coffee, cardamom, cinnamon, anise, ginger, shallots, saffron, mint, parsley, eggs, and wine.

Pre-Columbian Indians used vanilla to flavor chocolate and savory foods such as corn, annatto, and chile peppers. It can be used to round off stronger flavors in salsas, chutneys, and curries and can provide sweetness to lemon chicken, baked fish, or mole negro. In West Africa, it is used to flavor stews.

Spice Blends: corn chowder blend, mole negro blend, sambal blend, and ice cream blend.

Therapeutic Uses and Folklore: in ancient times, vanilla was used as a stimulant and an aphrodisiac. It was sometimes used as a digestive stimulant, antidote to poisons, and as a treatment for nervous disorders.

Chocolate and Cocoa

The word cocoa comes from the Greek words *theobromo cacao*, meaning "food of the Gods." The Aztecs first drank *chocolatl* (chocolate), mixing *cacauatl* (cacao beans) with chile peppers, corn kernels, annatto, and sugar. Aztecs toasted the cacao beans, crushed them between stones, and formed the paste into cakes, which they diluted with water and spiced with annatto and chile peppers. This mixture was beaten and stirred slowly over fire until it became a frothy liquid. The Aztec King Montezuma offered this chocolate drink to the Spanish explorer Cortez who brought its manufacture to Spain. Both cocoa powder (for beverages) and chocolate are made from the cacao beans.

Scientific Name(s): Theobromo cacao. Family: Sterculiaceae

Origin and Varieties: there are two main varieties of cacao beans, Criollo and Forastero, both cultivated in South America. Cacao is indigenous to South America and Mexico. The commercial Venezuelan types are Caracas, Carupano, and Porto Cabello; and Criollo types include Arriba, Machala, Bahia, and Guayaguil. Their flavors and aromas vary from mild and sweet to bitter. There are other varieties from Mexico, West Africa, Sri Lanka, Java, Malaysia, and the Caribbean (Trinidad and Jamaica).

Common Names: sjokolade (Afrikaan), cokolada (Czech), chocolade (Dutch), chocolat (French), suklaa (Finnish), cschokolade (German), shokolad (Hebrew), csokolade (Hungarian), cioccolata (Italian), coklat (Malaysian, Indonesian), sjiokolade (Norwegian), chocolate (Portuguese, Spanish), chakeleti (Swahili), choklad (Swedish), and chokolet (Thai).

Form: the ripe pods are 7 to 12 inches long, reddish dark brown to purple, depending on the variety. They are filled with a pale pink to whitish soft pulp that contains the seeds. The Criollo varieties are sweet with less bitterness, while the Forastero variety is very bitter.

Properties: cacao seeds are bitter because of the tannins. After fermentation, they become more reddish and less bitter. When cacao is dried, it becomes less astringent, and bitterness almost disappears. Roasting develops the bean's aroma and flavor.

Cacao beans contain an outer shell and an inner kernel, called the nib. The nib is ground into a fine paste called chocolate liquor, while the fat is separated and made into cocoa butter. The chocolate liquor is blended to give the desired, fat free, cocoa powder or chocolate. Chocolate products are white/milk chocolate (with whole milk and sugar), sweet chocolate, and semisweet or bitter chocolate.

Chemical Components: cocoa contains caffeine (230 mg/100 g) and theobromine (2057 mg/100 g), tannic acid, and oxalic acid. Cocoa butter contains 59.7% saturated fat, 32.9% monounsaturated fat, and 3% polyunsaturated fat. Chocolate is rich in phenylethylamine, which is also produced by the brain when one is "in love." Cocoa powder contains tannins (which give cocoa its color and flavor), purine, theobromine, caffeine, and starch.

How Prepared and Consumed: chocolate and cocoa powder are used in confectionaries, baked goods, and beverages. Europeans have traditionally used cocoa powder and chocolate in beverages and sweet products, but it is now being used more frequently in savory applications.

Chocolate is an essential ingredient in the popular mole negro from Oaxaca, Mexico. Chocolate pairs well with chile peppers, cinnamon, annatto, curry, fennel, ginger, clove, nutmeg, lemongrass, mints, pandan leaf, and many flower essences.

It can add a new dimension with its chocolatey bitter notes to sofritos, recados, fish soups, corn chowders, or meat roasts. In Spain and Italy, it is used with garlic, onion, tomato, and parsley in meat and fish dishes.

Spice Blends: mole negro blend, sofrito blend, recado negro blend, chicken curry blend, and corn chowder blend.

Therapeutic Uses and Folklore: the Aztecs and Mayans used cacao beans as currency. Chocolate was forbidden to the ladies during those ancient times, but today in Europe, the United States, and all over the world, it has become the symbol of "amor."

Coffee

Coffee comes from the Arabic word "qahwah," and began in ancient times as a food rather than a beverage. It was first used by the Galla tribe in Ethiopia, around AD 575 to 850, who mixed dried coffee berries with fat to make food balls. Later, in AD 1000, wine was made from the dried beans, including the skin. Coffee was introduced to Turkey in the thirteenth century, and from then on the Arabs roasted and ground coffee to produce their favorite beverage. It was spread by Muslims to Italy, the Netherlands, France, and England. The first coffeehouse was opened in Italy in 1645 and in England in 1650. Coffee came to the United States in 1668, and the first coffeehouse opened in New York in 1696.

Scientific Name(s): there are many types of coffees, the two main varieties being *Coffee arabica* and *Coffee robusta.* Family: Rubiaceae.

Origin and Varieties: differences in coffee flavors depend on the type of beans, soil, altitude, and climate conditions. *Coffee arabica* beans, grown at higher altitudes, are the most highly valued because of their delicate flavor.

C. robusta is less delicate with an earthy flavor but has higher caffeine than the former. Other types, such as *C. liberica* and *C. stenophylla,* are less fragrant than the former two. Coffee is grown in India, Indonesia, Brazil, Uganda, Ethiopia, Mexico, and Columbia.

Common Names: the names of coffee originate from the country, region, or quality grade of coffee the beans come from, such as Java, Costa Rican, Kona, or Kenya AA, Arabian coffee, and mocha (Yemen coffee, also the name for coffee with added chocolate). It is also called café (French, Portuguese, Spanish), kaffee (German, Danish, Swedish, Norwegian), caffe (Italian), koffie (Dutch, Afrikaan), kahvi (Finnish), kave (Hungarian), kahawa (Swahili), ko pyi (Korean), kopi (Malaysian, Indonesian), and kopee (Tamil).

Form: the coffee bean or fruit contains two seeds with adjacent sides that are flat and contained in the fleshy pulp. They are separated by a central groove. The seeds are greenish yellow in color and are separated from the rest of the pulpy material. Coffee's color changes as it ages. Coffee is ground to different degrees of fineness depending on the method of brewing.

Coffee is sold as whole roasted beans, coarsely or finely ground, decaffeinated (97% caffeine free), as blends, or as flavored coffees. Instant (soluble) coffee is roasted coffee extracted with water under pressure, and spray-dried or freeze-dried, then finely ground. Roasted coffee ranges from light to very dark colors and varies in flavor. Examples are American roast (usually medium color), French roast (very dark brown), and Italian roast (almost black). Flavors such as almond, cinnamon, chicory, or chocolate are added after roasting to give flavored or regional coffees.

Properties: green coffee does not have any aroma. The chemical composition of coffee beans changes as it matures and ages. Green coffee beans are fermented, dried and washed, thus separating the pulpy material. Then, seeds are roasted at about 400°F, anywhere from about 3 to 17 minutes, during which chemical changes called pyrolysis occur. Chemical substances are released that characterize coffee's taste and aroma. During roasting, the beans also become a brown color, lose moisture, and swell (from 50% to 100%). Its moisture level decreases from 7% to 2%, and its acidity also decreases. After roasting, the beans are quickly washed and cooled. The longer coffee is roasted, the darker it becomes and the higher its acidity, with lower tannins. Over-roasting gives rise to bitterness and burnt and smoky aromas.

When the beans are roasted to their peak flavor, the beans are cooled quickly (by air cooling, water quenching, or a combination of both) to stop the roasting process.

Chemical Components: coffee contains chlorogenic acid, caffeine, cafetannic acid, tannin, caffeic acid, trigonelline, starch pentosans, cellulose, hemicellulose, sucrose, proteins, and oil. Roasting decreases chlorogenic acid, tannin, and some trigonelline.

How Prepared and Consumed: globally, coffee is used principally as a beverage. It is also used to flavor desserts and ice creams. Sugar is usually added to decrease its bitterness. Spices have been used with coffee to create regional coffee favorites. In Saudi Arabia, dark-roasted coffee, mixed with ground cardamom, called gahira or gahwa, is served to guests as a symbol of hospitality. In the United States, cinnamon and chicory coffees are popular. Fondues and other sauces can be flavored with coffee blended with spices or fruits.

In certain cultures, coffee's bitterness balances savory and other taste profiles. Coffee can enhance or give roasted notes to savory foods, such as red salsas, dark curries, sambals with shrimp paste, and even roast turkey or chicken. It tends to pair well with tamarind, kaffir lime leaf, lime, soy sauce, galangal, cardamom, allspice, or mango.

Spice Blends: curry blends, sambal blends, and roast chicken blend.

Therapeutic Uses and Folklore: traditionally, Ethiopians and Middle Easterners chewed on coffee to increase energy or appetite. Today, it serves the same purpose.

FLAVORINGS FROM PREPARATION AND COOKING TECHNIQUES

Many spices and ingredients are common to different ethnic cuisines, yet they assume distinct flavor profiles and colors. The reason being these spices and ingredients are prepared and cooked in different ways using different techniques. How spices are cooked (refer to Chapter 3), including roasting coriander seeds in oil, dry roasting chilies, or popping mustard seeds in hot oil, produces distinct flavors and colors.

Cooking techniques also remove raw notes, intensify flavors, and sometimes modify colors. For example, dry roasting cumin seeds mellows the bitterness and gives a different characteristic flavor profile. Different techniques give differing flavor profiles with the same spice. So cumin seeds when boiled, braised, dry roasted or roasted in oil give various flavors. With certain bitter spices, such as fenugreek, clove, or mace, roasting in oil tends to take out some of their bitterness.

North Americans traditionally added spices in boiling water, simmering stews, or pickling solutions. These preparation methods did not necessarily give spices their optimum flavor characteristics. Today, North Americans are learning some newer techniques that can be used with spices to obtain more enhanced and intense flavor profiles. Thus, we see a growing demand for smoked chile peppers, roasted spices, or grilled tomatoes.

Cooking techniques fall into three basic categories: dry heat (air, fat), moist heat (water, steam, or other liquids except fat), and a combination of the two (Tables 28 and 29). Globally, these three techniques are called by many names. The equipment the food is cooked in creates distinct flavors, colors, and textures. The cooking utensil's size, shape, and most importantly, the material from which it is constructed, affect the food's ultimate taste.

Typical materials of cooking utensils include stainless steel, cast iron, enameled aluminum, anodized copper, ceramics, and claypot, sandpot, or crock-pot.

TABLE 28
Types of Cooking Techniques with Ethnic Cuisines

Type of Cooking Techniques	Ethnic Cooking in General
DryHeat	Grill, broil, roast, saute/stir-fry, barbecue, pan-fry, deep-fry, bake
Moist Heat	Steam, boil, poach, simmer, pressure cook
Dry and Moist Combination	Braise, stew, claypot, cocido, tagine

---

TABLE 29
Specific Global Cooking Techniques

Type of Cooking Techniques	Global Names
Grill, roast, bake, broil, smoke, barbecue	tandoor, bakar, pibil, barbacoa, jerk
Stir-fry	tumis, balti, bargar, bhoona, kwali, caldero, tarkar
Steam	couscousiere, tagine
Braise	korma, daubiere, pokkum
Stew	kallia, refogado, claypot, tagine
Pickle, ferment	escabeche, chutney, kimchee

The order in which ingredients are added to a recipe is crucial in creating a well-balanced flavor profile. For example, in making Southeast Asian sauces, garlic is sauteed first, then ginger and onions are added, followed by spices and chilies and tomatoes. Spices that add bitterness, such as fenugreek or clove, are generally added toward the end of cooking.

As discussed earlier, corn husk, banana leaf, and pandan leaf are used to wrap chicken, fish, rice, or meat which are then steamed, smoked, grilled, or barbecued. These wrappers enhance flavor and provide moistness, color, texture, and visual appeal to foods. They are discussed earlier in this chapter. Though commercially we may not be able to present food in these wrappers, we can simulate the flavors of these cooked wrappers.

7 Emerging Spice Blends and Seasonings

INTRODUCTION

North American tastes have become culturally diverse. Consumers, whether traditional or adventurous, are looking for something new and flavorful. Some consumers want familiar foods but with new twists. Other consumers are not afraid to try something totally different, so unique ingredients and bolder flavors appeal to them. Both types of consumers want more flavor, good visual appeal, and distinct textures. They have little time to prepare meals and want products that are convenient. Spice blends and seasoning blends can help create the foods that these consumers desire.

Asian curries or teriyaki sauce, Latin American adobos or moles, and Mediterranean harissas or pestos can provide desired flavors to poultry, meats, seafood, vegetables, rice, or pasta. Many ethnic seasonings have spices such as ginger, cilantro, cinnamon, or sweet basil that are recognizable to mainstream consumers and can safely "spice up" traditional foods. Others contain unique spices such as fenugreek, black cumin, galangal, or recao leaf that will appeal to the adventurous consumer.

Spice blends and seasonings are mixtures of spices or other ingredients, such as vinegar, pomegranate juice, soy sauce, wine, or fish sauce. Their possible combinations are almost endless, and left to the imagination of the creator or spice blender.

When combining different ingredients to create a seasoning blend, the food designer must understand that not all generic ethnic blends are the same. That is, not all salsas, curry powders, dukkahs, masalas, pestos, or sambals taste the same. Among different cultures or regions, variations in flavor (sourness, sweetness, or heat), color, or texture (creamy or thin) can occur.

For example, a basic curry blend from India predominantly contains cumin, turmeric, coriander, and ground red pepper. Other spices such as cinnamon, clove, fenugreek, and cardamom are added, depending on the application. Curry blends vary in flavor and color, depending on the countries or cultures of origin, the applications in which the blends are used, how their constituent ingredients are treated, and the order in which they are added to the applications. The Punjabis in the north of India eat milder curries with cardamom, anise, yogurt, nuts, and raisins, while Tamils in the south eat very hot, spicy curries with mustard seeds, tamarind, and curry leaves. Other variations in curry blends occur wherever they are enjoyed. Thai curry blends are pungent, fiery, hot, and sweet, with galangal, lemongrass, kaffir lime leaf, Thai basil, fish sauce, cilantro, hot chile peppers, and coconut milk. The fragrant Nonya curries of Malaysia contain pandan leaf, spearmint, candlenuts, dried shrimp pastes, and coconut milk. On the Caribbean island of Guadeloupe, curries have ground mustard, black pepper, habaneros, vinegar, and chives.

The food creator must also understand spices and flavorings in all of their varieties, because spices are the fundamental building blocks of these seasoning blends. But again, with many spices, there are regional variations with varying flavors. There is more than one type of basil, sesame seed, chile pepper, or citrus flavor. Some cultures or regions have specific preferences for particular varieties. By understanding the similarities and differences among the spice varieties, we can create authentic spice blends or appropriate fusion blends. Additionally, spices and seasoning blends interact in a system based on other ingredients (such as starch, fat, and proteins), pH, and processing techniques. The method of preparation of the spices (braising, roasting, or *tumising)* and how the foods are presented in a meal contribute to the overall flavor and texture of the foods.

Seasoning blends in western countries are usually a simple combination of spices that are generally mild. In Asia, Latin America, and the Caribbean, spices are more complex in their combinations, and they are more intense with varying heat. Malaysia, Singapore, Indonesia, Thailand, India, and Sri Lanka traditionally have the most varied, intense, and complex spice blends in the world. Some are already well known in the west. Others are less familiar but are growing in popularity.

Seasoning blends come as freshly made paste, liquid, dried, whole, or ground forms that are added to cooked or uncooked products. They can be added "as is" or roasted, fried or smoked. Some blends have fresh pureed (or blended) spices, while others have dried whole or ground spices. The wet mixtures are ground with, or added to, water, oil, vinegar, or coconut milk and are usually refrigerated after use. Seasoning blends can be created for snacks, soups, marinades, rice mixes, pasta sauces, desserts, and beverages. The emerging trend for spice blends is to provide the diverse flavors, intense aromas, and a perception of freshness

Spice blends can be authentic or fusion. Generational differences and acculturation or adaptation to the United States by immigrant communities need to be considered when developing spice blends. The first generation will have eating patterns and buying behaviors that are different from the second, third, and later generations who have adjusted to a new culture and language. For the latter, fusion blending gives a "taste of home" with the "new' flavors from North America and other flavors. Spice blends can also be created for regional United States preferences based on the residing ethnic groups, as discussed in Chapter 2.

In this chapter, we will look at some of the popular global seasoning blends categorized according to region. The focus will be on Latin American, Asian, Mediterranean, and Caribbean seasonings because they are the hot trends in the United States. African seasonings will also be discussed briefly to give the creator some idea of its regional flavors.

In addition, the chapter will examine popular regional North American blends and two types of seasoning blends that are popular globally, green blends and chile blends.

Northern European seasoning blends will not be discussed in detail because they are not emerging "new flavors" in North America unless it has become part of the fusion craze. Many of these seasonings are found in traditional U.S. cooking. Several popular French, Italian, and Spanish spice blends are discussed under the Mediterranean blends.

As in the United States, European cities are becoming more culturally diverse and, as a consequence, their traditional flavors are also slowly changing. They are fusing ingredients from newer immigrants, with unique seasonings emerging from the old favorites. For example, in England, Indian and Asian flavors have become mainstream. Curry to an Englishman is like salsa to a North American. Shepherd's pie, lamb roasts, and other traditional English dishes now often contain Asian Indian or Thai flavor. Tandoori, *nasi goreng*, biryani, and chicken *tikka masala* have become mainstream items in the United Kingdom. "Foreign" flavors are also infiltrating French, German, and many other European cuisines, such as Indian, Thai, or Tex-Mex. The discussion of emerging seasoning blends contained in this chapter will help the food product designer also understand emerging European flavors.

As noted above, the subsections of this chapter introduce the foods and ingredients of Latin America, Asia, the Mediterranean, the Caribbean, Africa, and the United States. The chapter also describes the many regional variations of these complex cuisines.

Included in each subsection are examples of blends that are commonly sold as prepared dry spice mixtures or pastes, such as five-spice blend, curry blends, adobos, and pestos. Also described are the spices and other ingredients that provide each region's characteristic flavors. They can be captured and made into spice blends or seasonings that provide the flavor profiles for many products. Spice blends in this section are dry, liquid, or as wet pastes. Whether pastes, liquids, or dry mixtures, they can be tailormade for specific applications once the characterizing spice blend is created.

LATIN AMERICAN SPICE BLENDS

The Latin American category includes the Spanish-speaking peoples and indigenous Indian populations of Mexico, the Caribbean islands, Central and South America, and Portuguese-speaking Brazil. In this book, seasonings from Spain will appear under the Mediterranean category.

Some seasonings have mass appeal throughout the Latin American market. Seasonings such as achiote, sazon, and sofrito add traditional flavors and colors to foods from many Latin American regions. However, flavor preferences vary among Latin American countries, based on the country of origin and other socio-cultural differences. For example, sofritos are used to flavor beans, rice, fish, and stews in many Latin American regions. A basic sofrito blend is made from chopped onions, garlic, tomato, and green bell peppers, and varies depending upon its origins. The Puerto Rican sofrito is pungent with culantro, the Yucatan sofrito is hot with habanero peppers, and the Cuban sofrito is mild with parsley. Likewise, adobo is a popular all-purpose seasoning that is used to marinate meat, chicken, or fish. It also comes in different flavors, *adobo con limon* for Puerto Ricans, *adobo con naranja agria* for the Dominicans, and *adobo con pimienta* for those who desire a little heat.

Not all Latin American food tastes hot. Latin American food is milder where there is greater European influence and spicier and hotter where it is more influenced by the indigenous population. Cubans, Puerto Ricans, and Dominicans use oregano, tomato, garlic, and black pepper with mild chilies to flavor their foods. In these Caribbean regions, flavors tend to be generally milder. In contrast, hot chile peppers are used extensively in the Andes areas of South America and in the Yucatan and Oaxacan regions of Mexico.

Latin American foods, especially Mexican foods, give intense flavors through dry-roasting and smoking of chile peppers, onion, garlic, tomatoes, and other vegetables. In the United States, our taste in food has been greatly influenced by Latin American flavors, particularly Mexican flavors because they are the largest Latin American group in the United States. Mexican food has been one of the three most popular ethnic foods for the last fifteen years, and today consumers seek the more authentic regional Mexican profiles.

Authentic Mexican foods differ in their flavors and heat levels. In northern Mexico, the food is milder, but farther south, toward Oaxaca and the Yucatan regions, spicier and hotter foods predominate with the influences of the Caribbean and the Mexican Mayans. The Oaxaca region has the most diverse cuisine of Mexico. It's unique moles vary in color, flavor, and texture depending on where it is prepared.

The next largest Latino groups in the United States are Puerto Ricans and Cubans. Together with Mexicans, these three groups comprise three-quarters of the total U.S. Hispanic population. The other Latino groups that constitute the remaining one-quarter of the growing Hispanic consumer market are Central and South Americans. South Americans are becoming a large and integral part of North America's Latino population. Colombians, Ecuadorians, Peruvians, and Brazilians constitute a major portion of this growing U.S. Latino category. Their seasonings for meats, poultry, desserts, and cocktail beverages are becoming increasingly popular.

Salsas are very popular in the United States, and now there emerge more varied salsa flavors, using ingredients from different Hispanic groups: *salsa de aji* from Peru, *pico de gallo nortena* from Mexico, *pebre* from Chile, *salsa Criollo* from Puerto Rico, or chimichurri from Argentina,. These sweet, spicy hot, tangy sauces are served fresh (uncooked) or cooked, with tomatoes, spices, chilies, or avocados.

Similarly, snacks such as empanadas, pastels, and saltenas are becoming better known in the United States. Like other Latin American foods, their fillings vary in flavor, based on their ingredients and how they are seasoned, with chile peppers, cilantro, fish, garlic, eggs, meat, and potatoes.

Let's look into the regions of Latin America and see how they are different and how they influence the overall Latin American flavor.

REGIONAL CUISINES OF LATIN AMERICA

Mexico

Mexican cooking has been influenced by many cultures, through trade or conquest. The Aztecs and Mayans used chile peppers, corn, tomato, and chocolate; the Spanish brought wheat, rice, citrus, and coriander; the French offered their cooking techniques and bread making; the Chinese, with rice, noodles, and sweet and sour sauces;

the Lebanese love for stuffed sweet pastries, cinnamon, and cumin; and the Germans with their cheeses and beer making.

Mexican flavors vary from region to region. In the north, called el Norte (Monterrey region), the food is influenced by the Mennonites (German) and North American cowboys. The food is mild with cheeses, beef, *caldillos* (beef stews), burritos, guacamoles, frijoles charros, tacos, and New Mexican chilies. In the North Pacific region (Jalisco, Sinaloa, Colima regions), seafood, tomatoes, poblanos, pozoles, ceviches, and pico de gallo predominate. The South Pacific region, which includes Oaxaca, is influenced by the indigenous groups (Toltecs, Zapotecs, and others). Moles, especially mole negro, *birria* (mutton or goat stews), *hierba sante*, salsa verde, chipotle, chile de arbol, jalapenos, guajillo, ancho, ginger, corn husk, tortilla, tamale, and pipian stews are popular. Chiapas has milder flavors than the Oaxaca region.

In the Yucatan region, the foods are traditional (from Mayan influence) as well as fusion, influenced by the Caribbean and Florida, as well as Asia and Europe. Yucatan flavors include recados, habaneros, sapote, mango, achiotes, epazote, *cochinata pibil*, tamales, rice and beans, banana leaf, *chiltomates*, misantle paste, and chili xcatik. The Veracruz region (Gulf area), influenced by Spanish and Creole cooking, has garlic, olives, mariscos, paella, coconut, sofritos, frijoles, tomato sauces, and *al ajillo*.

The area around Mexico D.F. provides old and new flavors such as nopales, squash blossoms, moles, recados, huitlacoche, mango, fish tamale, *barbacoa*, *tlacoyos*, fried rice, *taco pastor*, tortilla soup, and guacamole. The Bajio region, which includes Michoacán, Queretero, San Miguel, and Guanajuato, has influences from the Spanish and indigenous Mexicans. Cactus, cheese, enchiladas, anchos, moles, pinto beans, corn tamales, thyme, cumin, avocado, beef, pork, fish, and jalapenos abound. Spanish, Iberian, Muslim, and indigenous Mexicans exert their influences in Central Mexico. Its foods include mole poblano, *chile en nogada*, ancho, mulato, pasillas, chipotle, *gusanos*, maguey, *taco pastor*, *chalupas*, and the famous mixiotes.

Caribbean

Caribbean cooking blends indigenous Amerindian ingredients with others who settled there—Africans, Asian Indians, Arabs, Chinese, Spanish, French, Dutch, and British. This cultural diversity has given rise to a fusion-style cooking with jerked meats, pepperpot stew, rice and beans, cho-cho, cassava, curries, Creole-style sauces, seafood escabeches, or banana flambé. Their cooking is based on hot and fruity sauces, seasoned rice, coconutty stews, tubers, and marinated meats and seafood. Soups and stews including *sopito*, *bebele*, conch chowder, *caldo gallego*, *sancocho*, pepperpot, and *quimbomba* are staples of the Caribbean diet. Rice, cassava, and habaneros have a mass appeal in all the islands, but there are also distinct flavor preferences in the islands. East Indians have brought their curries and rotis, Africans yam, okra, and oxtail stews, French their bouillbaise and rouille, Spanish their sofritos and escabeches, while the Chinese, their fried rice and noodles.

Flavors and cooking styles differ from island to island with each island boasting its special dishes. Criollo cooking, from the Spanish-speaking regions of Cuba,

Puerto Rico, and the Dominican Republic (influenced by the Spanish, Taino Indians, Arawaks, Caribs, and Africans) use cilantro, annattos, mojos, and adobos abundantly. Cuban flavors can be mild or hot depending on the region and feature pork, seafood, black beans, mojos, parsley, *cachucha*, annatto, plantain, *congri* (rice and beans), squash, tamale, *ajaico* (vegetable stew), and citrus. Puerto Rican and Dominican foods are mild to slightly spicy. They feature adobos, sofritos, achiotes, culantro, cilantro, oregano, paprika, *aji dulce*, *picadillos*, *mofongo*, rice and beans, platanos, and paellas. Creole cooking, from the French-speaking islands of Martinique, Guadeloupe, and Haiti flavor seafood stews and soups with bouquet garni and chives. The Dutch islands of St Maarten and Curacao add Indonesian touches with satays and *nasi goreng*. The British islands of Jamaica, Trinidad, and Barbados enjoy rice and peas, jerk meats and seafood, and curry dishes.

Central America

The flavors of Central America differ depending on what cultures influenced their cooking (Mayans, Spanish, Germans, Italians, North Americans, or Africans) and what ingredients are abundant in the region, whether rice, corn, seafood, beef, pork, or chile peppers. Central Americans who live along the east coast use a more Caribbean style of cooking, which is very different from the cooking of those who live inland. This Criollo cooking is influenced by the various cultures from the Caribbean and Africa, with rice and beans, coconut, seafood, bananas, and curries. Guatemalan cooking has strong indigenous Indian influence and influence from Mexican cooking with corn, yuca, tacos, pork, and beef.

Smoky-tasting tortillas form the staple meal with grilled (*parillas*) or roasted meats (*asados*). They are served with spicy condiments of cilantro, onions, and chile peppers. Rice and beans and plantains, *pacaya*, *chile rellenos*, ceviches, *recados*, *bistec*, stews (*caldos*), *chuletas* (pork), tamales, frijoles, *refritos* (boiled mashed beans refried in lard), and *enteros* (boiled beans with onions) are commonly eaten.

South America

South America's culture and "foodways" are as diverse as its geography and people. Largely dominated by the Inca civilization, later colonized by the Portuguese and Spanish who brought in African slaves, and eventually settled by other Europeans, South America is home to the mestizos, mulattos, and the Zambos. Brazilians and Guyanese and other Caribbean South Americans largely constitute the indigenous, non-Hispanic South American population. Japanese, Chinese, and Jewish emigrants are prominent in Argentina, Peru, and Brazil. Spanish is spoken all over, with Amerindian dialects, specifically from the Quechuas, Guaranis, Aymaras, Mapuches, and Portuguese in Brazil.

South America is home to potatoes, chilies, beans, and corn, and meals are generally served with white rice or potatoes. Potatoes are daily foods of the indigenous peoples of the Andes highlands with maize (corn), yuca (manioc or cassava), squash, sweet potato, and hardy grains—quinoa, *kañiwa*, and *kiwicha*. Guava, mango, oranges, pineapple, passion fruit, apples, peaches *camu camu*, guarana, acerola, *açai* (berrylike), or *caja* and pears are abundant. Bananas, avocados, limes, and coconut milk are also plentiful to season dishes or make into snacks, bakery items, or beverages.

Chimichurris, *pebre*, *salsa de mani*, *aji molido*, salsa Criolla, *pickapeppa* sauce, or *molho apimentado*, made with chilies, vinegar, onions, cilantro, parsley, and fruits, become marinades and dips for barbecued or grilled meats, like steak (biftec), as sides for seafood stews and soups, or to flavor cooked potatoes, manioc, corn, or beans. Sofrito, a mixture of onions, garlic, bell peppers or tomatoes, sauteed in olive oil, forms a foundation for many Spanish-style sauces. Some sofritos call for *aceite de achiote* (annatto-infused oil), which the indigenous population also uses to color and flavor dishes.

Soups and stews (*caldos*, *cocidas*, *locros*, *ajiacos*, *chupes*, *carbonadas*, or *guisados*), prepared with potatoes, avocado, chickpeas, squash, corn, rice, seafood, chicken, meats, tomatoes, fruits, and grains, and seasoned with ajis, garlic, cilantro, and other spices, lime juice, and coconut milk are sustaining meals for the indigenous population. Empanadas, also called pastels, empadas, *salteñas*, or *empanitas* with fillings of ground meat, seafood, potatoes or corn, chopped hard-boiled eggs, chopped olives, ham, cheese, and generally seasoned with chilies, dried shrimp, capers, hearts of palm, and raisins. Of Amerindian origin, tamales, also called *humitas* in Peru, Bolivia, and Argentina, or *hallacas* in Venezuela, consist of corn-meal dough filled with chicken, meat, seafood, potato, puréed young corn, and cheese, and then wrapped and cooked in corn husk, banana leaf, or other vegetable leaves. Seviches or cebiches are enjoyed throughout South America prepared with raw fish or shellfish in citrus juice, red onions, garlic, black pepper, ajis, and garlic. Desserts and pastries—with Moorish and Portuguese influences—are rich and sweet, made with coconut, bananas, pumpkin, avocado, mango, and many other local fruits.

Regions that have more European influence (Germans, Italians, or Spanish) have milder flavors, while those with a strong Inca and other indigenous Amerindian (especially Andes regions), and African influences (Bahia, northeast Brazil), have spicier and hotter flavors. Colombian and Venezuelan foods with strong Spanish influences are generally mild, with corn, rice, arepas, tortillas, chicken, seafood, casseroles, creamy sauces, plantains, *apio* (celery root), *pabellon criolla*, black beans, and empenadas. Brazil, with a diverse culture—indigenous Indians, Portuguese, French, Lebanese, Chinese, Japanese, Africans, Dutch and Italians—have flavors that vary from hot to mild. Where there is a strong Amerindian or African presence, foods are spicy and hot. Brazilian favorites include fish sauces, meats, peanuts, malagueta peppers, rice and beans, annatto, dende oil, dried shrimp, cassareep, coconut, couscous, lemon, *vatapas*, *feijoada completa*, empenadas, and hearts of palm. Argentina and Uruguay have similar flavors with grilled steaks (*matambre*, churrascos, and *parillas*) served with chimmichurri, salsas (parsley based), empenadas, *humitas*, corn, and fruits.

Paraguay is very much influenced by the indigenous population (Guarani Indians) and Europeans. Its cooking uses beef, cheese, oregano, onion, parsley, poblano, and olive oil for soups and stews. Bolivia has picante flavors and uses rocotos, ajis, saltena, potatoes, and eggs. Chileans enjoy less meat, with more seafood, cooked with hot chilies. Its foods include fruits, potatoes, squash, shrimp chowders, fish stews, or empenadas seasoned with garlic, *aji molido*, *pebre*, and cilantro. Chile's national dish is *porotos granados*, which has seafood, beans, and corn, served with hot *pebre* sauce. Peru's staples are corn, quinoa, and potatoes, which were the staples of the Incas. Potatoes are seasoned in a variety of ways with cheese, milk, chile peppers, onions, peanuts, jalapenos, or olives. Ecuador is well known for its hot, sour, and slightly bitter ceviches. Potatoes, corn, cheese, *naranga agria*, ajis, paprika, fish, cilantro, plantains, and peanut sauce are commonly eaten.

The Caribbean South America, including Suriname (Dutch), Guyana (English), and French Guiana follow the foodways of the Caribbean, their European colonizers, as well as the Asian and African workers brought to their colonies. Pepper-pot stew, cow-foot soup, curries, Indonesian satay and fried rice, and Vietnamese noodles and soups abound. In Guyana, pickapeppa sauce, made with lime juice, chilies, ketchup, and brown sugar, adds a punch to many dishes. Salsa Criolla, with tomatoes, black pepper, and onions, is added to fried fish and meat dishes.

Following, find the basic Latin American spice and seasoning blends that need to be known in order to develop authentic Latino and increasingly popular Latin-influenced foods.

POPULAR LATIN AMERICAN SPICE BLENDS

Adobos

Description

Adobo is a popular, all-purpose seasoning that has a savory, garlicky flavor. It is an indispensable item in Puerto Rican, Cuban, and Dominican kitchens. The earliest adobos were salt and vinegar mixtures used to preserve meat. The word adobo comes from the Spanish, meaning pickling sauce of olives, vinegar, and spices, or from adobado, which means pork pickled in olives, vinegar, or wine and spices. Today, the dry spice blend, adobo, is used with vinegar and olive oil to marinate beef, pork, poultry, and fish, or to zip up rice, stews, and sauces.

Ingredients

Adobo consists of salt, garlic, black pepper, oregano, and turmeric. Optional ingredients include onions, olive oil, lime juice, sour orange juice, or vinegar, which are added to create marinades for meat, fish, or chicken.

The basic adobo has mass appeal among Latin Americans, but specific ingredients, such as onion, seville orange juice, lemon or lime juice, parsley, chipotle, or cumin are added to appeal to specific Hispanic groups. Cubans enjoy garlic, cumin, and sour orange juice; Puerto Ricans prefer vinegar and oregano; Mexicans use a fiery hot adobo made with habanero or chipotle peppers and sour orange juice; Dominicans use sour orange juice, while Nicaraguans prefer it with annatto. Adobo without black pepper is prepared for Hispanics who do not want any heat. Accordingly, the adobos in Table 30 can be tailored for each Latin American region.

Applications

Adobo provides flavor to flank steaks, chicken, pork, fish, and shellfish before grilling, deep-frying, or sauteing. In Latin America, this preparation is so essential

TABLE 30
Types of Latin American Adobos

Latin America	Types of Adobos
Basic adobo	Types of Latin American Adobos.
	Garlic, oregano, black pepper, turmeric
Puerto Rican	Adobo con limon (adobo with lemon)
Cuban	Adobo con cumin (adobo with cumin)
Dominican Republic	Adobo con naranga agria (adobo with sour orange)
Mexican	Adobo con chipotle (adobo with chipotle)
Nicaraguan	Adobo con annatto (adobo with annatto)

that the verb "*adobar*" (to adobo) is commonly used during food preparation. Adobo can be used as a base seasoning for stews, sauces, sour cream, chicken stock, vinaigrette, baked potatoes, steamed vegetables, and as a rub on chicken, fish, or meats before they are barbecued or roasted.

Mole Blends

Description

Moles are the "royal" sauces of regional Mexican cuisine. The word mole comes from the Nahuatl word *molli*, which means sauce or mixture. They were originally created for the rulers of pre-Hispanic Mexico, using chocolate (made from cacao), an ingredient reserved for royalty. Nowadays, these smooth, rich sauces, with their sweet, nutty, roasted, and slightly bitter flavors are used with chicken and seafood dishes.

The most popular moles, and perhaps the best, come from the Oaxaca and Puebla regions. Oaxaca is known as the "land of seven moles." However, every Oaxacan seems to have his or her own idea of what the seven moles are. In fact, there seems to be at least eight types that are popular in Oaxaca: mole Negro, mole rojo, mole verde, *mole coloradito*, *mole amarillo*, *mole de almendra*, *mole manteles*, and *mole chichilo*. The uncontested king of moles is Oaxaca's mole Negro. Mole poblano is the most popular mole from the Puebla region of Mexico.

Ingredients

There are hundreds of mole recipes in Mexico. In fact, each family seems to have their own variation, depending upon their preferred ingredients and cooking techniques. That is why moles seem to come in so many different colors: black, brick red, yellow, and green. However, most moles contain dried chilies, roasted tomatoes, pureed nuts and seeds, spices, herbs, and chocolate. However, not all moles have chocolate. The chile peppers provide the mole's characteristic color and flavor.

It takes hours to prepare a good mole. Preparation and cooking techniques, such as grinding, pureeing, grilling, browning, roasting, and toasting are essential for intensifying ingredient flavors and obtaining the right consistency. Most ingredients are roasted on a comal (a large, lime-coated ceramic pan). After roasting, the ingredients are either hand ground on a heavy, granite metate or commercially ground at the local market in a machine called a *molino*. Finally, once the ingredients are prepared, the sauce mixture is cooked with lard in a *cazuela*, or mole-pot, and stirred continuously.

Mole negro is a complex, fiery, black sauce made from numerous ingredients: toasted and ground dried chile peppers (*chilhuacle negro, pasilla*, mulato, guajillo and chipotle), blackened chili seeds and stems, almonds, tomatilla, spices (garlic, onions, sesame seeds, black peppercorns, canela, cloves, thyme, marjoram, Mexican oregano), and chocolate.

Mole poblano has several types of dried chilies (ancho, mulato, pasilla, chipotle), with anise, clove, canela, black pepper, fried garlic, tomatillas, sesame seeds, ground almonds, peanuts, and bitter chocolate. Table 31 lists the ingredients of nine mole blends. However, there are many interpretations of these nine blends and their ingredients.

Applications

Moles are generally eaten with white rice or tortillas, the latter used to scoop them. The thick sauce is poured over chicken, turkey, seafood, pork, or beef. Mole sauce can also be used as a filling for empanadas and tamales or poured over enchiladas.

A few popular moles are not sauces at all. In pre-Hispanic time, moles were more like soups and were served as bowl meals. Some of these ancient moles such as *mole de maiz quebrado* (corn mole), bean mole, and shrimp mole are still consumed today. A popular one, *mole de lla*, is a hearty soup full of beef, onions, vegetables, cilantro, lime, and epazote, with added rice and corn tortillas.

TABLE 31
Types of Latin American Moles

Mole Blends	Ingredients
Mole negro	Chipotle, chilhuacle, negro, pasilla, mulatto, guajillo, chili seeds and stems, almonds, onions, tomatillas, garlic, sesame seeds, black peppercorns, canela, clove, thyme, Mexican oregano, chocolate
Mole poblano	Ancho, mulatto, pasilla, chipotle, anise, clove, canela, black pepper, sesame seeds, almonds, tomatillas, garlic, peanuts, chocolate
Mole rojo	Red chilhuacle chili, ancho, pasilla, and peanuts
Mole coloradito	Ancho, pasilla, guajillo, canela, raisins, sesame seeds, almonds, chocolate, and fried bread
Mole verde	Green jalapeno, epazote, cilantro, parsley hoja santa, tomatillas, and masa
Mole amarillo	Guajillo, chiluacle amarillo, ancho, cloves, cumin, chepil, zucchini, chayote, and tomatillas
Mole manches manteles	Anchos, pasillas, chilcostle chilis, sesame seeds, fresh pineapple, ripe plaintains, tiny apples
Mole de almendra	Almonds, ancho, cloves, sesame seeds, and fried bread
Mole chichilo	Chihuacle negro, pasilla, guajillo, blackened chili seeds, roasted avocado leaves, and masa

Mojo Blends

Description

Mojo is to Cuban cuisine as vinaigrette is to French cuisine. Also called Cuban garlic citrus sauce, it is tart, tangy, and garlicky. It can be used to marinade meats or as a condiment to add zest to meals. It is Cuba's national table condiment.

Ingredients

Mojo's ingredients include garlic, olive oil, sour orange juice (narangia agria), and black pepper. Ground cumin is optional. In Cuba, naranga agria, a green, bumpy orange that tastes like lime juice, is the acid used. In other Caribbean regions, including Puerto Rico and the Dominican Republic, vinegar is substituted for the naranga agria. In the Miami area, lime juice or other fruit juices, such as pineapple, carambola, and grapefruit are generally substituted for naranga agria.

Application

Mojo is a vinaigrette-like Cuban sauce used as a marinade, especially for poultry and meat. Mojo is also served on Cuban sandwiches, grilled seafood and meats, boiled yuca, and many other foods. Rosemary, cilantro, or mint can be added to create new flavors.

Salsa Blends

Description

Salsa means sauce in Spanish, and it comes in many forms and flavors. Some Latin American sauces are called salsas, while others have specific names, but all are added to provide zest or heat to foods. Latin American regions have many versions of salsas that range in flavor from sweet and mild to hot, pungent, and slightly tangy. They vary depending upon ingredients and preparation techniques used.

Salsas can be fresh or uncooked (salsa cruda, salsa fresca, salsa de molcajete) or cooked, based on roasted chilies, tomatoes, and other ingredients. Salsas can be smooth or coarsely textured, thick or thin. Salsa verde (green salsa) is uncooked and is tart and hot. Salsa rojo (red salsa) is cooked and is sweet and spicy.

Popular salsas include red (rojo) and green (verde) salsas of Mexico, *xni pec* from the Yucatan, pico de Gallo from northern Mexico, *salsa di aji* from Peru, *pebre* from Chile, salsa *criollo* from Puerto Rico, and chimmichurri from Argentina.

Ingredients

When North Americans refer to salsa, they generally mean a condiment of tomatoes, onions, chilies, and bell peppers. Spices such as garlic, cumin, oregano, and cilantro are also added. Whole tomatoes, diced tomatoes, and tomato pastes form the basis of commercial salsas. Lime, lemon, sour orange juice, or vinegar are often added to provide acidity. Depending on regional preferences, tomato, bell peppers, habaneros, ajis, jalapenos, corn, or fruits are added.

The basic red salsa is made from red roasted tomatoes, red bell peppers, and roasted red chile peppers. Green salsa is made from tomatillas, cilantro, and green

TABLE 32
Types of Latin American Salsas

Salsa Blends	Ingredients
Salsa rojo	Chipotle, red jalapeno, red bell peppers, garlic, tomato, cilantro
Salsa verde	Green jalapeno/serrano, cilantro, tomatilla, garlic, onion, lime juice Pico de gallo, Red serranos/jalapenos, cilantro, oregano, onions, tomato, lime, orange, jicama, melon, cucumber, olive oil, shredded coconut
Chimichurri	Garlic, parsley, cilantro, vinegar, black pepper, oregano
Salsa de aji	Aji, onion, tomato, cilantro
Molho Malagueta	Malagueta pepper, olive oil, vinegar, onion
Ajilimojili	Garlic, sweet chile peppers, black pepper, olive oil, lemon
Pebre	Onion, cilantro, parsley, green hot peppers, lemon juice, olive oil
Aji molido	Aji, olive oil, garlic, vinegar
Salsa criolla	Onion, tomato, garlic, black pepper, parsley, olive oil, vinegar
Xni pec	Tomato, onion, cilantro, habaneros, sour orange juice

chile peppers with green leafy spices. Nowadays, the emerging trend in the United States is to incorporate fruit flavors with salsa blends to create new flavors.

Salsa fresca, salsa cruda (or fresh salsa) uses basic seasonings—chopped jalapenos, serranos or habaneros, tomatoes, white onions, and cilantro. These salsas are eaten raw or are cooked briefly in lard or oil. Cooked salsas generally use dried, roasted chilies, such as chipotle or ancho, with roasted tomatoes and garlic.

Some examples of Latin salsa blends and their ingredients are given in Table 32.

Applications

Salsas are used as dips for snack chips or tortillas, or poured over eggs, fajitas, and enchiladas. Salsa crudas (uncooked salsas) appear on Mexican tables like salt and pepper. They are used as condiments to add zest to rice and beans, meats, fish, eggs, and beans, and also serve as toppings or garnishes on enchiladas, tacos, quesadillas, and for *antojitos*. Puerto Ricans serve salsa *ajilimojili* with meats, tostones, and boiled vegetables. South Americans serve chimmichurri sauce as a dip or condiment for grilled meats (or churrascos).

Salsas add a flavorful topping for grilled foods, baked fish, fillings for crepes and wraps, and as exciting condiments to add extra flavor to foods. They are also being used as marinades, salad dressings, and even as a topping for baked potatoes.

Recado Blends

Description

Recados are ground spice mixtures or wet pastes that are essential to traditional Yucatecan, Central American, and Puerto Rican cooking. They come in different flavors and colors, depending upon their ingredients (Table 33). Most widely used is the recado rojo (sometimes referred to as achiote). Others are recado negro (or chilmole), recado cumin de *bistek*, recado de adobo, and *recaitos*. They can be mild, pungent, smoky, and hot with red, black, or olive colors. The *chilmole* is black and

TABLE 33
Types of Latin American Recados

Recado Blends	Ingredients
Recado rojo	Ground annatto seeds (achiote), garlic, peppercorn, onion, cilantro, oregano, allspice, and sour orange juice
Recado de bistek	Oregano, canela, garlic, onion, black pepper corns, cumin, cloves, chili de arbol, and sour orange juice
Recado de negro (Also called chilmole)	Annatto paste, toasted chili xcatik, allspice, roasted onions, roasted garlic, cumin and burned totillas
Recado de adobo	Red chilies, oregano, garlic, clove, cumin, cilantro, saffron, canela, Seville orange

pungent, while the recaito is greenish. Recados are used to zip up sauces, stews, meats, seafood, and beans. They can be served over cooked poultry or rubbed on meats, chicken, and pork, before grilling or barbecuing.

Ingredients

Recados' ingredients include garlic, chile peppers, oregano, onions, annatto, and black peppercorns. Other ingredients are cumin, allspice, cilantro, canela, culantro, or cloves depending on regional preferences. There are many variations, even in the Yucatan.

Applications

Most Yucatecan dishes use recados as a base flavoring for many dishes. Turkey is roasted in recado Negro and fish is grilled in recado rojo. To create the famous pibils of the Yucatan, recado is rubbed on pork or chicken that is then wrapped in banana leaves and pit roasted. *Chilmole* is used to season meatballs or ground meat.

In Puerto Rico, an olive green spice mixture called recaito is used as a base seasoning in the local sofritos. Recaito contains onion, cilantro, recao leaves, garlic, and aji dulce.

Sofrito Blends

Description

Sofrito is an aromatic mixture of spices used to flavor meats, rice, beans, stews, and fish. It originated as tomato sauce from the Catalan region of Spain. Now, it is an essential seasoning in Puerto Rican, Spanish, Cuban, and Mexican cooking. The word sofrito comes from the Spanish word sofrier, which means lightly cooked. Sofrito blends range in color from bright orange to red, and their flavors are mild or pungent. Preferences and ingredients vary with country of origin (Table 34).

Ingredients

Typical sofrito ingredients are green and red bell peppers, onions, annatto, garlic, oregano, tomatoes, cilantro, or parsley. Puerto Rican sofrito is pungent, and its ingredients include recaito leaves, culantro stems and roots, aji dulce (also called

TABLE 34
Types of Latin American Sofritos

Sofrito Blends	Ingredients
Basic sofrito	Sweet and mild with garlic, tomato, sweet bell peppers, mild chile pepper
Spanish sofrito	Sweet, mild with garlic, red or green bell peppers, onions, olive oil
Puerto Rican sofrito	Pungent, slightly spicy with recao leaves, culantro root and stem, aji dulce, tocino, alcaparrado, annatto
Cuban sofrito	Mild with garlic, onion, cachucha, cured ham, parsley
Mexican/Yucatan sofrito	Peppery hot with roasted garlic, black pepper, cumin, habaneros

rocotillos or sweet chile peppers), annatto, garlic, onions, oregano, cilantro, and tomato sauce. In Puerto Rico, it is added with ham, *tocino* (salt pork), or *alcapaardo* (salted capers, olives, and pimentos). Cuban sofrito is mild, with garlic, onion, cachucha (or aji dulce), and cured ham. Spanish sofrito is sweet, with bell peppers, onions, tomatoes, and olive oil. The Yucatecan sofrito is peppery and spicy with habaneros, roasted garlic, black pepper, and cumin.

Applications

Sofrito is used as a starting base seasoning or added toward the end of cooking to give a finishing touch to many Latin American and Spanish dishes. Cubans flavor soups, stews, and casseroles with sofrito or use it as a topping sauce over grilled chicken, beef, and pork. In Puerto Rico, sofrito is lightly sauteed in lard or olive oil and added to stewed beans, braised chicken, or fish. In Spain, sofrito is made with toasted almonds and is an essential ingredient in a seafood stew called *zarzuela de pescado*.

When sofrito is made with annatto, the annatto is first heated in oil or lard until it becomes red, then the other ingredients are added to complete the sofrito blend. Annatto gives a wonderful orange yellow hue to the foods in which the sofrito blend is used.

ASIAN SPICE BLENDS

The abundance and variety of Asian spices and spice blends provide the well-balanced and intense flavors found in no other cuisine. Asian foods have spice and seasoning blends that mainstream consumers are comfortable with, such as ginger, sesame, hoisin, or teriyaki sauce. They also have the aromatic sensations and unique tastes that excite adventurous consumers, such as sambals, curry powders, or kimchis. Authentic Asian blends have the well-balanced flavors and textures, great visual appeal, and variety that today's consumers are seeking.

The growing demand for foods that promote a healthier lifestyle is also fueling consumer interest in the Asian way of flavoring and eating. Asian seasonings provide both the flavor and a healthy meal. Many seasonings also contain ingredients such as lentils and soy products that are perceived as natural and healthy.

North Americans are combining Asian food concepts, including ingredients, cooking techniques, and presentation styles, with traditional foods to create new American foods, including Pan Asian, Asian Latino, New Californian, or Pacific Rim.

Some Asian ingredients and seasoning blends, such as ginger, soy sauce, fish sauce, garlic, star anise, and curry may have mass appeal throughout Asia, but Asians come from many ethnic backgrounds—including Indians, Pakistanis, Chinese, Japanese, Koreans, Thais, and Vietnamese—and each Asian region has its preferences. There are also differences in flavor profiles within each region. For example, in Southeast Asia, Malaysian foods are distinct from Thai or Vietnamese foods. Even within Malaysia, flavor differences occur because of its diverse cultures. The Malay dishes are seasoned differently from the Chinese, Indian, or Nonya foods.

Religion also adds to the variety and unique differences in seasonings found in Asian foods. Hindus, Muslims, Buddhists, Christians, and Jews of Asia have specific preferences and prohibitions for ingredients. Certain ingredients and preparation techniques are specific to different religious groups. Rice wine, widely used in Chinese and Japanese cooking is Haram (not Halal), so regions of Muslim predominance, such as Malaysia, Indonesia, and Pakistan, do not use it in their cooking. Garlic and onions are not used by the Buddhists. As a consequence, there are endless varieties of seasoning blends in Asia for use with fish, chicken, beef, or vegetables.

Asian concepts of meals and food preparation differ from North American and European cooking in several basic ways. In traditional Western meals, there is a tendency to segregate tastes on a plate. Meats, fish, pasta, potatoes, and or vegetables are not mixed together and eaten, and sweet flavors are left for dessert. In contrast, Asian cuisines, including Chinese, Indian, and Southeast Asian foods, combine tastes and textures in a meal, often even within a single dish to create balance and harmony in a meal. And sweet and savory flavors are all interacting in a dish.

For example, Indian cooking categorizes foods into six tastes: sweet, sour, salty, spicy, bitter, and astringent. The proper, well-balanced Indian meal contains all six tastes. This principle explains the complex spice combinations and depth of flavor in Indian foods. Similarly, in Chinese cooking, there are also six different tastes: sweet, salty, bitter, sour, spicy, and umami. The proper balance of these different flavors with appropriate textures and colors creates good taste in Chinese cooking.

Asian cooking principles go beyond the balancing of tastes. Foods can be hot, cold, moist, dry, heavy, or light. Every meal should be well balanced between these sensations to promote digestion and well-being (see Chapter 3).

Asian presentation techniques are designed to carry out this philosophy of eating. For example, Asian foods redefine the concept of an entree. In traditional western meals, the entree typically consists of a large piece of dry unseasoned or lightly seasoned meat or fish. In contrast, Asian meals feature smaller servings of meat or fish that are more seasoned and sauced, cut thin, and served with steaming white rice. In Malaysia and Indonesia, the word "nasi" means cooked rice but is also the word for "meal." Many Asians believe that rice nourishes body and soul. Without rice, food is only a snack, not a meal. Thus, rice is the centerpiece of every Asian meal, accompanied by a variety of crunchy stir-fries, pungent condiments, and aromatic curries. Rice is also a comfort food for Asians. Rice porridge with shredded chicken and pickled vegetables, called *congee*, is often taken when one is sick.

In Southeast Asian cuisines, the concept of a "main entree" and side dishes sometimes blend together to form a "one-pot," "one-dish" meal or bowl meal. A one-pot meal may be a stew, soup, sauced noodles, or fried rice containing onions, tofu, chilies, leftover meat, vegetables, or anything available in the pantry, served with a spicy condiment. However, the addition of these ingredients is not simply random. They are chosen and added to create a balanced "system" of taste and texture. A variety of seasoned rices and noodles are commonly eaten all over Southeast Asia, for lunch, dinner, breakfast, and even as "pick me ups."

Side dishes and condiments contribute variety to an Asian meal. The hot, sour, and crunchy flavors of side dishes and condiments, whether chutneys, curries, sambals, or dals, enhance or provide balance to the overall flavor and texture of the main entree. Condiments are an integral part of Asian meals, whether teriyaki, sambal trasi, mango pickle, or koch'ujang, that perk up a dish and provide added appeal to individual tastes on a meal table. One of the most notable aspects of Thai and Vietnamese cooking is the emphasis on condiments based on fish sauce, garlic, tamarind juice, leafy spices, and chile peppers.

Aromatic hot sauce to an Asian is like ketchup to a North American or salsa to a Mexican. Every region has its own hot sauce or condiment, fresh, pickled, or cooked. Southeast Asia is noted for its pungent fruity, sweet, sour, and hot condiments. There are a variety of condiments—*sambal olek*, chili garlic, Sriracha, *sambal trassi*, and *nuoc cham*—which contain vinegar, tomato, brown sugar, shallots, tamarind, coconut milk, garlic, lemongrass, turmeric, and black pepper.

Artfully presented garnishes and toppings also provide a final note of contrasting textures, flavors, and colors. Malaysian and Indonesian seasoned rices and noodles are served with nutty roasted peanuts, caramelized crispy shallots, or crunchy bean sprouts. Fresh leafy spices are important Southeast Asian garnishes that provide fresh aromas, appealing colors, and many textural variations.

The way spices and spice blends are prepared in Asian cuisines is of significance in providing distinct flavors, colors, and textures to Asian foods. Chile peppers, coriander seeds, or sesame seeds are smoked, pickled, or toasted before being added to spice blends to give more intense flavors. Steaming or grilling meats, fish, or rice in pandan or papaya leaves gives them unique flavors and visual appeal.

In Asian cuisines, texture is an important element, especially in vegetable entrees. Consumers perceive crispy, crunchy vegetables as fresh and natural. In traditional American cooking, vegetables are often boiled soft and become unappetizing. Cantonese cooking, which uses steaming and stir-frying techniques with light seasonings, brings out the vegetables' fresh colors, flavors, and textures.

Appearance is also an important factor in Asian cooking. Thai and Japanese foods add sculptured fruits, chilies, flowers, and vegetables to rice dishes, noodles, grilled chicken, and fish.

Spice blends provide to Asian foods the balance of hot, sweet, sour, savory, and aromatic sensations all in one meal.

The spices commonly used in Asian cooking are ginger, cumin, cassia, coriander, star anise, galangal, chile peppers, coriander leaf, basils, spearmint, turmeric, clove, and garlic. A variety of fresh leafy spices, including mint leaf, cilantro, basil leaf, lemongrass, and scallions, is important in flavoring and garnishing foods and drinks.

Seasonings such as hoisin, masalas, *sambals*, and *panchphoron* create magic with fish, steamed chicken, stews, soups, or sauces.

Certain spice blends have a "cross-over" appeal among different Asian ethnic groups, such as curry, five-spice blend, or teriyaki. However, with other blends, such as nam pla, kimchi, or chat masala, only specific preferences exist.

Also, not all generic spice blends have the same flavor such as curry, sambal, or five-spice. They vary depending on where they originate, what ingredients are added, and what end product they go into.

Asians have mastered the art of preparing spice blends—as dry roasted, roasted in oil, or braised—all of which enhance the flavor of the finished product. These cooking techniques remove bitter notes, intensify flavors, or add fragrant aromas. Different preparation techniques suit specific spices and these create a multitude of flavor dimensions and colors in Asian cuisines.

In the next section, some of the Asian spice blends from different regions will be discussed to provide an understanding of the similarities and the distinct differences in creating authentic Asian foods or Americanized Asian flavors.

Regional Cuisines of Asia

South Asia

South Asians have mastered the art of combining different spices to create numerous blends with unique depths of flavors. A basic bouillon seasoning used in India and Pakistan is an example of regional variations in South Asia. Yakhni or akni is the South Asian seasoning equivalent to the bouillon cubes of the western world. It is made from mutton or lamb with basic spices such as onions, garlic, ginger, green and red cayenne, bay leaf, and parsley. A North Indian version is made from onion, coriander seed, cloves, green ginger, fennel seed, and cayenne. Black cumin, mustard seed, fenugreek leaf, cardamom, cinnamon, kari leaf, and parsley are added to create regional tastes. This blend is used as a base flavoring for many Indian dishes, such as curries, sauces, snacks, soups, and grilled foods.

Pakistani foods have Arabic, Iranian, Turkish, North Indian, and Afghan influences. They are noted for their festive biryanis, wheat-based breads called naan, marinated mutton, and lamb kebabs. Pork and alcohol are rarely consumed because of Pakistan's Muslim population.

Nepal and Tibet are influenced by Chinese, Indians, and Europeans, and they use plenty of dried and fermented foods derived from buffalo meat, mutton, vegetables, fish, fruits, and legumes. Corn, rice, millet, and wheat are eaten, along with soups, chutneys, and dumplings. The seasonings are mild, and curries are sweeter.

Bangladesh, formerly called East Pakistan, has flavors similar to India and Pakistan. Because the population is mostly Muslim, pork is prohibited. Rice, fish, legumes, chutneys, and vegetables flavored with mustard, poppy seeds, and nigella are popular foods.

Indian foods have tremendous variety because India is influenced by many cultures and religions. The major religious influences are Hindus, Muslims, Buddhists, Jains, Sikhs, and Parsis. Indian cuisines have great aromas and in-depth taste profiles that are derived from a complex combination of spices and preparation techniques of spices. Rice and wheat are popular ingredients, and consumed based on regions; for example, wheat predominates in the north and rice in the south. Other commonly used ingredients are legumes, vegetables, fruits, nuts, fish, mutton, lamb, dairy products, coconut, and numerous spices. In the north, milder and yogurt-based curries are popular, while the southerners enjoy fiery tamarind and coconut-based curries.

There is no one type of Indian spice blend or flavor. Indian cuisine possesses distinct flavors that vary, depending on regional and cultural preferences. In the northern regions (around Punjab and Kashmir with Mogul, Pakistani, and Afghan influences), ghee, nigella seeds, fenugreek leaves, almonds, dill seed, mint, bay leaf, pomegranate, lotus seeds, lemon juice, fennel seeds, saffron, rose petals, and garam masala flavor *parathas*, lamb kebabs, pilafs, tandooris, and kormas. In the south (Kerala and Tamil Nadu, influenced by Hindus, Muslim Keralites/Moplas, Arabs, Christians, and Jews), kari leaf, tamarind, kodampoli, turmeric, hot chilies, coconut, fenugreek seeds, black pepper, and mustard seeds flavor fish curries, sambars, urad dal, sambars, and sour dosais. In the southwest (Goa), the Portuguese-influenced vinegary vindaloos and tomato-based flavors are popular. In the east (Bengal, with British influence), mustard oil, mustard seed, nigella, chilies, raisins, and panch-phoron flavor chutneys, kedgeree, and crabs. And, in the northwest, around Gujerat and Bombay (with Parsi, Jain, Muslim/Bohris, and Hindu influence), dried apricots, green *chili*, coriander leaf, black cumin, tamarind, asafetida, heavy cream, and kala masala flavor sweet and sour sauces, spicy sour curries, *bhel poori*, channa dal, and aromatic biryanis.

Sri Lankan cooking is influenced by Hindus (Tamils), Buddhists (Singhalese), Portuguese, Arabs, British, Malaysians, and Dutch. It has fiery hot curries of fish, beef, chicken, and vegetables, *sambols*, chutneys, savory rices, and fermented breads. Their cuisine is flavored by darkly roasted spices, hot chile peppers, coconut milk, dried fish, kari leaf, tamarind, green jackfruit, and mango.

Southeast Asia

The countries of Southeast Asia are geographically close and have common ingredients and similar preparation techniques. Yet each region has its own unique flavors. It has flavors that are a fusion of many cultures. Ginger, chile peppers, fish sauce, shrimp paste, tubers, leafy greens, noodles, and legumes are widely used in all of these regions, yet each region prepares and uses them to their taste preferences, resulting in unique flavors.

For centuries, Southeast Asia has been the crossroad for trade and religious exchange between the East and West. Conquerors and traders have had a significant impact on their languages, religions, and foods. Some of the cultures that have influenced Southeast Asia include Chinese, Indians, Sri Lankans, Arabs, Jews, Portuguese, Dutch, English, French, Spanish, and Americans. The Malays enjoy their pungent shrimp pastes, sambals, and coconut milk; the Indians introduced curries, fenugreek, and dals; the Chinese introduced soy sauce, taucheos, and tofu; the Spanish introduced tomatoes, beans, and adobos; the Portuguese introduced their vinegar, tomatoes, and mustard seeds; while the Arabs introduced kebabs, flatbreads, and pastries.

As a result, the cuisines of Southeast Asia are some of the most varied in the world. When we taste Southeast Asian foods, we experience a burst of differing flavors and textures. Simple or complex combinations of many well-balanced sensory perceptions can be tasted at the same time, including hot, sweet, bitter, sour, crunchy, or astringent. Their foods are fragrant, peppery, spicy, sweet, and sour and have wonderful textures and colors.

A typical Southeast Asian meal consists of white rice, surrounded by one or two meat, poultry, or fish dishes, a vegetable dish, and several condiments. The meal ends with fresh cut fruits. Soups are typically served in Chinese meals.

One-dish or bowl meals are a local favorite for lunches, quick-fix dinners, as snacks, and pick-me-ups. They are either stir-fries or soup- and sauce-based dishes that consist of rice or noodles, meats or seafood, vegetables, and an accompanying condiment.

In Thailand, soups and noodles, a Chinese influence, curries from India and Sri Lanka, and pungent fish and coconut sauces from North Malaysia. Flavors vary among its regions. Mild and hot flavors predominate in the mountainous regions of the north, with wild game, mushrooms, steamed glutinous rice, and chili-lime dips. Fiery, tangy salads (*laab*) are popular in the northeast, with minced pork and chicken, lime, chile peppers, and pungent shrimp paste. Spicy, sweet, coconut-based curries, satays, and lemongrass soup characterize the foods of the south. In the central regions, hybrid flavors with elegant presentations of rice, noodles, seafood, and desserts are enjoyed.

Vietnamese foods have fragrant aromas and vivid textures. Its charcoal-fired grilling techniques and crispy salads combine with Chinese cooking styles and sauces, lighter and mild versions of Indian and Thai-style curries and French baquettes, crepes, and pates. All this is balanced with pungent fish sauce, aromatic leafy spices, and crunchy, crispy, al dente textures derived from vegetables, fruits, and cooked rice. Foods in North Vietnam have more Chinese flavors, including soy sauce, ginger, and black pepper with stir-fries and claypot dishes. In the Northwest, foods are seasoned with galangal and lime, and grilled over charcoal. The central region, the home of former royalty, has lemongrass, mints, basil, shrimp paste, and fish sauce. The south, bordered by Cambodia and closer to Thailand and Malaysia, has spicier flavors with fish sauce, chilies, fruits and curries, and rice.

Indonesian flavors in cooking vary widely among the different islands, with influences from Malaysia, Singapore, China, and the Arab world. Most islands have a Muslim background, except for Bali, which has Hindu influence. Sumatra has fiery hot rendangs and spicy root vegetables, Java has stir-fried noodles and hot salads, and Bali has pungent curries and fish sambals.

Foods from Malaysia and Singapore have similar flavors, with pungent Malay sambals and tamarind-laced fish, Chinese stir-fried noodles and soups, Nonya chili-based laksas, Indian curries and dals, Cristiang (Portuguese influenced) hot condiments and seafood, and European-style cutlets and salads. They have unique regional and cultural flavor variations derived from fermented bean pastes, shrimp pastes, wrappers, sour fruits, leafy vegetables, coconut milk, nuts, and a variety of spices.

Kampuchean (or Cambodian) foods have a strong Chinese and Indian influence with rice, fish, peanuts, ginger, and black and white peppers as essential flavorings. Laotian foods have flavors similar to Cambodian and Thai foods as well as origins from China. Sticky rice, beans, tubers, pork, chicken, chilies, coriander, mint, coconut milk, shrimp, and fish sauce are important ingredients. Philippine foods have Malay, Chinese, Spanish, and North American influences with pungent fish pastes, coconut, adobos, and fruit juices. Foods from Myanmar (or Burma) are influenced by neighboring India, China, Cambodia, and Thailand. Rice, noodles, duck, chicken and fish curries, fermented fish pastes, garlic, turmeric, tomatoes, and onions are popular ingredients.

East Asia

East Asian flavors such as Japanese and Chinese have become mainstream foods with North Americans. Chinese foods have some influences from Japanese, Mongolians, and Indians, depending on the regions. In South China, people eat mainly rice or noodles, while the northerners eat wheat, millet, and sorghum. The most popular meats are pork, beef, and lamb, the latter two being preferred by the Muslims. Vegetables are eaten in abundance.

Regional cooking of China has become better known in the United States. Cantonese foods have mild seasonings and, like French cuisine and rely on the subtle juices of ingredients and preparation techniques. Their seasonings are soy sauce, ginger, garlic, and oyster sauce. Beijing (or Peking) cuisine incorporates pancakes, steamed buns, stir-fried or steamed dishes, pickled vegetables, tofu, scallions, roasted or barbecued pork, and lamb. Shanghai cuisine on the east coast has sweeter flavors and uses plenty of noodles, sesame seeds, fish, dark red sauces, sesame oil, and black bean sauce. Szechwan and Hunan cuisines in the west are hotter, using chile peppers, preserved vegetables, fagara, and cloves.

Korean foods have been influenced by neighboring Japan and China. Rice is the staple with noodles, seaweeds, pickled vegetables, pork, and kimchis. Essential flavorings include sesame, ginger root, ginseng, gingko nuts, garlic, soy sauce, fish sauce, and chile peppers. The three essential sauces are *kanjang* (soysauce), *toenjang* (fermented soybean paste), and *koch'ujang* (hot red pepper paste). The south has spicier and saltier flavor profiles than the north. A typical Korean meal, called *panchan*, consists of steamed white rice with side dishes of kimchi, soup, spinach "namul," toasted seaweed, pickled vegetables, meat, or pickled seafood.

Japanese foods have aesthetic appeal and simplicity. They have some influence from China and Korea. Japanese foods are lightly seasoned, focusing more on preparation techniques, textural and visual appeal. If seasonings are used, such as gomasio or shichimi, they are used as garnishes and condiments. Their foods delight the senses and are based on cooking techniques using fresh ingredients.

Their staple is glutinous rice, with a variety of noodles, fish and other seafood, pickled vegetables, mushrooms, seaweeds, soy sauce, wasabi, sansho, ginger, miso, mirin, and sesame. Japanese flavors are obtained mainly from the way ingredients are prepared and cooked (such as grilling, steaming, simmering, or deep frying).

POPULAR EAST ASIAN SPICE BLENDS

Fermented Soybean Blends

Soybeans have inspired many dishes in China, Japan, Korea, and Southeast Asia. Soybeans are frequently fermented to achieve end products that are richer in appearance, flavors, proteins, and vitamins. These fermented bean pastes are called *jiang* (Mandarin), *jang* (Korea), *taotsi* (Philippines), *taucheo* (Malaysia), and miso (Japan).

The sensory properties of fermented soybean pastes vary with fermentation time, aging, and added ingredients—chile peppers, seafood, peanuts, sesame, onions, or garlic. They can be smooth or chunky, salty, sweet, or hot. They are commonly used as seasonings to flavor the meats, seafood, vegetables, soups, dips, and dressings of China, Korea, and Japan. They are a major ingredient in Szechwan sauces, marinades, hoisin, and oyster sauces. They are also important in Singaporean braised meats and stir-fries and Indonesian sambals.

Fermented bean pastes are made to complement or enhance particular dishes. Some of the fermented soybean products are the Japanese and Korean misos, the salty Malaysian *taucheo*, pungent Korean *toenjang*, and the salty, sweet, dark black and brown Chinese bean pastes. Miso is a common ingredient in Japanese and Korean soups, sauteed dishes, and marinades. They are sweet, nutty, or salty and white, brown, or dark red colored. Their flavor and color depend on whether they are made from soybeans alone or with rice or barley, their fermentation time, and other added ingredients. The yellow or brown bean paste is slightly sweet and flavors stir-fries of Peking origin. The black, salted and fermented, pungent soybean paste is made with garlic and is used to flavor seafood and pork dishes of Shanghai origin. The red, sweet bean paste made from adjuki beans is used in pastry and bun fillings and in the puddings of the Guandong (Cantonese) region.

Fermented Bean Seasonings
Fermented bean seasonings include soybean sauce/paste (*jiangs, taucheo*), sweet bean sauce (*timcheo*), oyster sauce, misos, and *toenjang*.

Description
Asians ferment, salt, and pickle soybeans and other beans to create sauce blends with intense flavors, colors, and textures. These blends are rich in flavor and have high levels of B vitamins and minerals. Each region uses its own distinct aging process and raw materials to create unique, intense flavors. They are red, black, white, or yellow and are pungent, spicy, sweet, nutty, or salty, depending on the added ingredients (rice, barley, wheat, salt, chile peppers, garlic, and sesame oil) and their aging times.

Ingredients
Soybeans or rice, barley, and buckwheat are the starting ingredients used in creating fermented bean pastes. Salt, sugar, garlic, red chile peppers, seaweed, sesame oil, ginger, rice wine, or scallions are some of the typically added ingredients.

Applications

Bean pastes enhance or complement many dishes, such as soups, stir-fries, cooking sauces, or marinades. They also thicken and provide consistency to these products. They are mixed with hoisin, sweet and sour, or tomato sauce to give a wonderful blend of sweet, pungent, and savory tastes to many Chinese dishes.

Miso blend, a must in Japanese cooking, is a fermented product made from soybeans with sea salt and *koji*, together with other ingredients such as rice, wheat, barley, or buckwheat. *Aka, hacho, mugi, shiro, shinso, soba,* or *kome* are different misos that are mild, salty, intense, sweet, or hearty, depending on the raw ingredients and fermentation time. The three most popular misos are the *aka miso* (all soybean), which is dark reddish black and salty; the *shiro miso* (soybean and rice), which is sweet and light yellow; and the *shinso miso* (soybean and barley), which is semisweet and dark brown. Misos' applications vary by region, depending on the raw materials available, climatic conditions, and eating customs. For example, *shiro miso* is typically used in the Kyoto region of Japan, *aka miso* in northern Japan, and *shinshu miso* in Nagano region. Spreads, marinades, salad dressings, stir-fries, and soups use different types of misos. Meat and seafood go well with *aka miso*, while *shiro miso* pairs well with soups and salad dressings.

Soybeans are fermented to create salted whole beans (*taucheo*), pungent chunky black bean sauce, and the slightly sweet pungent brown bean sauce (*timcheo*). They are used throughout East Asia and Southeast Asia, wherever there is Chinese influence. The sauces are thick, pungent, salty, or sweet and provide consistency and flavor to pork, fish, or stir-fries. They are also used as spreads for fresh spring rolls or moo shu pork skins.

Black bean sauce is thick with a deep reddish brown color and a rich flavor, whereas the yellow bean sauce is sweeter and less pungent. *Taucheo*, the salted whole soybean, is used whole or pounded to flavor fish, noodles, vegetables, and pork dishes.

Oyster sauce (also called *hou yu jiang* in Mandarin, *nam man hoi* in Thai, and *kicap tiram* in Malaysian, Indonesian, and Singaporean) is a thick, reddish black seasoning that is commonly used for Cantonese-style cooking, marinades, and stir-fries. It contains oyster extract or juice combined with fermented soybeans, sugar, starch, and other ingredients. It is added to fish sauce, hoisin, and soy sauce to intensify the flavor of stir-fries, pork, fish, and meat dishes and is used with spices as a topping sauce for fried tofu. It adds a sweet, meaty, and slightly smoky flavor to beef, seafood, tofu, and vegetables.

Dwenjang/toenjang, a fermented soybean paste, is flavored with garlic for a spicy pungent flavor which is a staple flavoring in Korea. *Koch'ujang*, which contains red pepper paste and glutinuous rice powder, is an essential ingredient for stews, vegetable dressings, soups, and dips. Garlic, sesame oil, vinegar, sugar, soy sauce, or sesame seeds are added to *koch'ujang* to create a hot pungent table condiment.

Chinese Five-Spice Blend

Many Asian regions, such as China, Thailand, and India, have their own five-spice blend that is commonly used to season many dishes, or used as a "sprinkle on" seasoning at the meal table. Each one contains different ingredients based on regional preferences. The Chinese five-spice is the most well-known five-spice blend in the United States while the Bengali *panchporon* is an emerging blend.

Description

Chinese five-spice blend is a ground mixture that is popular in Chinese cooking, especially Szechwan, Hunan, and Peking dishes. It is also called *wu xiang fen* in Mandarin and *phong pha lo* in Thai. It is used in the Chinese-influenced dishes of Vietnam, Thailand, Malaysia, and Singapore. It has a pungent, anisey taste with slight camphorlike notes, and its color ranges from a tan to gingery brown. Foods get bitter if too much of the spice blend is used. The Chinese believe that the number five has curative powers, so this blend started its use as a medicine.

Ingredients

Chinese five-spice blend is made from star anise, fennel, clove, cassia, and Szechwan pepper. Black peppercorns can be substituted for Szechwan pepper. Today, commercially sold blends also contain ginger, licorice, cardamom, or ground dried tangerine peel.

Applications

Chinese five-spice blend is an important flavoring in hoisin sauce and other Chinese marinades for deep-fried, roasted, grilled, braised, or barbecued foods, such as pork, duck, chicken, and other meats. It is mixed with red color, soy sauce, and sesame oil to create the popular Chinese roast pork called char siew. Five-spice is mixed with salt and used as a dipping sauce for roast meats, poultry, and raw or deep-fried vegetables. The blend pairs well with rice wine, dry sherry, soy sauce, shallots, and garlic.

Japanese Seven-Spice Blend/Shichimi Togarashi

Description

Shichimi togarashi is a finely ground mixture of spices that is essential to Japanese cooking. It has a hot spicy taste with a citrus aroma. Its texture is gritty.

Ingredients

It contains sansho, chile peppers, dried orange/tangerine peel flakes, dried nori/seaweed flakes, white and black sesame seeds, and white poppy seeds.

Application

In Japan, this seasoning is sprinkled on prepared foods. It is generally used to season grilled meats, soups, and noodle and rice meals. It is also used as a popular table condiment.

Teriyaki Blend

Description

Teriyaki adds flavor and color to raw and cooked foods in Japanese cuisine. It is slightly sweet and soy-sauce-like and is sold in dried, paste, and liquid forms.

Ingredients

There are many versions of teriyaki sauce. The basic teriyaki ingredients are soy sauce, sugar, vinegar, and sweet rice wine. Ginger, garlic, chives, sesame oil, and fish stocks can be added to give extra edge or different flavor notes to enhance specific applications.

Application

Teriyaki complements broiled and steamed dishes such as steaks, chicken, and spinach. It flavors dipping sauces that are used for wontons, dumplings, sashimi, deep-fried and battered seafood, or vegetable tempuras. It is commonly used as a Japanese-style marinade for grilled or roast beef, chicken, seafood, and vegetables. Pineapples, oranges, lychees, tomatoes, onions, and bell peppers are chopped or sliced and added to teriyaki sauce.

Hoisin Sauce Blend

Description

Hoisin sauce is one of the most important seasonings in Szechwan, Peking, and Cantonese cooking. It is a thick, reddish brown or dark brown sauce with a slightly sweet, spicy, and tangy flavor. Traditionally, it is made with wheat flour and sugar. It originated in northern China and is called *tianmian jiang* in Mandarin, which means "sweet wheat paste." In Cantonese, hoisin means "sea fresh sauce."

Ingredients

Hoisin sauce blend is made from soybeans, wheat, sugar, caramel, vinegar, and five-spice blend. Sometimes, other ingredients are used, such as ground sesame seeds, red beans, or garlic. Plum puree, sesame oil, or chile peppers are sometimes added to create differing flavor variations in applications.

Applications

Hoisin is used as a dipping sauce, a marinade for ribs, shellfish, and duck and as an accompaniment to Peking duck and moo shu pork. It is combined with chilies, garlic, plum sauce, and vinegar to create char siew or roast pork (a Cantonese specialty), double-cooked pork (a Szechwan specialty), and stir-fry dishes. Hoisin provides a deep, rich flavor and color to stir-fries, roasts, and barbecues. It is also used as a table condiment and dipping sauce. Hoisin pairs well with chicken, duck, ribs, shrimp, scallions, sesame oil, rice wine, soy sauce, vegetables, and ginger. Hoisin sauce is widely used in Cantonese-style cooking to season meats, tofu, vegetables, and seafood dishes.

POPULAR SOUTH ASIAN SPICE BLENDS

Curry Blends/Kari Podis/Masalas

Description

Curry originated in India. Through trade, colonization, and immigration, curry spread to other parts of Asia, Europe, Africa, and the Caribbean, with resulting variations, such as *rendang, w'et, gulai, gaeng,* or *colombo.*

While spice mixtures called *kari podis* and masalas for flavoring sauces have been made for thousands of years in India, the term "curry" is derived from the Indian word "Kari." Curry or curry powder is an English term for a commercialized spiced sauce or spice mixture from India. There are many versions of how the word "curry" originated. The English coined the term curry from the word "kaari" or "kaaree" in the South Indian Tamil language, which means "sauce" or "gravy." Other stories say that the British took the word curry from "kari leaf," a pungent leafy spice used in sauces of South India. Another origin of the word curry is derived from "karahi," a woklike vessel used to cook curries in North India and Pakistan. Yet another source traces curry from the Hindustani word "turcurri," which was shortened to "turri" and mispronounced as curry by the Westerners.

Curry blends are used not only to flavor sauces but also to flavor stews, snacks, or rices. Generally, curries are a mixture of many different spices (whole, ground, or crushed) with fruits, vegetables, nuts, seeds, and other ingredients. There are hundreds of curry blends in South Asia. Curry blends differ in flavor and color, depending on who creates them, the types and proportions of ingredients used, their preparation, and the end application. They are becoming "hot" in the United States because curries satisfy consumers' increasing appetite for spicier ingredients, low fat and low sodium dishes. Whether called curry powder, garam masala, *panch-phoron,* chat masala, vindaloo paste, Madras curry powder, korma paste, or *sambar podi,* all contain mixtures of spices blended for specific applications and to suit regional preferences.

Ingredients

A good curry blend or masala contains a well-balanced proportion of spices. It would be difficult to develop an industry-wide standard for curry blends because of the numerous types of curries. A basic curry blend from South Asia consists primarily of cumin, coriander, black or red pepper, and turmeric. Many versions exist depending on regional preferences, cultural influences, the availability of ingredients, and the application—meat, fish, vegetable, or poultry. Spices such as cardamom, clove, cinnamon, mustard seeds, fenugreek, kari leaf, mint, coriander leaf, and celery seeds are added to create these variations. South Asians toast the whole spices then grind them to obtain more fragrant aromas and crunchiness and to remove raw tastes of spices so proper digestion of food is achieved.

Curry blends pair well with garlic, ginger, shallots, coconut milk, yogurt, lime juice, lemon juice, tamarind, lemongrass, mango, tomato, other vegetables, and vinegar.

South India

South Indian curries, or kari, are hot, spicy, coconutty, and pungent with some sour notes. They can be dry, saucy, soupy, or thin and are generally made with fish, goat, chicken, lentils, or vegetables. Essential ingredients in the curry blend are chile peppers, coconut milk, tamarind, turmeric, black pepper, fenugreek, and kari leaf. "Popped" mustard seeds are usually added, along with shallots, ginger, and kokum. Optional spices are clove, cinnamon, and green cardamom.

These curry blends characterize vindaloo, Madras curry, *konju meen*, *sorpotel*, *keera kootu*, *meen kolambu*, *keema bafat*, *erachi kotu* (mutton), and *xacuti curry*.

North India, Pakistan, and Bangladesh

Curry spice mixtures in the north are generally called masalas (plain masala or garam masalas), which comes from the Arabic language. Curries from North India, Pakistan, and Bangladesh are influenced by Mogul and Persian cooking and often have yogurt, dried fruits, and nuts to characterize them. Other ingredients are coriander leaves, methi leaves, paprika, ground mustard, saffron, and ghee. They are mild, sweet, rich, nutty, fruity, and aromatic, creamy, or dry. They are generally made with lamb, chicken, vegetables, and legumes.

Some popular curry sauce blends are *saag* blend, *jalfrezi* blend, *rogan josh* blend, korma blend, *channa makhani* blend, *masala murgh* blend, *matar keema* blend, chicken masala blend, Kashmir curry blend, and *kofta* curry blend.

Sri Lanka

In Sri Lanka, most curries are chile-based, pungent, and extremely hot. The milder curries contain coconut milk and tamarind. The "black" curries contain kari leaves, darkly roasted spices such as coriander, cumin, fennel, and fenugreek, and other ground spices including cloves, cinnamon, and cardamom. Tiny bird's eye peppers provide the heat. Some curries have dried and powdered Maldive fish, eggplant, mango, and unripe jackfruit.

Some popular Sri Lankan curries are fish *pindun*, beef *smoore*, pork *padam*, *frikkadel*, and duck *padre*.

Specific Curry Blends

Chat masala is a popular green curry mixture used in North India. It comes as a wet paste or a dry mix, and there are many versions of it. A typical blend uses green chilis, mint leaves, coriander leaves, coriander seeds, cumin seeds, ajwain seeds, black peppercorns, cloves, ginger, asafetida, and dried mango powder (amchur). Others contain pomegranate seeds, lime juice, bay leaf, cinnamon, and cardamon. Chat masala has a tart, salty, and hot flavor. It is commonly used to flavor sauces and snacks.

Garam masala is the most popular masala used in North Indian cooking. It is generally used with curry blends to add zest to sauces. Garam masala means "hot or warm spice mixture" containing spices that create heat in the body. These spices are ground for use in North Indian curries. Its ingredients vary, but, traditionally, garam masala contains the "warmer," more pungent spices such as brown cardamom, black peppercorn, clove, cinnamon, mace, and nutmeg. Today, it includes the cooler spices, such as fennel, green cardamon, and bay or cinnamon leaf. Based on regional preferences, cassia, fennel, or celery seeds are added. It usually lacks turmeric.

Garam masala is dry-roasted and used to enhance many curry dishes. It contributes to the aroma of a finished sauce and is usually sprinkled on just before serving or toward the end of cooking to avoid bitterness caused by prolonged cooking. It is generally added to meat dishes, less in poultry and biryanis (layered meats and rice), and rarely in vegetable or fish dishes.

Korma is a braised, dry curry from North India. It is brown, aromatic, and mild, with a creamy consistency. *Korma blend* has almonds, cardamom, cumin, coriander, turmeric, yogurt, and onion. It goes well with lamb, chicken, and certain vegetables. In Muslim-style biryanis, korma sauce is layered between cooked basmati rice and is baked to give biryanis their characteristic aromatic flavor.

Madras curry blend is a hot, pungent curry from Tamil Nadu in south India. It contains roasted dry spices such as dried ground red chilies, coriander seeds, cumin, mustard seeds, black pepper, fenugreek, turmeric, asafetida, and kari leaves. It can also contain ground, toasted lentils and rice. This spice blend is added to chicken, fish, mutton, or vegetables.

Kashmiri masala is mild and aromatic with green cardamon as the predominating spice. Other spices are similar to the garam masala. It is used for baked dishes. *Xacuti curry blend* is from Goa and contains cumin, coriander, fenugreek, black peppercorn, chile peppers, and shredded coconut, which are roasted until a dark color is obtained.

Char masala, a four-spice mixture used in Afghanistan, contains any combination of cumin, coriander, cinnamon and cloves, green cardamon, or black pepper.

Kala masala, a black spice mixture popular in Sri Lanka, consists of roasted and ground clove, cinnamon, cumin, coriander, red chile peppers, and black pepper. Maharashtra, in north India, has *kala masala* in many versions, a typical one having cumin, coriander, cinnamon, cardamom, sesame seeds, chile peppers, and shredded toasted coconut.

Samanthi podi is another popular dark brown spice mixture in Kerala. It contains dried toasted coconut, kari leaf, tamarind, onion, and ground red chili. Toasted and ground lentils are optional. It flavors condiments and chutneys that are served with local breads such as puttu, thosai, or idli, especially at breakfast.

Tandoori blend, an aromatic, spicy marinade, was created by the Moguls of North India who baked food in a "tandoor" or clay-type oven heated with charcoals. Prior to baking, chicken, lamb, shrimp, or fish are marinated in a blend of yogurt, lemon juice, and ground spices, such as ground chile pepper, ginger, cumin, coriander, clove, cinnamon, cardamom, nutmeg, mace, black peppercorn, and bay leaf.

Panchphoron is a five-spice mixture from Bengal that contains fennel, mustard, cumin, fenugreek seeds, and *kalonji*. It generally complements fish, dals, and vegetable dishes.

Applications

Commercial curry blends come as powders, pastes, sauces, or oils that offer convenience and consistency for a recipe. They also are available hot, medium, or mild depending on the amount of chile pepper added. While in the United States, curry blends contain ingredients that have mass appeal, such as cumin and turmeric, but they often lack the flavor-enhancing effect required for a specific application. To create the authentic products that consumers demand, curries must include specific or specially prepared ingredients that will enhance the flavor, texture, and color profiles of fish, meat, poultry, or vegetables. It is also important to maintain the aroma and taste of freshly ground spices.

When creating a sauce or curry, ginger, garlic, onions, potatoes, vegetables, nuts, or seeds provide the "body." The consistency of a finished curry varies and can be thick or thin, dry or wet. Color is also an important sensory attribute of curry. Color is contributed by turmeric, saffron, paprika, green leafy spices (cilantro, basil, mint), certain vegetables, nuts, or chile peppers.

Ingredient preparation and cooking techniques create numerous flavor, texture, and color profiles in a curry sauce blend. By using ground or whole spices and cooking techniques, such as dry-roasting, toasting, braising, or sauteing, different flavor dimensions can be achieved.

The order and time when ingredients are added in the cooking process are also crucial in obtaining the flavor desired. Some spices are more volatile than others. Cumin, coriander, and fennel are dry-roasted and added first. The more volatile spices, such as mace, cinnamon, and cardamom, are added later to retain their aromas.

Sambar Podi, Rasam Podi, and Dal Podi

Description
South India vegetarian dry spice mixtures often contain ground or whole "dals" (lentils) and are used in sauces, soups, and curries and to enhance flavor and texture of chutneys, dips, and snacks. These spicy lentil mixtures include *sambar podi*, *dal podi*, and *rasam podi*. They are added directly into cooking "as is" or they are dry-roasted or "popped" whole in hot oil to render them fragrant, then ground into the "podis" or mixtures.

Ingredients
Indian "dals" or lentils are fried or dry-roasted, then added whole or ground with spices and other flavorings. The *podis* usually contain *urad dal*, *channa dal*, or *toor/toovar dal*. Many other lentils or beans are also used in combination with spices to add zest or texture to foods.

Sambar podi contains *channa dal* (*Bengal gram*), *toovar dal*, and *urad dal* (black lentil/black gram) that are mixed with spices, such as coriander seeds, cumin seeds, mustard seeds, fenugreek seeds, black peppercorns, asafetida, kari leaves, and ground red chilies. The spices are roasted for a few minutes, then cooled, ground, and used for stews and soups.

Rasam podi contains *toor dal*, channa dal, coriander, cayenne, black peppercorns, cumin, turmeric, and kari leaves. This blend is added with tamarind juice to make a spicy, sour soup.

Dal podi contains *toor* dal, *channa* dal, *urad* dal, turmeric, cumin, coriander, asafetida, cayenne, black peppercorn, and onions. It is used in stews, sauces, and dips.

Applications

In Indian cooking, dals are used in many different applications, and an understanding of their usage will help create appropriate lentil spice mixtures. *Channa dal*, popular in all of India, is used raw in chutneys, roasted whole as a spicy snack, ground for sweets, or whole with vegetables and chutneys. *Urad* dal is widely used in South India where it is fermented with rice and mixed with spices and curry leaves to create *dosai*, steamed *idli*, and snacks such as *vadai* or *pappadum*. *Toovar* dal exhibits a thick and more gelatinous consistency and is combined with *channa* dal, spices, and chilies to create *sambars*. *Kala channa*, a smaller, darker chickpea, is popularly used in the Punjab region and is braised for use in tangy, spicy cumin and coriander-based sauces. The leafy spices are usually added crushed or added whole to the ground spices.

These *podis* are added to chutneys, pickles, soups, and sauces of south India. They are usually eaten with local breads or rice.

Pickle and Chutney Blends

Description

Pickles and chutneys are essential accompaniments to Indian meals, snacks, soups, breads, and appetizers. Chutneys originated in India and were introduced into England and Europe by British colonials. Chutneys or *chatnis*, which translates to "licking good" in Sanskrit, and pickles, are essential condiments and side dishes in every Indian meal. They add zest and texture to the main meal with their hot, salty, sour, and sweet flavors. They consist of spices, fruits, vegetables, vinegar, and sugar. There are unlimited variations of these condiments, depending on regional preference as well as on the type of food it is eaten with—meat, fish, rice, bread, or vegetables.

Ingredients

The main ingredients used in pickles and chutneys are mango, lime, coconut, apples, chile peppers, ginger, garlic, mustard oil, mustard seeds, turmeric, sesame seeds, mint, cilantro, and tamarind. These ingredients are usually fermented or pickled to achieve intense, characterizing flavors. They can also be freshly prepared (uncooked) or cooked.

Pickles contain discrete pieces of mangoes, green chile peppers, or limes, whereas chutneys contain smooth or coarsely textured ingredients such as coconut, mint, or tomato. There are many variations of chutneys and pickles, and some are included below.

The British chutneys tend to be sweet and tart, incorporating fresh or dried fruits such as citrus, pineapples, cranberries, apples, plums, figs, peaches, apricots, and dates. They are served alongside hot curries and cold foods. Today, fresh vegetables such as corn, squash, bell peppers, cucumbers, carrots, and green tomatoes are added. These chutneys can be compared to the freshly made or cooked salsas of Latin America.

Mint/*pudina* chutney is fragrant, uncooked, smooth-textured chutney that goes well with roasted/grilled meats, tandooris, and fried foods. It contains mint, ginger, coriander leaves, black pepper, green chile peppers with vinegar, lemon juice, and sugar. It is also commonly served as dips with samosas, pakoras, and other appetizers.

Coconut/*nariel* chutney is an uncooked, coarsely textured, and spicy sweet specialty condiment of South India. It usually accompanies fermented breads, such as dosai or idli, and fish and chicken. It has toasted coconut flakes, roasted cumin seeds, "popped" mustard seeds, garlic, ginger, green chilies, and roasted curry leaves.

Tomato/*thakali* chutney is an uncooked or cooked, sweet, sour, and spicy condiment, commonly used in South and North India. It has roasted tomatoes, onions, ginger, garlic, roasted, ground cumin seeds, black mustard seeds, cayenne, coriander leaves, and lemon juice. It complements seafood dishes, barbecues, and fried foods.

Mango pickle is a spicy, tart pickle with green/ripe mango with mustard seeds, mustard oil, turmeric, ginger, and garlic.

Green chile achar is a pungent spicy pickle with green cayenne, mustard oil, ginger, sesame oil, ground mustard, fenugreek seed, and turmeric.

Mixed pickle/*sabzi* achar is a spicy, sour, and hot mixed vegetable pickle commonly eaten all over India with any meal. It contains garam masala, onions, ginger, cayenne, turmeric, black mustard seeds, nigella, fennel, mustard oil, carrots, green mangoes, green chilies, and lotus root.

Applications

Pickles containing mango, lime, shredded carrot, or chilies with spices and yogurt are generally used to provide bite to vegetarian meals as well as to balance the flavors in curries or dals. They are used for breakfast, lunch, snacks, or dinners. Chutneys usually accompany snacks and tone down fiery sauces. They are also used as dips for breads, such as *idli*, *thosai*, or *chappatis*. They can be spicy, sour, or slightly sweet and include mango, mint, tomato, cauliflower, white radish/*mooli*, carrot, eggplant, onion, coconut, and tamarind.

POPULAR SOUTHEAST ASIAN SPICE BLENDS

Sambal Blends

Description

Sambals are the hot sauces of Southeast Asia and are used the same way Tabasco sauce is used in the United States or salsa is used in Mexico. They are essential table condiments used to perk up rices, breads, mild curries, soups, and noodles. Sambals are also used as base sauces for cooking shrimp, fish, chicken, meat, and vegetables and for spicing fruit and vegetable salads.

Sambals are hot, sweet, and very fragrant. The fragrant aroma is developed when the sambal sauce blend is "tumised" or sauteed slowly with constant stirring until the oil seeps out from the mixture. Indonesia and Malaysia have many varieties of sambals with their meals. Some examples include *sambal belacan* (Malaysia, Singapore), *sambal trasi* (Indonesia), *sambal olek* (Southeast Asia), *sambal badjak* (Indonesia), cucumber and pineapple *sambal* (Malaysia), *sambal tomat* and *sambal matah* (Bali), and *nam prik* (Thailand).

Ingredients

The basis for most sambals is chile peppers, shallots, tamarind, brown sugar, and lime juice. It can be as simple as shallots with *cili boh* (pureed chilies) and lime juice or more complex with other added ingredients, such as lemongrass, nutmeg, shrimp paste, palm sugar, fish paste, ginger, galangal, kaffir lime leaves, coconut milk, candlenuts, and tomatoes.

Each ethnic group adds its special flavoring to these sambals, such as ginger and soy sauce (for less heat) by the Chinese; bird peppers, *belacan* (fermented shrimp pasta), and coconut milk by the Malays; candlenuts, mint, or "bunga kantan" by the Nonyas; or a little curry powder and coriander leaf by the Indians.

Applications

Southeast Asians add sambals to almost anything, even as spreads to sandwiches and as fillings to buns and pastries. A typical breakfast in Malaysia and Singapore, called "nasi lemak," is a coconut milk and lemongrass-based rice dish that is topped or served with sambals made from prawns or *ikan bilis* (an anchovy-like fish).

Fruits and vegetables, such as pineapple, mango, cucumber, green papaya, *petis* (fermented shrimp pasta), *belimbing wuluh*, and carrot are added to sambals. Salads, laksas, rices, or noodles that are enjoyed by Malaysians and Singaporeans are served with fiery pungent sambals.

Malays add fish, prawns, *ikan bilis*, cucumber, cuttlefish, tofu, mango, *bengkuang* (jicama), tomatoes, peanuts, or belimbing in their *sambals* and serve them as side dishes with their "nasi" or boiled white rice. Every meal has a choice of one or more *sambals* to suit the accompanying dish. Nonyas, from Malaysia, generally serve *sambals* with laksas, fried rice, or fish dishes.

Curry Blends

Description

Indians have influenced Southeast Asian with their curry blends, which have become important flavorings in Southeast Asian cooking. Like sambals, they are used in sauces (wet or dry), stir-fried noodles, soups, and condiments. Each region has developed its own distinct curry blends.

Ingredients and Applications

Burma/Myanmar

Their dishes are influenced by the Chinese, Indians, Thais, and Cambodians. Their curry blends are mild or medium hot as compared to South Indian curries. Burma's curry blends include basic spices such as onion, garlic, ginger, turmeric, fish sauce, and chile peppers, with other added vegetables, tamarind, and the bark of the banana tree. Their curries are oily, with sesame, tomatoes, and peanut oils. Curry blends are used in duck, pork, fish, chicken, and vegetable dishes.

Examples are *kyethar peinathar hin*, *bairather sepiyan*, *mohinga*, *Chettiar* or monsoon curry, and *pazun hin*.

Cambodia/Kampuchea

Cambodian dishes have strong Chinese and Indian influences. The essential ingredients are fish sauce, lemongrass, galangal, garlic, coconut milk, and chile peppers. The lemongrass provides their curry blends with a fibrous texture, and the coconut milk gives them a creamy consistency. Curry blends are used with meats, such as pork and water buffalo. Bean sprouts, potatoes, peanuts, and eggs are other added ingredients.

Examples include lemongrass curry, egg and potato curry, and *Cambogee*.

Laos

Laos is greatly influenced by southern Chinese and northern Thai cooking. Their flavors are similar to Cambodian cooking and have influences from India. The southern region of Laos uses fish sauce, coconut milk, mint, and chile peppers as essential ingredients to spice up their sauces. The Hmong cuisine of Laos, influenced by Chinese cooking, uses soy sauce instead of fish sauce to flavor their foods. Laotian curry blends' essential ingredients are ginger, shallots, chile peppers, coconut milk, yams, taro, and sweet potatoes. Curry blends are used with pork and chicken, catfish, and eggs but rarely with vegetables, which are usually stir-fried. Laotian curries are very smooth and starchy.

Examples are catfish curry, *kalee peel*, and *mok pla*.

Vietnam

Vietnamese curries are influenced by Chinese, Indians, and French. North Vietnam cuisine is influenced most by the Chinese. South Vietnam has been influenced by Indian traders who introduced curries, which are called caris. Their neighboring Thais have also influenced with their curries. The curry blends include lemongrass, mint, coconut milk, vinegar, sugar, turmeric, garlic, chile peppers, lime, fish sauce, and coriander leaf. They are milder than Thai curries, and less intense than Indian curries. They are slightly sour and sweet with a strong turmeric color. Curry blends are added to ground meats, chicken, frog legs, mung bean noodles, taro, and sweet potatoes.

Some examples include Saigon *cari*, *cari ga*, and *Echnau cari*.

Thailand

Thai curry blends are hot, pungent, sweet, sour, and aromatic and some of them have a strong, coconutty flavor. Thai curries, called *gaengs* or *kaengs*, are hot and fiery with plenty of fresh or dried chile peppers that include cayenne or bird peppers. Thai red curry is pungent and intense with coriander root, shrimp paste, galangal, and dried, red chilies; green curry has coriander leaves, green chilies, lemongrass, and green eggplant; yellow curry includes fresh turmeric and yellow chilies; Mussaman curry, which is influenced by Indian Muslims, contains more spices, including clove, nutmeg, cinnamon, and cardamom; and Panang, which has influences from Penang (in North Malaysia), has coconut milk shallots, peanuts, and tamarind. Thai curry blends pair well with pork, seafood, and chicken. Kaffir lime leaves, Thai basil, lemongrass, fish sauce, shrimp paste, coconut milk, bamboo shoots, eggplant, and green beans are some essential ingredients in their local curries. Thai curry pastes are added to pork, chicken, or beef and are used to perk up soups, dips, salads, and stir-fries.

Red (*nam prik gaeng ped*), green (*kreung gaeng kiow wahn*), yellow, Mussaman, Panang, and spicy basil are some of the popular Thai curry blends.

Malaysia/Singapore

Malaysian and Singaporean curry blends are influenced by South Indians, Malays, Nonyas, Chinese, and the Portuguese. Essentially, their curry blends contain Indian spicing with added Malay and Nonya ingredients. They can be dry or soupy. They are generally hot, pungent, and very aromatic. Curries are very popular as side dishes to rice, as dips for flat breads (naan, *chappati, dosai, roti canai*), to flavor rice (called biryanis), or as bases for soups, pickles, sandwich fillings, and spicy salads.

In Singapore and Malaysian curries, lemongrass, star anise, galangal, pandan leaf, ginger flower, lime, coconut, tamarind, candlenuts, mint, coriander leaf, turmeric, shallots, shrimp paste, or sambals are some of the ingredients that are added for variety.

The Nonyas add tamarind, fish sauce, turmeric, candlenuts, and lemongrass to make pungent, tart, and fiery laksa curries. The Malays add bird peppers, mint, belacan, galangal, and coconut milk to create hot and aromatic fish curries. Indians use many spices, including kari leaves, yogurt, tomatoes, garam masala, ginger, coconut milk, and tamarind to make vindaloos, mutton curry, *sothis, sambars*, tandooris, biryanis, or *keemas*. Their curries vary depending on their South Asian ancestry: Punjabi, Gujerati, Bengali, Keralan, or Sri Lankan.

Singapore, which has a stronger Chinese influence, adds soy sauce, fish balls, seafood, and local vegetables to create their famous fish head curry, dry mutton curry, *masak lemak*, and *rendangs*.

Examples are *rendang* blend, laksa blends, *kari ayam* blend, *debil kari* blend, *kapitan* curry blend, kurma blend, fish head curry blend, dry mutton curry blend, *asam pedas* blend, *keema* blend, *Mughlai* blend, and *sayur lodeh* blend.

Indonesia

Arabs, Indians, Chinese, Dutch, and Portuguese influenced Indonesian curries. They are similar to Malay curries, but vary regionally. The Minangkabaus in East Sumatra eat fiery curries with beef and coconut milk; the Javanese enjoy sweet and hot curries with kecap manis and ginger; and the Balinese, with a heavy Hindu influence, use more aromatic spices with the local sambals. In Bali, curry pastes are rubbed on fish, chicken, pork, or beef, wrapped in banana leaves and then grilled or baked.

Essential ingredients in curries are galangal, coconut milk, tamarind, *trasi, salam* leaf, and chile peppers. Other ingredients are eggplant, okra, potatoes, manioc, and lime leaves. Fried shallots and sliced boiled eggs are common toppings for curries. Some Indonesian curry sauce blends include beef *rendangs, gulai kambing, terong kari, ubi kentang kari*, and *gulai putih*.

MEDITERRANEAN SPICE BLENDS

Mediterranean foods are tasty, trendy, and part of a healthful lifestyle. They use plenty of fruits, vegetables, nuts, olive oil, roasted flavors, and spices that provide great variety and visual appeal. Mediterranean blends range from the familiar Italian pasta blends to the more exotic North African harissas. Mediterranean foods take their unique flavors from many regions and encompass all of the varied foods of southern European, northern African, and the Middle Eastern countries that surround the Mediterranean.

Certain ingredients are commonly used throughout the Mediterranean. Olive oil, garlic, parsley, beans, lamb, dried fruits, and pimientos have mass appeal in all of the regions, but each region has its own unique flavors. Northern Mediterranean cuisine (Provence, Italy, Basque regions) is characterized by sweet leafy spices (basil, thyme, rosemary), tomatoes, and wine; the west (Spain) by bacalao, sofrito, and saffron; the east (Middle East) by mint, sesame, pomegranate juice, and yogurt; and the south (northern Africa) by coriander, cayenne pepper, pippali, and harissa.

Mediterranean foods will continue to grow in popularity because of their light sauces and healthy ingredients and cooking techniques. The Middle East, Turkey, and northern Africa (Ethiopia, Egypt, Morocco, and Tunisia) have stronger and spicier notes that will excite an adventurous consumer.

Religion adds to the variety and unique flavor differences found in Mediterranean foods. Muslims, Christians, Jews, and Arabs have specific preferences and prohibitions for ingredients such as lamb, fish, beef, or pork. Certain ingredients and preparation techniques are specific to different religious groups. Wine, widely used in southern French and Tuscan cooking, is not a Halal item, so regions of Muslim predominance do not use it in their cooking. Likewise, Muslims or Jews do not use pork flavoring.

There are no intricate flavor combinations with Mediterranean cooking. The cooking is simple and rooted in the soil. Antipasti, mezzes, and tapas using a variety of vegetables, fruits, breads, and olive oil are a way of life with Mediterraneans. The natural flavors and textures of the ingredients are kept intact using light sauces. The variety comes from the different ingredients. Spiced and preserved meats and fish, crystallized and dried fruits, pickled vegetables, and olives are some of the other unique flavors.

Seafood plays an integral part in the Mediterranean diet with many varieties of fish which can be used fresh, dried, grilled, pureed, as stocks for a main entree, or to flavor other dishes. Bouillabaisse, *zarzuela de pescado*, *bourride*, *soupe de poisson*, and *tagine del hout* are some of the popular fish dishes. Lamb and mutton are seasoned and grilled as kebabs in the Islamic regions, where eating of pork is forbidden. Pork and veal are popular in Spain, France, and Italy.

Cheese such as feta, mozzarella, romano, gruyere, domiatti, manchego, ricotta, pecorino, and Parmigiana (parmesan) are common in Mediterranean meals and are used for salads, with vegetables, in desserts, or as toppings for main entrees. They are typically derived from sheep, buffalo, or goat's milk. Yogurt is used abundantly in Middle Eastern and North African meals. It is served plain or with fruits and is used as a cooking medium, a thickener, or a beverage. It is sauced over fried eggplant, mixed into cucumber soup, and served alongside pollou.

Rice is served plain accompanied by stew or condiments, or as a one-dish meal (paella, pilaf, or risotto) using arborio, Patna, or Spanish rice. Each Mediterranean region has its own preference for rices, whether sticky, firm, soft, separate, fluffy, or creamed.

Couscous, a popular dish from Morocco and other regions of North Africa, is precooked pasta that takes on the flavors of other ingredients with which it is prepared. It is served as a side dish or a one-bowl meal and made into wonderful desserts with fruits and spices. It can be eaten with stews, fried chicken, meat kebabs, fish, or seasoned vegetables. The presentation and preparation of couscous varies depending upon the country. In traditional Moroccan cooking, it is steamed (covered or uncovered) in the aromatic vapors of a meaty stew or broth.

Mediterraneans enjoy a variety of flat breads that are fermented, baked, or fried and are used to soak up sauces, scoop up food, or as a base to spread olive oil, anchovy paste, or tapenade. The long French baguette, Italian focaccia, North African *baladi* and *injera*, and Middle Eastern pita, *barbari*, and *tanoor* are eaten with spicy condiments, yogurt dips, lentil salad, spicy lamb stew, or clarified butter.

Broad beans, chickpeas, fava beans, flageolets, Borlotti beans, black-eyed peas, green and yellow split peas, red lentils, and red kidney beans are the peasant foods of this region. They are prepared and seasoned in a variety of ways and used in salads, falafel, casseroles, antipastos, risottos, spreads, stews, soups, and as purees. They are combined with vegetables, rice, wheat, and meat to provide a complete meal, such as *pasta e fagioli*, cassoulet, *ful medamus*, and *cholent*. Chickpea flour is used in a hearty vegetable soup for breaking the daily fast during Ramadan, a holy time of the year for Muslims. In the Middle East, navy beans, great northern chickpeas, and black-eyed peas provide body and texture to dishes of rice, noodles, stews, and salads. In Ethiopia, lentils are seasoned with berbere seasoning and eaten with injera. Falafel, a street food in Israel, Egypt, and Lebanon is made with white fava beans or chickpeas with spices and chile peppers and is served at breakfast, lunch, or dinner. Hummus, made from pureed chickpeas and sesame, and *bissara*, a garlicky herb puree made from fava beans, are popular foods eaten in the Middle East.

Olives are ubiquitous in Mediterranean cooking, whether pickled, braised, roasted, pureed, to season foods, as garnishes, or as a cooking oil. There are many varieties with different colors, shapes, and sizes. Black kalamata, icon-ian, and nafplion olives from Greece, Spanish manzanilla and Spanish Queen, and the Italian green and Moroccan types have become regular items on our grocery shelves. Olives are served as appetizers with cheese, nuts, breads, and chorizos. They are pickled and seasoned with lemon, vinegar, wine, garlic, fresh leafy spices, or cayenne before they can be eaten. They can also be pitted or stuffed with anchovies, almonds, or capers. Tapenade, an aromatic olive and caper spread, salad nicoise, pizzas, tagines, and puttanesca sauce are many of the familiar items with olives.

Nuts, such as walnuts, almonds, or pine nuts add crunchiness to poached fish, lamb patties, and salads, thicken soups, and are ground into flour for desserts, spreads, and sauces or are steamed with couscous. They are compatible with many leafy spices, tomatoes, raisins, butter, lemon rind, and chile peppers.

Fruits are eaten fresh, preserved in syrups, or added to ice cream, candy, tarts, and sorbets. There are a variety of fresh fruits, such as citrus, grapes, apples, figs, nectarines, peaches, pears, melons, pomegranates, persimmons, quinces, and strawberries. Dried fruits such as figs, prunes, dates, apricots, and raisins are mixed with nuts, raisins, spices cooked with fish, pork, or duck and used as stuffings for lamb.

Mediterranean foods offer a variety of vegetables, such as aubergine, artichoke, asparagus, savoy cabbage, cardoon, zucchini, courgette, fennel, kohlrabi, leeks, mushrooms, pumpkin, potatoes, or radicchio. They are usually eaten raw, grilled, sauteed, deep-fried, marinated, baked, or steamed. Cucumber and yogurt salad, served as a mezze, is a popular item in eastern Mediterranean cooking and has many variations using garlic, mint, or dill. Eggplant is popular in soups and dips, relishes, or in Middle Eastern ratatouille. It is fried in oil, grilled, pickled, stuffed and baked, braised or pureed, and seasoned with garlic, lemon juice, cayenne, paprika, basil, mint, oregano, or yogurt.

There is great diversity not only with ingredients, but also with cooking styles, which vary among the Arabs, North Africans, Turkish, and southern Europeans. Marination, grilling, pickling, and braising are preparation techniques that create healthy, tasty side dishes with variety and excitement. Grilling techniques using aromatic woods are popular around the northern Mediterranean. Grilling creates wonderful aromas and flavors with vegetables, meats, and fish.

Mediterranean foods are aromatic with fresh leafy spices such as mint, thyme, parsley, dill, basil, tarragon, and chervil; peppery with cayenne, black pepper, clove, and ginger; savory with coriander, cinnamon, and caraway; nutty with almonds, pine nuts, pistachios, and walnuts; and fruity with citrus, figs, pomegranates, and nectarines.

There are numerous seasonings used with fish, meats, couscous, salads, and appetizers. Some of the popular seasoning blends are pesto, sofrito, tapenade, harissa, picada, rouille, romesco, salsa verde, aioli, *tarator*, and the many different tomato sauces. Garlic and olive oil are popular items used with a variety of ingredients such as cayenne pepper, parsley, cumin, paprika, eggs, walnuts, mint, and bread to create unique marinades, sauces, and dips.

Let's look into the regions of the Mediterranean, how they are different, and their influence on overall Mediterranean flavor.

REGIONAL CUISINES OF THE MEDITERRANEAN

The northern Mediterranean includes Spain, southern France, and the Italian regions.

Spain

Spain had influences from the Moors, Phoenicians, Arabs, Romans, Italians, and French. Spanish explorers also brought peppers, chocolate, vanilla, potatoes, and beans from the New World. The essential ingredients in Spanish cooking are olive oil, garlic, onions, olives, parsley, tomatoes, sweet red peppers, and sweet bell peppers. Other typical flavorings in Spain include saffron, paprika, piquillo peppers, nora peppers, chorizos, and mint, which are used in dishes such as paella Valenciana, stuffed piquillos, *zarzuela de pescado*, *gam-bas al ajillo*, and gazpacho.

Spanish cuisine has many varying flavors and ingredients, depending on its regions. On the north coast (Galicia and Asturia), seafood, anise, ham, and onions characterize its cooking; the northeast (Catalonia) with Italian and French influences has various sauces—sofrito, sanfaina (onion, tomato, pepper, and aubergine), romesco (dried red chile peppers, garlic, tomatoes, and almonds), *picada* (pureed almonds, crushed garlic, parsley, and bread crumbs) and *allioli* (garlic and olive oil). These sauces are popular with grilled meats and fish and are used as base sauces for rices and stews. Valencia has a Moorish influence with paellas, citrus, and almonds. The Basque region, which neighbors France, has garlic, onions, dairy products, and seafood. The south (Andalusia), with Arab, Roman, and Phoenician influences, has given Spain its olives, olive oil, and fried seafoods. Arabs brought saffron, nutmeg, mint, black pepper, almonds, citrus, honey, and sweet and sour sauces to Andalusia. In the central region (Extremadura), the hearty cooking of the local Spanish nomads has contributed roast lamb, sausages, cured meats, Manchego cheese (from sheep milk), and pork with tomato and olive oil.

Codfish, tomatoes, onions, garlic, and pimiento characterize the flavors of the Spanish Basque region; romesco, garlic, and parsley the Catalonia region; saffron, oranges, almonds, chocolate, lemon peel, and rice in Valencia; and olives, cinnamon, and Spanish paprika/pimenton are typical of Andalusia.

France

France has many diverse flavors that vary with climate, terrain, and regional history. The south and southwest of France, which includes the regions of Provence (with Roman, Greek, and Italian influences), Languedoc (with Roman and Arab influences), Corsica, and the Basque country, forms the northern boundary of the Mediterranean. These regions tend to have lighter sauces than the north and other regions of France. Tomatoes, beans, garlic, olives, olive oil, white wine, anchovies, saffron, clove, sweet and hot peppers, red wine, figs, citrus, and bouquet garni are popular flavorings. Many sweet leafy spices such as rosemary, thyme, marjoram, sage, juniper, oregano, fennel, and savory provide aromatic flavor to the foods.

In Provence, *soupe au pistou*, sauce provençale, *bourride*, *brandade*, bouillabaisse, *daube*, rouille, tapenade, *pissaldiere*, *soup de poisson*, ratatouille, and aioli are some of its popular dishes. In Languedoc, leeks, mutton, lamb, tomatoes, garlic, and sweet peppers flavor cassoulet (white beans and meat stew), *pot-au-feu* and polenta, while ham, mushrooms, garlic, charcuterie, and potatoes are popular foods in the French Basque region.

Corsica, with French and Italian influences, has hotter flavored versions of sauces and spicier tomato pastes than Provence. Their characteristic ingredients include hot peppers, pimientos, rosemary, and juniper to flavor anchoiade, hot sausages, polenta, game meats, raviolis, and coulis.

Italy

The flavors of Italy vary with its terrain, climate, and cultural heritage and with the effects of trade and conquest. Italian cooking is influenced by Greeks, Etruscans,

French, Saracens, Germans, Austrio-Hungarians, Spanish, Chinese, and Arabs. These cultures introduced tarragon, salted cod, rice, cream, butter, cayennes, chowders, and sausages. Early explorers also brought tomato, pimiento, corn, vanilla, peanut, chocolate, and potato from the New World.

In Italy, antipasto leads a typical meal. It is followed by the first course of rice, pasta, or soup. The second course is the main dish of meat, chicken, or seafood, accompanied by a vegetable dish (*contorno*). Condiments and sauces, such as red sauce, green sauce, and béchamel sauce, are served with pasta.

Italian flavors vary widely around the country, with tomato an essential ingredient in the central and south regions, butter in the north, olive oil and garlic in most regions, and hot peppers in the south. Pasta is eaten in all of the regions, but each region has its own flavors and methods of preparation. Olives (made into olive oil) and grapes (made into balsamic vinegar) are significant ingredients in Italian cooking.

While every Italian region does not border the Mediterranean, all of Italy's regional flavors have influenced Mediterranean cuisine.

In Liguria and coastal Tuscany, pesto is added to pasta, minestone soup, and trofie (Genoa dumpling). *Stoccafissa* (codfish stew), stewed veal, and seafood chowder are also popular foods. Leafy spices (sweet basil, thyme, and marjoram), citrus, olive oil, potatoes, and mussels are common ingredients.

In Lazio (Rome), marsala wine, ricotta, pecorino, porchetta, sage, and marjoram flavor carbonara sauce, *saltimbocca*, veal scaloppine, and chicken diavola. Typical foods from Emilia-Romagna, Veneto, and South Lombardy are *risi e bisi* (rice with peas), *osso bucco*, *risotto alla Milanese* flavored with mortadella, proscuitto, red wine, saffron, gorgonzola, and creamy mascarpone. Piedmont and North Lombardy, with German, Austro-Hungarian, and French influences, have simple hearty and rich foods such as potato dumplings, bollito (sauced meat and vegetable dish), stewed red cabbage, *carpacio*, *bagna cauda*, spinach dumpling, and barley soup flavored with bacon, cheese, bread, white truffle, paprika, gorgonzola, horseradish, and poppy seeds.

Inland Tuscany, Umbria, and Le Marche have simple, light cooking with soups, seafood stews, bean salads, stuffed pork chops, spit roasted pheasant, and lamb. Olive oil, ham, beans, sausages, tomatoes, olives, mushrooms, and plenty of meat and vegetables predominate. Onions and garlic are essential flavorings along with fresh leafy spices such as rosemary, sage, bay leaf, and fennel. Sardinia (also influenced by Greeks) has robust hearty foods such as stewed and roasted meats and game. Citrus fruits, figs, fennel, myrtle leaves, ham, salted pork, pecorino, ricotto, and feta cheese flavor stuffed calamari, tuna, or pork dishes.

In the Abruzzi and Apulia regions, meat, cheese, and spicy chile peppers (pepperoncini) are added to fish chowders, *orecchiete alla puglese*, and *fettucine alla abruzzese*. Campagnia (Naples) has spicy flavors with diverse pastas such as *spaghetti alla puttanesca* or *spaghetti alla vongole*. Basilicata has Greek-influenced pork and lamb sausages and vegetable dishes. Typical ingredients of these regions include basil, fennel, mozzarella, citrus fruits, vegetables, pork, sausages, onions, and clams.

Calabria and Sicily (which have Asian, northern African, Arab, Spanish, and Greek influences) use pickled vegetables, citrus fruits, sun-dried figs, capers, couscous, salt cod, cinnamon, nutmeg, raisin, and pepperoncini to flavor cheese, fish, lamb, tomato, and eggplant. Caponata, stoccafisso, and cabbage soups are popular.

Italians have basic stocks to flavor fish, meats, chicken, or duck. Generally, garlic, onion, black peppercorn, bay leaf, fennel, thyme and parsley, tomato paste, carrots, and celery are used with chicken bones, fish bones, or beef bones.

The Middle East

The Middle East includes Israel and Arab regions (Lebanon, Jordon, Syria, Palestine, Gulf States, and Yemen). The Middle East has been influenced by Arab, Iranian, Indian, and European cultures and thus has varied ingredients. Generally, garlic, saffron, nuts, yogurt, sumac, fenugreek, parsley, and mint are popular flavorings in all of these regions. Some regions prefer olive oil, while others use clarified butter (*smen/samneh*) from sheep or goat's milk. Other popular ingredients include sesame oil, olive oil, garlic, fenugreek, cinnamon, paprika, cayenne, coriander, cumin, cloves, nutmeg, fresh coriander leaf, mint, dill, cassia, allspice, mahleb cherry, sesame, caraway, almonds, pine nuts, grape, and vine leaves. Yogurt is a staple flavoring in all of these regions. *Torshi* or pickles, *hilbeh*, pickled lemon, tahini, kebabs, pilafs, garlic dips, *kibbeh*, *kofta*, *cholent*, and lentil soups are also popular foods.

A popular seasoning that is essential in Middle Eastern cooking is a stock called *yahni*, which is made from onion, garlic, cardamom, *baharat*, and clarified butter with lamb or mutton bones. These spice mixtures generally come dried. *Zhug*, *zahtar*, and *baharat* spice blends are popular in most regions. Aromatic salt, which is sea salt combined with spices, is used as an essential flavoring in many Middle Eastern dishes.

Throughout these regions, there are ingredients and foods that have a mass appeal, but each region has its own specific flavor preferences. The southern Gulf States like hotter and curry-type flavors. The northern regions of Iraq and Syria enjoy spicy, sweet, and pungent spices such as pomegranate, Aleppo pepper, and cumin. The west (Israel and Lebanon) has sesame paste, parsley, and sour, milder flavors. The east (Iran) prefers aromatic leafy spices and almonds.

Some typical dishes are muhammara, a Syrian spicy hot dip that contains Aleppo pepper, walnuts, cumin, lemon juice, and pomegranate juice; *kofta* which are meatballs made from lamb, *baharat*, aromatic salt, mint, cilantro, and garlic; *hilbeh*, ground fenugreek seeds combined with garlic, olive oil, green chilies, tomatoes, and lemon juice used as a spread on breads before it is baked; and *talia*, a garlic, onion, and chile pepper seasoning used in Yemen.

Greece, Turkey, Armenia, Iran, and Cyprus

Greece, Turkey, Armenia, Iran, and Cyprus have quite distinct cultures, yet certain seasonings have mass appeal throughout these countries. Turkey, predominantly Muslim, uses garlic, Aleppo peppers, bay leaf, paprika, mahlab cherry, pomegranate juice or concentrate (*pekmez*), yogurt, walnuts, turmeric, and vinegared chile pepper as essential flavorings for their kebabs, dips, soups, fish, or meats.

Greece, influenced by Ottomon Turks and West Asians, uses oregano, garlic, olive oil, almonds, black olives, goat cheese, raisins, yogurt, and capers as base flavorings for its simple cooking. Other ingredients include onions, pickled green chilies, Aleppo pepper, Boukovo pepper, vine leaf, bay leaf, sesame, mahleb cherry, cinnamon, anise, clove, parsley, dill, and saffron. These flavorings are used in many dishes such as dolmas, tarator soup, *skordalia* sauce, *taramosalata*, *moussakas*, pilafs, or *trahanas*.

Iran, with Arab, Aryan, Greek, and Ottoman influences, combines sweet, savory and sour tastes using molasses, honey, lemon, pistachio, yogurt, almonds, parsley, dill, marjoram, and dried fruits (figs, prunes, dates, apricots). Typical dishes are *koresh* (a creamy, sweet-sour, lamb sauce with cloves, sumac, turmeric, cinnamon, dried fruits, smen, and pickled lemon), aromatic *pollou* (which contains cardamon, cumin, cassia, saffron, clarified ghee, and many leafy spices), and *mahi now rooz* (stuffed fish with spinach, smen, lime, nuts, raisins, cinnamon, mint, and garlic).

North Africa

North Africa includes Tunisia, Morocco, Algeria, northern Egypt, Libya, and Ethiopia. It has been influenced by many cultures through trade and conquest, including the French, Moors, Berbers, Arabs, Phoenicians, Sephardic Jews, Romans, Greeks, and the Far East.

Most North Africans are Muslims, so pork and alcoholic beverages are taboo. Lamb and mutton are preferred meats, and coffees, spiced teas, and fruit- and flower-based drinks are enjoyed.

Tunisian blends have spicy and hot profiles with turmeric, dried chilies mixture, harissa, and dried mint. Moroccan food is not as spicy and is famous for its couscous, lemons, paprika, and tagine dishes. Egyptians enjoy falafel, white fava beans, *zahtar*, *dukkah*, sesame paste, cumin, and coriander, while Ethiopians prefer sesame seeds, pippali, *berbere*, and *injera*. The Berbers contributed the slow-cooked tagine dishes to North African cuisine. Pickled foods called *cisbech* are very common Algerian appetizers or accompaniments to foods and use spices, such as cayenne pepper, paprika, cumin, coriander, or many herbs.

Grilled chickpeas, marinated olives, *kibbeh, ful medamis* (bean with mint, cumin, garlic, and onion), and *labneh* (fried cheese) from Egypt; *sohleb* (millet porridge), brochettes, and pickled sardines from Algeria; nit'ir qibe (spiced butter), *kifto*, and *doro w'ets* of Ethiopia; and *bistilla* and tagines from Morocco are commonly eaten foods in north Africa.

Popular flavorings include saffron, cayenne, mint, rose petals, turmeric, roasted spices, paprika, black pepper, cinnamon, cilantro, nutmeg, cloves, and sumac. Typical North African seasonings are *tabil, chermoula, la kama, harissa, dukkah, ras-el-hanout, semit,* and *berbere*. Chermoula, popular in Morocco, is a paste made from garlic, olive oil, parsley or coriander leaf, red chile peppers, paprika, saffron, and onion and is used for marinating fish, meat, or poultry. *La kama*, which consists of ground black pepper, ginger, turmeric, nutmeg, and cinnamon is also a popular flavoring in Morocco for stews, soups, and salads.

Some of the typical Mediterranean spice blends are discussed below.

Popular Mediterranean Spice Blends

Spanish Spice Blends

Romesco Sauce Blend

Description
Called salsa romesco in Catalonia, Spain, it is a popular table condiment, dip, or sauce used to perk up fish, pasta, vegetables, and chicken. Its name comes from the mildly hot, dried, red peppers (romesco) that are the main ingredients in this sauce blend. It has a rich, smoky, nutty, and spicy taste.

Ingredients
Romesco sauce is made from dried romesco pepper, onions, tomatoes, olive oil, garlic, red wine vinegar, roasted almonds, or hazelnuts. Other ingredients that can be added are black pepper, salt, toasted white bread, and seafood broth. All ingredients are roasted to give romesco sauce its deep rich flavor. If more heat is desired, dried, hot, red chile peppers are added. There are many versions of romesco from the Catalan region in Spain.

To prepare romesco sauce, tomatoes, garlic (with the skin on), and onions (no skin) are roasted in olive oil. Then, chile peppers are boiled with vinegar. Tomato and garlic are peeled and deseeded. The vinegared chile peppers are added to the onions and roasted until the onions brown. All ingredients with their juices are blended together until a smooth puree is obtained.

Application
Romesco sauce accompanies fish stews, other seafood dishes, salads, and boiled vegetables. It can be used as a spread for breads, a dip for vegetables, or added to zip up pasta sauces.

Sofrito/Soffritto Blend

Description
Sofrito is an aromatic mixture of spices with tomato, sweet bell peppers, and pimientos that are browned in olive oil and added to soups, sauces, or stews. This principal sauce characterizes the Catalan cooking of Spain. The Spanish also introduced sofrito to the Caribbean islands, the Philippines, Italy, and Egypt, which have other flavor versions. (Latin American versions are discussed earlier.) Sofrito comes from the Spanish word "sofrier," which means to lightly fry. Called soffritto in Italy, its taste varies from slightly sweet to spicy and peppery. Soffritto is combined with pancetta (cured ham seasoned with black pepper, nutmeg, cinnamon, cloves, and juniper berries), and garlic and onions. All are browned in olive oil and are used as a flavor base for seafoods, soups, vegetables, and fricassees.

Ingredients
A basic sofrito blend has tomatoes, onions, garlic, olive oil, pimiento, and saffron (Table 35). Other Mediterranean versions can be created with bell peppers, chile peppers, annatto oil, butter, cilantro, celery, carrots, oregano, thyme, parsley, lemon, and black pepper.

TABLE 35
Sofritos of the Mediterranean and Their
Characterizing Ingredients

Sofrito Blends	Ingredients
Basic sofrito	Garlic, onion, tomato, olive oil
Spanish sofrito	Sweet and mild with sweet bell peppers, onions, pimiento, olive oil, saffron
Italian soffritto	Mild and savory with onions, celery, carrots, parsley, pancetta

Applications

To make sofrito sauce, onions are gently fried in oil until they become light brown. Then, deseeded tomatoes and/or other ingredients are added and gently stirred until the sauce becomes somewhat thick. Sofrito is used to flavor stews, rices, soups, and beans, as a rub for chicken and fish, and as a condiment. It is used as the starting base for many dishes and adds zest to dishes such as the famous Catalan seafood stew called zarzuela de pescado, paellas, soups, or sauces.

French Spice Blends

Tapenade Blend

Description

Tapenade is a caper and olive paste from Provence, France. Its name is derived from the French word tapeno for capers. There are many versions of tapenade. Generally, it comes as a smooth and dark paste, with a salty, pungent, sharp taste.

Ingredients

Ingredients in tapenade include capers, garlic, black olives, ground black pepper, Dijon mustard, lemon juice, and dark rum. Other ingredients are anchovies, bay leaf, and other leafy spices. Tomato paste, sugar, or mashed cooked vegetables are added to tone down its bold flavor. To prepare tapenade, olives, capers, or anchovies are soaked to remove excess salt. They are then rinsed, drained, and blended with the rest of the ingredients until a smooth, thick sauce is obtained.

Application

Tapenade is used as an appetizer or spread on toasted bread and topped with leafy spices and cheese. It is also served with boiled potatoes and cold fish and can be added to perk up blander foods.

Herbs de Provence

Description

Herbs de Provence is a mixture of fresh or dried French and Italian leafy spices. It has its origins in southern France. The blend has sweet, floral, and slightly spicy notes. The Romans had a great influence in Provence, and, thus, the presence of Italian sweet basil and fennel seed.

Ingredients

Herbs de Provence has sweet spices such as savory, thyme, rosemary, fennel seed, oregano, and marjoram with lavender. Sweet basil and tarragon are the other spices added.

Application

Herbs de Provence is used as a rub to add a sweet, herblike flavor to roast chicken, beef, cornish hen, and pork tenderloins. They also perk up stews, baked vegetables, sauces, and salads.

Rouille Blend

Description

Rouille is a garlicky, spicy, mayonnaise-like condiment that is used to accompany fish-based sauces and stews in Provence, France. In French, rouille means "rusty," describing its reddish coloring. It is thick, smooth, and spicy.

Ingredients

Rouille's ingredients include garlic, roasted red bell pepper, dried hot red pepper, bread, olive oil, black pepper, and fish stock. Sometimes saffron, thyme, parsley, chervil or sage, paprika, cheese, and tomato paste are added. One version of rouille uses mayonnaise as a starting base, then adds spices and other ingredients.

Application

To create rouille, garlic, peppers, and bread pieces (soaked in water and squeezed dry) are blended with stock and water in a food processor or blender. The olive oil is slowly added until the mixture gets a thick consistency. Black pepper and salt are added, followed by the warm fish stock. The rouille mixture flavors fish stews and soups, especially the noted bouillabaisse. Rouille is also served in the French islands of the Caribbean, where it is spread over toasted bread, which is then topped with grated cheese and eaten with fish stew or soup.

Quatre Epices

Description

Quatre epices is a mixture of dried spices that is also called Parisian/Parisienne spice or epices fines. It is the most popular seasoning used in France, other parts of Europe and the French islands in the Caribbean. (In France, quatre epices is also the name for allspice.)

Ingredients

The basic ingredients in quatre epices are white or black peppercorns, nutmeg, ginger, cloves, mace, cinnamon, and allspice. Leafy spices such as bay leaf, sage, marjoram, and rosemary are also added. All of these spices are then blended and pounded with the leafy spices added at the end.

Application

Quatre epices is used to flavor French charcuterie, slow-cooked meats, and red cabbage dishes.

Bouquet Garni

Description

Bouquet garni is a mixture of leafy spices, especially the sweeter types. It flavors many European dishes, especially those requiring long simmering. Bouquet garni is a French term for a bundle of herbs. The types of herbs used and their proportions vary to complement particular dishes, such as poultry, fish, game meats, or lamb.

Ingredients

In a bouquet garni blend, the leafy spices that are added need to complement each other. The classic mix is parsley, thyme, and bay leaf. Crushed garlic is sometimes included. In Provencal cooking, dried orange or lemon peel is also used. Other spices are added to complement specific applications. Thyme, tarragon, parsley, rosemary, and bay are added for poultry; rosemary, mint, parsley, and thyme are added for lamb; thyme, basil, marjoram, and bay are added for pork; juniper berries, bay, and savory are added for game meats; and sage, lemon verbena, dill, and chives are added for fish.

Application

Fresh leafy spices are tied in bundles with a string, while the dried herbs are tied in a cheesecloth bag. These bundles are removed at the end of cooking or when sufficient flavor has been imparted but before bitterness is added to the application. Bouquet garni is used to flavor soups, sauces, stews, and meat dishes of Europe, especially France for *pot-au-feu*, bouillabaisse, and *daubes*.

Italian Spice Blends

Pesto Blends

Description

Pesto is a green blend of spices that is traditionally used in the cooking of Genoa, in Liguria (Italy). The traditional blend, *pesto alla Genovese* (or basil pesto), is sharp, pungent, and fresh tasting. The French version, called *pistou* (in the French Riviera, adjoining Liguria), is influenced by the Genovese pesto but is much milder in flavor.

Ingredients

There are many versions of pesto. The essential ingredient is sweet basil. Pesto alla Genovese contains sweet basil, garlic, olive oil, pine nuts, and cheese (pecorino or parmesan). Other versions add sun-dried tomatoes, parsley, or sweet red bell peppers. The more recent versions add walnuts, almonds, green chile peppers, mint, or cilantro.

Applications

In Italy, basil leaves are deveined and ground in a mortar and pestle with garlic, salt, and pine nuts. Olive oil is added drop by drop until the desired consistency is obtained. Finally, cheeses are ground and added. Nowadays, a food processor is used to make pesto. Pesto is typically served with gnocchi, tossed with pasta, or stirred into soups, such as minestrone. Today, pestos are used as dips, flavoring for rice, or as a topping sauce for chicken and seafood.

Pasta Sauce Blends

Description
Italian sauces give their foods character. Endless varieties of these sauces are used for flavoring bland pastas. Each region of Italy has its own characteristic sauce. The heavier tomato-based sauces are more popular in the south, while some creamy sauces are common around Rome and the north. Emerging pasta sauces contain fire-roasted garlic, sun-dried tomato, sweet and sour flavors, saffron, or Burgundy wine flavor. These sauces can be made into seasonings that can contain a tomato base, garlic, onion, vinegar, sugar, olive oil, and many spices such as basil, thyme, parsley, or sage.

Ingredients
A basic cooked pasta sauce blend has tomatoes, olive oil, garlic, black pepper, and basil. Sugar and vinegar are added for a sweeter or more acidic preference. Other sauces contain ingredients that are generally preferred on a regional level, such as cayenne pepper, capers, anchovies, mushrooms, cream, meats, clams, squid, spinach, carrots, peas, or pancetta.

Applications
These sauces are generally applied over differently shaped and sized pasta, such as spaghetti, penne, fusilli, orecchiete, tagliatelle, angel hair, fettucine, linguine, or pappardelle. They are added fresh or cooked. Most pasta sauces are cooked, but there are freshly made uncooked sauces that are most frequently eaten during the summer.

Generally, in making a pasta sauce, garlic is sauteed first in olive oil, then tomatoes are added, and finally, the other ingredients are added. Cheese is usually added before serving. The seasonings are simmered. These sauce blends can come as pastes or dried mixtures that are then added with vinegar, wine, or cream to give varying flavors for a finished bottled sauce.

Examples of Classic Pasta Sauce Blends

Aglio e oilo, the most basic pasta sauce blend, contains olive oil, garlic, and parmesan cheese. Other ingredients such as parsley and red pepper flakes are optional.

Bolognese sauce blend is a creamy meat sauce from Emilia-Romagna, containing ground beef, meat broth, sausage, tomatoes (for color and flavor), olive oil, butter, white wine, porcini mushroom, onion, garlic, parsley, sage, bay leaf, celery, carrots, and Parmigiano reggiano (parmesan cheese).

Puttanesca sauce blend is called the "style of the prostitute," because Italians say that the sauce is spicy, like the ladies of the night. Another says that these ladies attracted their clients with the sauce's enticing aromas. A third mentions that the sauce was quickly prepared as a fast food and eaten in between clients. It is an aromatic, pungent sauce that contains olive oil, tomatoes, garlic, black olives, anchovies, capers, parsley, fresh oregano, crushed red peppers, and black peppers.

Marinara sauce, popular in the south, is a sweet aromatic sauce that contains tomatoes, basil, oregano, garlic, black pepper, and olive oil. Olives and capers are optional ingredients.

Arrabbiata sauce is a very hot, spicy, deep reddish orange sauce, from the far south, that has tomatoes, olive oil, red cayenne pepper, pancetta, garlic, basil, and pecorino romano. It is usually tossed over penne.

Alfredo sauce is creamy with butter, heavy cream, Parmigiano reggiano, garlic, nutmeg, and black pepper.

Carbonara sauce is a mild and creamy sauce that contains pecorino romano (or Parmigiano reggiano), white wine, pancetta, olive oil, butter, eggs, parsley, and black pepper.

Vongole sauce contains clams, olive oil, white wine, garlic, parsley, and red pepper flakes. Tomatoes are added if red wine is used instead of white wine.

Middle East, Greece, Turkey, Armenia, Iran, and Cyprus Spice Blends

Za'atar/zahtar Blend

Description
Za'atar is a blend of spices that varies in flavor, whether it comes from Turkey, North Africa, Lebanon, Jordan or Syria. It is also the term for wild thyme in Arabic. In Morocco, it is referred to as a leafy spice that is a hybrid of oregano, thyme, hysop, and marjoram. Za'atar spice blend can be mildly tart or very tangy depending on the amount of sumac added.

Ingredients
Za'atar is usually a blend of ground thyme, marjoram, sumac, and roasted sesame seeds. Pistachios, hazelnuts, olive oil, mint, hyssop, or garlic are added to create varying flavors. Sesame seeds are dry-roasted for about one minute until aromatic and are then added to the rest of the ingredients and blended well.

Application
Za'atar is used as a seasoning for dips, to flavor breads and salads, and is eaten with cheese. In Syria, it is mixed with olive oil and is used as a dip for pita bread. Za'atar is also used as a dressing in salads with other leafy spices such as parsley and mint. It goes well with lemon juice, garlic, black pepper, tomatoes, and olive oil.

Zhug/Zh oug/S'µg

Description
Zhug is a spicy, aromatic, rich, green blend of spices used in Middle Eastern cuisines. It can be called the "pesto" of the Middle East. Its flavor varies from country to country. Zhug originated in Yemen where it has a hotter flavor profile than other regions.

Ingredients

Ingredients include coriander leaf, green chilies, ground cumin, cardamom, and garlic. Jalapenos or serranos, mint, lemon, or vinegar can be added to give varying flavor profiles. The ground spices and garlic are fried in oil until they are aromatic and are then cooled, combined with other ingredients, and blended into a paste.

Application

Zhug is a table condiment and is used to add zest to meats, vegetables, breads, and salads or is used as a base sauce in stews and sauces. Zhug can be tossed over salads, used as a rub for barbecues, and a spread or filling for sandwiches, bagels, pitas, or dumplings, or as a sauce over pasta or noodles.

Tahini Blend

Description

Tahini is a base blend or paste that adds zest to many cold and hot dishes of the Middle East, Turkey, and Northern Africa, such as hummus, *baba ghanoush*, salad dressings, creamy sauces, soups, cooked meats, and fish. Tahini is an oily, shiny paste with a very strong, roasted sesame flavor. It provides a strong, nutty flavor to soups, salads, and sauces.

Ingredients

Tahini is made from sesame seeds, sesame oil, garlic, and lemon juice. Depending on regional preferences, ground chile pepper, paprika, cumin or clove, chopped parsley, mint, almonds or walnuts, yogurt, or olive or hazelnut oil can be added with tahini. Tahini blends well with yogurt, olives, tomatoes, nuts, and pomegranate.

In preparing tahini, sesame seeds are lightly roasted, then ground and mixed with the sesame or olive oil and other ingredients to make a smooth paste. Boiled chickpeas with water and the tahini paste are pureed in a food processor or blender to achieve a smooth or coarsely textured puree.

Application

Hummus is made from pureed chickpeas (garbanzo) and tahini paste, while baba ghanoush is made from pureed roasted eggplant and tahini. They are favorites throughout the Middle Eastern regions, from Israel to Syria to Egypt. Hummus is a very smooth to coarsely textured, thin or very thick puree. It is creamy, with a slightly sweet, nutty, and roasted sesame taste. Depending on the type of sesame seeds used (white or brown) or how long seeds are roasted, hummus has different colors and flavor strengths. Hummus and *baba ghanoush* are typically eaten as dips with pita bread, as spreads, or as appetizers. They accompany shish kebabs, vegetables, or fish. They are served chilled or at room temperature, usually with a little paprika oil and chopped parsley.

North African Spice Blends

Harissa Blend

Description

Harissa is a fiery chile-based sauce that resembles sambals and is a fiery North African version of ketchup. It accompanies appetizers and many other dishes in North Africa. It has a garlicky, slightly sweet, fresh, chile flavor. Tomato and red bell pepper tone the heat down. There are many different versions of harissas, depending on the region, Algeria, Tunisia, or Morocco. Its other versions are called *felfel sudani* in Morocco and dersa in Algeria. The Tunisian version is the hottest of all three. It is often sold as pastes.

Ingredients

Harissa has a deep red color with a smooth, thick consistency. It has dried chile peppers, tomatoes, garlic, ground cumin, ground coriander, vinegar, and olive oil. Hotter versions omit the red bell peppers and increase the chile peppers. Other added ingredients are crushed mint leaves, caraway, lemon juice, cilantro, or olives. For more intense flavor, bell peppers and chile peppers are roasted.

Preparation

Dried chile peppers are soaked in water until soft, and are then blended in a food processor or blender with all of the other ingredients to achieve a smooth texture. Enough olive oil is added to form a smooth, thick paste. Traditionally, a mortar and pestle are used to grind all of the ingredients into a smooth puree.

Applications

As a table condiment, harissa enhances the flavor of entrees and side dishes. In Tunisia, it accompanies meat and fish and adds zest to salads and soups. It is scooped over couscous, rice, or noodle dishes. Harissa goes well with grilled meats and is used as a rub for barbecues, as a topping for hot dogs, or to add zest to stews, soups, and sauces. In the Middle East, harissa accompanies mezzeh (called *aadou* in Tunisia and kemia in Algeria) or finger foods and dips for breads. In Tunisia, harissa is placed over a variety of foods and is even used with bread and butter for breakfast.

Ras-El-Hanout

Description

Ras-el-hanout is a Moroccan classic mixture of twenty or more spices. The word in Moroccan means "top of the shop." The shop owner mixes and creates a blend that is perceived as his or her ultimate creation. These vary in ingredients depending on the shopkeeper. All of these blends contain spices, dried flowers, and some aphrodisiacs (cantharides).

Ingredients

A typical blend includes cardamom, cinnamon, cumin, fenugreek, lovage, mace, cubeb, long pepper, allspice, nutmeg, ginger, rose petals, lavender blossoms, orange blossoms, Spanish fly, dung beetle, belladonna berries, monks pepper, grains of paradise, berries, black pepper, chile pepper, turmeric, nigella, ash rhizome, orris root, lesser galangal, chufa nuts, thyme, and rosemary. Whole spices are generally mixed with the rose petals and leafy spices.

Application

It is used to enhance many dishes in Morocco and other neighboring regions. Ras-el-hanout is used on game meats, rices, couscous, lamb tagines, and desserts.

CARIBBEAN SPICE BLENDS

Caribbean cooking is a beautiful collage of tropical ingredients with varied cultural influences. It blends indigenous ingredients grown by the Caribs and Arawak Indians with ones imported by the different peoples who settled there, including Africans, Asian Indians, Arabs, Chinese, Spanish, French, Dutch, and British. This cultural diversity has resulted in a popular fusion cooking with a wide range of tastes, textures, and colors that differ from island to island. Some Caribbean products that are becoming more familiar in the United States are jerk seasoning, hot pepper sauces (Solomon Gundy, Jonkanoo hot pepper sauce, coconut run down sauce), and curry *colombos* (also called *poudre de colombo*).

REGIONAL CUISINES OF THE CARIBBEAN

Because of the proximity of the Caribbean islands, they draw ingredients from spice, fruit, and vegetable sources common to all of the islands, with certain ingredients and seasonings having mass appeal. At the same time, distinct taste preferences reign on each island. What makes each island's flavors distinct is the way food is seasoned and prepared. Cassava, habaneros, and rice and beans are widely eaten in all of the regions, but diverse preparation techniques create variations on the theme. For example, rice and beans vary in flavor from Puerto Rico to Guadeloupe to Jamaica to Trinidad, depending on the type of beans, seasonings, and added ingredients.

Through the years, authentic profiles have been modified to suit changing tastes. The Africans have influenced the cooking with groundnuts, cornmeal, and oxtail; the Asian Indians have influenced it with curries, rotis, and pigeon peas; and the Amerindians have influenced it with habaneros, cassava, and annatto. Also, each island's flavors reflect the tastes of its European colonizers. For example, on the French-speaking islands of Guadeloupe and Martinique, bouquet garni, chives, and cream-based sauces predominate; in the Spanish-speaking islands of Puerto Rico, the Dominican Republic, and Cuba, cilantro, olives, and garlic are commonly used; while in the British-speaking islands of Jamaica, Trinidad, and Barbados, curries, chutneys, and jerk foods are enjoyed; and in the Dutch islands of Curacao and St. Maarten, nasi goreng, satays, meat with Edam cheese fillings, and ristaffels are popular.

The eastern Caribbean—Trinidad, Barbados, and the French islands—generally eat hotter foods than the western Caribbean—Puerto Rico, Cuba, and the Dominican Republic—except for Jamaica. Appearance is an important factor in Caribbean cooking. Colorful, sculptured fruits, chilies, flowers, and vegetables sit on a backdrop of rice and beans, grilled chicken, fish, and lobster and add visual pizazz to our plates.

Caribbean cooking is based on hot, fruity sauces, coconutty flavors, seasoned rices, and marinated meats. Fruits, vegetables, seafoods, and chile peppers are in abundance throughout the year and are easily available as fresh ingredients that add flavor, texture, and color to foods.

Soups and stews such as *sopito*, conch chowder, *bebele*, pepperpot, *quimbomba*, *caldo gallego*, *sancocho*, and pumpkin soup are staples of Caribbean diets. They are often made with meats and seafood (including leftovers) and are spiked with spices and hot peppers. Tubers or vegetables lend bulk, frequently making the soup or stew into a one-pot meal that is typically eaten with rice or breads.

Dasheen, pumpkin, breadfruit, yuca, plantains, callaloo, okra, and chayote are added to create body and textures in stews and soups. Chayote, cho cho, or christophene, a mild-flavored vegetable found in most regions, can be stuffed, steamed, sauced, or added to soups and stews. Okra, also called gumbo or *quimbombo*, is traditionally used in dishes of African or Amerindian influence. Breadfruit has a starchy pulp that is boiled, grilled, or roasted and added to soups and stews. Beans, such as pigeon peas (also called gandules or gungo peas), are curried and used in fillings for rotis or pastries, added to rice and peas, or used in stews and soups. Tubers play a prominent role in Caribbean cooking, such as yuca (also called manioc or cassava), boniato, sweet potatoes, and yams that are added to provide bulk and texture to soups and stews.

There is an abundance of fruits in the Caribbean. Tropical fruits such as mango, papaya, guava, soursop, citrus, cherimoya, passion fruit, tamarind, pineapple, and *ackee* flavor nectars, preserves, cordials, relishes, hot sauces, and stews. *Ackee*, which is often served with saltfish for breakfast in Jamaica, has the texture of scrambled eggs. Papaya, called paw paw, is used in stews, soups, and salads when green. Coconut is a versatile ingredient in these islands' cooking. It is made into snacks, seasoned with allspice, clove, nutmeg, and cinnamon, turned into a refreshing beverage, or used as cream in curries, soups, stews, breads, and desserts. Plantains (sometimes called platano macho), which are larger than their banana cousins, are popularly eaten as a side dish with rice and beans. When ripe and sweet, they are thinly sliced and fried, or when green and mealy, cut thin for tostones. When ground, they thicken sauces and are used in breads and cakes. Plantain leaves are used to wrap meats, rice, cassava, and desserts before cooking or serving.

Seafood, whether pickled, salted, sauced, or fried is a common part of the diet in the Caribbean. Each island offers its own seafood specialty, curried, grilled, stewed, fried, or served with hot condiments, such as conch fritters, fried flying fish, feroce de morue (salt cod with Scotch bonnet and lime juice), or blaff du poisson (with bouquet garni, white wine, habaneros, and allspice). Pork is a popular meat, fried, stewed, baked, or processed into a sausage called boudin. Creole pork chops, Jug jug (a Barbadian dish with pigeon peas, thyme, and habaneros), and a Caribbean take on Scottish haggis are popular dishes. Goat, especially curried, is frequently eaten in Trinidad and Jamaica. Beef is popular in the Spanish-speaking islands and is added to *ropa vieja*, *beef al ajillo*, and *picadillos*. Chicken is made into *poulet colombo*, jerk chicken, calypso chicken, lemon chicken, or Bajan chicken.

Cornmeal is made into many products such as *funchi* (dumplings with spices), *coucou* (with okra and spices) and tutu with black-eyed peas, or is wrapped in banana leaves and boiled. Rice is popular on all of the islands, but flavors vary depending on the island. It is cooked with onions, salt cod, ginger, soy sauce, sofrito, adobo, curry seasoning, beans, or peas to make Cuban Moros y Christianos (rice and black beans), Jamaican peas and rice, *nasi goreng*, Chinese fried rice, or paella.

The spices commonly used in Caribbean cooking are allspice, thyme, black pepper, clove, nutmeg, cinnamon sticks, ground mustard, celery, turmeric, clove, and garlic. Bay leaf, cilantro, recao leaf, lemongrass, chive, and a variety of other fresh leafy spices season foods and local drinks. Seasonings such as adobo, sofrito, mojo, curry, rouille, and jerk create magic with fish, fried chicken, pork chops, stews, soups, or sauces.

Indian curry powders and garam masalas found their way to most Caribbean islands. They are available in powder and paste forms and are mostly consumed in Trinidad, Jamaica, and other islands where there is a greater Asian Indian population. Their flavors are generally mild and, depending on the islands' preference, come with added green mangoes, hot peppers, vinegar, vegetables, or coconut milk. Curry seasonings often flavor dishes made with goat, lamb, fish, shrimp, conch, pork, or chicken, and they are served with white rice or as fillings in rotis, the flat breads of the Caribbean. Curries called colombos in the French islands of Martinique and Guadeloupe are heavily laden with turmeric, ground mustard, and vinegar that add a sweet and slightly sour taste.

Jerk barbecue, a classic Caribbean item that has captured the U.S. market, is from Jamaica. Escovitch (or escabeche), a tart, hot, and spicy marinade for fish, is derived from the Spanish and is an integral part of Puerto Rican and Jamaican cooking. Another popular sauce is Creole sauce, which comes from the French-speaking islands but is also popular elsewhere. It is added to steaks, hamburgers, grilled chicken, and fried fish.

Hot pepper sauces are integral to Caribbean cooking. The first hot sauce was created by the Arawak and Carib Indians, the original settlers of these islands. Today, many varieties are becoming better known, with tangy, fruity, sweet, sour, and fiery notes enhancing barbecued steaks, grilled fish, chicken wings, stews, curries, soups, and vegetables. Each island has its own preference for its hot sauce, using habanero peppers, bird peppers, cachuchas, rocotillos, or pimientos as the principal ingredients. Hot sauces such as salsa *ajillo*, sauce *chien*, sauce Creole, rouille, mango chutneys, red pepper remoulade, sauce *ti-malice*, sauce picante, or *ajilimojili* are some of the local popular sauces. Pickled and chopped Scotch bonnets with sliced onions and spices are important table condiments.

POPULAR CARIBBEAN SPICE BLENDS

Jerk Seasoning

Description

Jerk seasoning is a hot, spicy marinade for pork, chicken, and fish and is part of a Jamaican tradition of marinating and slowly smoking seasoned meat, pork, or fish over a fire pit of pimiento wood. The technique of jerking came from the Maroons, the runaway Jamaican slaves who escaped the British in the 1600s. They hunted wild boar and pigs, which they preserved and cooked in deep pits lined with pimiento wood and leaves. This smoking was a way of preserving the meat. Jerking the meat from side to side was a method of texturizing it.

Ingredients

Jerk marinade consist of allspice, thyme, nutmeg, cloves, scallions, and Scotch bonnet peppers.

Applications

Jerk seasoning is used as a marinade for meats, chicken, or seafood or as a sauce to perk up boiled or grilled dishes. It is sold as a dry mix or a wet paste.

Escovitch/Escabeche Blend

Description

Escabeche is the Spanish word for a pickled product and has its origins in the Spanish-speaking islands. It is commonly used in the Spanish-speaking islands of Puerto Rico, Dominican Republic, and Cuba and is also an integral part of Jamaican cooking. Escovitch is the Jamaican version of the blend that is tart, hot, and spicy. This seasoning is generally used as a marinade for fish, chicken, and pork.

Ingredients

Escabeche consists of pimientos, black pepper, onions, garlic, vinegar, and Scotch bonnet peppers.

Applications

Fish, chicken, pork, and other seafood or game meat are marinated with escovitch and are also simmered in this marinade. Sometimes, the whole fish is deep-fried before it is simmered in the marinade. It also seasons tamale fillings and vegetables.

Creole/Criollo Sauce Blend

Description

Another popular Caribbean sauce is Creole or Criollo sauce, which is a cooked tomato-based sauce. It is popular in the English-, French- and Spanish-speaking Caribbean Islands and has flavor contributions from many ethnic groups, including Africans, Spanish, Chinese, French, and North American Creoles. Where the islands have Spanish origins, this sauce is called Criollo sauce, and in the French-speaking islands it is called Creole sauce. It has a spicy and slightly sour and sweet flavor. It is a spicy version of the Chinese sweet and sour sauce. Its heat varies with different regions.

Ingredients

Creole or Criollo sauce contains onions, tomatoes, hot pepper, vinegar, and black pepper. Other ingredients such as bell peppers, tomato puree, ketchup, Worcestershire sauce, white wine, chicken stock, scallions, celery, and lime juice are added, depending on the island preferences.

Applications

To make the sauce, onions are sliced and sauteed, then sliced tomatoes, hot peppers, and spices are added and cooked until the sauce becomes slightly thick. Then vinegar is added and the whole mixture is simmered. The finished sauce is spooned over steaks, grilled chicken, pork, lobsters, crayfish, hamburgers, and fried fish or is served as an accompanying condiment.

Hot Sauce Blends

Description

The Caribbean has a reputation for its tangy, fruity, sweet, sour, and fiery hot sauces. Hot sauce to a Carib is like ketchup to a North American or sambal to a Southeast Asian. Hot peppers have been relished in the Caribbean throughout history and come in many varieties. Hot sauce can be aromatic with hints of apricot, peach, leafy spices, and citrus or pungent with garlic and mustard.

The Carib and Arawak Indians, the original settlers of these islands, used hot pepper juice with cassava and spices to season their foods and developed the first Caribbean hot sauce.

Ingredients

The popular hot peppers in the Caribbean are the Scotch bonnets, bird peppers, rocotillos, pimientos, and cachuchas. Scotch bonnets (also called Congo peppers, Bahama mama, Bonney peppers, Jamaican hots, and pimient) form the principal ingredients of the hot sauces. They come in many colors and sizes, are aromatic, and have hints of apricot, peach, and citrus. Other ingredients typically used in the Caribbean hot sauces are mango, papaya, vinegar, chives, or tomatoes with allspice, mustard, turmeric, or black pepper.

Each island has developed its own preference for hot sauce. Barbados has mustard-based sauces, Guadeloupe makes a sauce with chive and garlic, while Trinidad is well known for its Scotch bonnet and papaya-based sauces.

Pickapeppa sauce, Matouk's calypso hot sauce, West Indian hot sauce, flambeau sauce, and Busha Browne's hot sauce, each claiming to be hotter and tastier than the other, are popular condiments in the United States. Pickapeppa is a hot steak sauce with tomatoes, onions, sugar, vinegar, peppers, mangoes, and spices. It is a spicier version of A-1 sauce used in the United States.

Applications

Caribbean hot sauces can enhance barbecued steaks, grilled fish, buffalo wings, stews, soups, pastas, and vegetables. Pickled, chopped, chile peppers, sliced onions, and spices are important local table condiments. *Pickapeppa* sauce is used for basting, grilling, sauteing, and stewing meats, seafood, and vegetables. There are other numerous varieties of hot sauces on the market that include West Indian hot sauce, Matouk's, Solomon Gundy, Jonkanoo hot pepper sauce, and calypso sauce, each one claiming to be hotter than the other.

Curry Blends/Colombo Blends/Garam Masala

Description

Asian Indians have had a significant influence in the Caribbean, especially in Trinidad, Jamaica, Guadeloupe, and Martinique. When they emigrated to the Caribbean, they brought with them the curry blends of India, which were then modified using locally available ingredients. Garam masalas in the Caribbean are mild, do not usually contain turmeric or chili powder, and are added with curry blends to curries. Colombos or poudre de colombo are fiery aromatic local curries in the French islands of Martinique and Guadeloupe.

Curries can be mild or hot and spicy, depending on the islands' preference. They include spices familiar to the islands, such as allspice, thyme, black pepper, onions, mangril, ginger, Scotch bonnets, green mangoes, hot peppers, vinegar, vegetables, or coconut milk. A general curry powder contains cumin, coriander, turmeric, fenugreek, celery seed, and fennel seed.

Poudre de colombos are heavily laden with mustard, turmeric, and vinegar, yielding a sweet and slightly sour taste. They are generally sold in powder or paste forms.

Ingredients

A typical Caribbean curry blend contains coriander seeds, cumin seeds, poppy seeds, cloves, brown mustard seeds, anise seed, fenugreek, black peppercorns, ground turmeric, and ground ginger. Scotch bonnets or Congo peppers are added for heat. Generally, the masalas do not have ground chilies. They are added to the curries toward the end of cooking or are served as a condiment. Spices can be in whole or ground form. Colombos have mustard seeds, toasted rice, hot peppers, coriander, turmeric or saffron, black pepper, garlic, onions, tamarind, and vinegar. White wine, stock, coconut milk, eggplant, chayote, papaya, pumpkin, mangoes, pumpkins, okra, green tomatoes, taro, yam, and lime juice are optional ingredients. Coconut milk is not added to fish or shellfish *colombos*.

Most whole spices are roasted and then ground. Chicken pieces are seasoned with curry blend and browned. Then water, onions, tomatoes, and other vegetables are added and simmered until they are cooked. Caribbeans usually add vegetables to curries.

Most Trinidadian curries are mild and contain ghee, turmeric, and mangril but usually lack chile peppers. Those who want heat in their curries use hot sauces. In Jamaica, where curry goat is a specialty, curries are generally very hot and contain a good amount of black pepper, Scotch bonnets, and allspice. *Roucou* (annatto oil), cilantro, and vinegar mustard flour are some of the other added ingredients.

Applications

Caribbean islands are noted for their goat curries. Other ingredients cooked with curry are conch (lambi), fish, shellfish, chicken, lamb, pork, shrimp, beans, and many vegetables. These dishes accompany boiled white rice or are used as fillings for the famous Caribbean flat breads, called "rotis," that are popular street foods. Curry blends are also added to barbecues, soups, and stews. Curries are accompanied by mango chutney, grated coconut, salad, or fried plantains.

AFRICAN SPICE BLENDS

Many groups, including Arabs, Europeans, Asian Indians, and other Asians, through wars, colonization, slavery, and exploration have influenced the indigenous cuisines of Africa. For purposes of discussion, this section has divided this vast land into four geographic sectors: North, East, West, and South Africa, each with foods that have quite distinct flavors. Curry blends, *egusi* seeds, bitter leaf, pumpkin leaf, cassava, beans, okra, fenugreek, mace, cloves, groundnuts, rose petals, grains of paradise, cornmeal, cassava, palm oil, and waterblommetjie are some typical ingredients used in African cuisine. Starch-based meals and stews are the staples of most regions of Africa.

REGIONAL CUISINES OF AFRICA

North Africa

North African cuisine is influenced by the Berbers, Coptics, Greeks, Arabs, Jews, southern Europeans, and Middle Easterners. This book classifies North African cooking under Mediterranean. Because North African cuisine follows a Mediterranean-style diet, olives, spicy sauces, fish, leafy spices, legumes, and couscous are used abundantly. Tagines, stews, breads, couscous, and condiments characterize the cuisine.

West Africa

West African cuisine is influenced by Africans, English, Arabs, and French. Starch-based meals made from rice, breads, cassava (*gari*), potato, yams, millet, and plantains predominate. Ground cornmeal (*foofoo*), grated roasted cassava porridge (*gari*), roasted corn flour (*ablemanu*), or moinmoin are served with spicy stews and sauces. Stews form the foundation of the meal. They are mostly made from beans and root or leafy vegetables, with little meat, poultry, or fish. They are flavored with peanuts, red cayenne, peanut or palm oil, black pepper, or cubebs. Except in South Africa, meats are used more as flavorings. Soups are thickened with okra and nuts and are also eaten with a starchy base. Sauces containing hot peppers, groundnuts, tomatoes, or seafood are served with boiled roots, plantains, and vegetables.

East Africa

East African cuisine is influenced by Africans, Asian Indians, Middle Easterners, Persians, Arabs, Dutch, English, and Portuguese. The Indian influence is predominant in the local cuisine and comes mainly from Ismaili Muslims and Gujeratis, who are mostly vegetarians. Curries, kababs, pilaus, samosas, flat breads, stews, mashed root vegetables, and rice are essential foods. Typical ingredients are lentils, ghee, mint, coconut milk, masalas, and hot peppers. *Ugali* (cornmeal porridge), okra stew, groundnut stew, *matoke*, or yams are typical foods of the indigenous population.

South Africa

South African cuisine is influenced by Europeans, Africans, Arabs, Indonesians, Malays, and Asian Indians. A popular style of cooking is Cape Malay cuisine, a melting pot of flavors from Malays, Dutch, French Huguenots, indigenous Khoisan, Asian Indians, and Portuguese. Cape Malay dishes include *bobotie* (sweet curried meat casserole), *sosaties* (sweet and pungent grilled lamb marinated with curry powder and coconut milk), *bredies* (vegetable/legume stews), *beeriani* (rice-lamb dish flavored with ginger, tomato, fennel seed, saffron, cumin, and onion), and *atjars* (spicy, sweet, pickled condiments). Curried dishes are served with chutneys and sambals. Cornmeal is made into *pap* or *putu* and served with seasoned stews. Common flavorings are curry powders, garam masalas, clove, allspice, black pepper, coriander, tomatoes, chile peppers, and bay leaf.

POPULAR AFRICAN SPICE BLENDS

Curry Blends

Description

Curries in Africa are influenced by Muslim Indians, Arabs, Europeans, Malays, and vegetarians from India. Different regions have their own curry blends influenced by different cultural groups. They can be hot, medium, or mild and dry or soupy. Some of the different types of curry blends include *frikkadel* in South Africa, *kimasin* in East Africa, and *kefta tagine* in North Africa. South African curries, influenced by Malays and Indians, are hot, spicy, and sweet with dried fruits and apples. *Berbere*, Cape masalas, *qalat dagga*, *ras-el hanout*, and Malawi curry powder are some examples of African curry blends.

Ingredients

There are many types of curry blends, especially in East and South Africa where they are popular. Usually, these curry blends contain cumin, coriander, clove, cinnamon, turmeric, ginger, fennel, cardamom, mustard, black peppercorn, and fenugreek, with or without ground chile pepper. East African curries, such as *kima*, have chopped beef, red chile pepper, curry spices, eggplant, and coconut. Mozambique's curries use piri-piri peppers and nuts to provide flavor and consistency. Ethiopians have their own versions of curries called *w'ets* with ajowan, pippali, mint, sweet basil, paprika, and tomato. In West Africa, groundnuts are added to give some crunchiness to their curries, and in South Africa, dried fruits, ginger, and bird chilies flavor kormas, *murghi kalya*, and *atjars*.

Applications

In Africa, curries accompany cooked rice, couscous, or breads. They are used with lamb, poultry, fish, casseroles, vegetables, pastries, samosas, soups, stews, or chutneys.

Berbere

Description

Berbere is a spicy, classic blend that resembles the curry blends of India. It flavors many Ethiopian and Eritrean mutton and lamb dishes, especially traditional stews called wats or *w'ets*. It is pungent and aromatic, with slightly sweet and peppery to very peppery notes.

Ingredients

Berbere has ajowan, cardamom, black cumin, fenugreek, cardamom, cinnamon stick, cloves, ground ginger, garlic, onion, and ground, dried chile powder. Other ingredients such as black pepper, long pepper, coriander leaf, tomato, mint, or olive oil can be added. Basil, paprika, mustard, turmeric, nutmeg, and allspice go well with it.

Applications

To make a berbere paste, the spices are first dry-roasted for a few minutes until they become aromatic. Then, chile powder is added, and the mixture is dry-roasted again for about a minute. Garlic and onions are blended with the other ingredients until smooth.

Berbere is an indispensable seasoning of Ethiopian dishes, especially the currylike, stewed dishes called *w'ets*, which are the mainstay of their meals. Berbere is used to flavor lamb stews, chicken stew (*doro wat*), and vegetable or lentil stews (*shiro wat*). Muslims eat *shiro wat* during their fasting month of Ramadan.

Bebere primarily enhances fish and lentils. They are eaten with local sourdough flat bread called *injera*, which is made from a grain called teff. Berbere is also used in a condiment that accompanies grilled meats or poultry and is also added to soups. Berbere pairs well with salted fish, tomatoes, rice, potatoes, yams, butter, lemon juice, and peanuts.

Piri-Piri/Pili-Pili/Peri-Peri/Peli-Peli

Description

Piri-piri is a slightly sweet, hot mixture that characterizes Mozambique's cuisines. There are several versions of it, depending on whether it is made in West Africa, Angola in South Africa, or Mozambique. Piri-piri is also the name of a small, red, fiery capsicum pepper, which resembles bird pepper. Today, piri-piri refers to any hot spicy blend made from fresh or dried, ground piripiri pepper. It is used as a wet marinade, a dry rub, a table condiment, or a hot dipping sauce.

Ingredients

Piri-piri is made from garlic, onions, ground red piri-piri peppers, hot paprika, bay leaf, parsley, black pepper, olive oil, and lemon juice. Other versions have added cardamom, basil, ginger, butter, and coconut milk. It pairs well with peanuts and shrimp paste. The West African version has tomatoes, onion, garlic, habanero pepper, and prepared or grated horseradish.

Applications

Piri-piri is used as a marinade or rub for cooked, broiled, charcoal-grilled, or barbecued chicken, fish, prawns, and shrimp. To make a piri-piri marinade or rub, simply blend all of the ingredients. Piri-piri is also used to add extra flavor at the meal table. It gives added zest to stews, vegetables, and cooking sauces. For use as a table condiment, blend all ingredients and cook gently for about five minutes. Coconut milk can be substituted for lemon juice.

In West Africa, piri-piri is used as a condiment and cocktail sauce and as a flavoring for grilled meats or sandwiches. Mayonnaise flavored with piri-piri sauce is used for cold chicken salad and coleslaw.

Atjar Blends

Description

The South African term "atjar" came from the Indonesian Malay word "acar" that originated from the Indian word "achar." Achar translates to spicy pickled vegetables or unripe fruits in oil or sauce. It is crisply textured, and its flavor varies depending on its ingredients. Generally, it is spicy and sweet. The South African favorite is a lime atjar, which is a sweet, spicy type that contains olives, limes, mangoes, and chile peppers. It is an essential flavoring in the Cape Malay cuisine of South Africa. It is also used as a table condiment for the Indian-influenced cuisine of East Africa.

Ingredients

Atjar was originally made with fish oil and brine, but today olive oil and peanut or other vegetable oils are added with brine. A general atjar marinade has anchovy fillets, garlic, ground fenugreek, turmeric, chile peppers, curry powder, and vegetable oil. To prepare a lime atjar, limes, salt, mustard seeds, bay leaves, and vinegar are combined. Other variations come with green mango, onions, cauliflower, or green beans.

Application

Atjars are used to perk up entrees of fish, meats, or rices. Vegetables such as green beans, okra, carrots, cucumber, and brinjal can be atjared. They are served as accompaniments to rice, curries, fish, and vegetable or meat dishes.

Dukkah (Also Called Za'Tar Or Do'Ah)

Description

Dukkah is a blend of spices, roasted chickpeas, and nuts and is a staple flavoring in Egypt since ancient times. It has a spicy and nutty taste with a coarsely ground or crumbly texture. It usually accompanies breads on the mezze (or tapaslike) table. There are many versions of dukkah in the Middle East.

Ingredients

Dukkah is a dry mixture of coriander seeds, cumin seeds, lightly roasted sesame seeds, black peppercorns, mint leaves, unsalted roasted hazelnuts or peanuts, and dried, roasted chickpeas. Other ingredients added are thyme, savory, oregano, marjoram, or savory to give varying tastes.

Application

To make dukkah, whole spices are roasted, cooled, mixed with the remaining ingredients and pounded in a mortar and pestle or blended in a food processor. It is eaten for breakfast with breads, as an appetizer, or on the streets as a snack. It is sprinkled over many main entrees to add texture and flavor. In Egypt, a bread called aish is dipped into olive oil and then into this blend before it is consumed.

West African Hot Pepper Blends

Description

West African hot pepper blends are a fiery mix of hot peppers with other ingredients and are used commonly to add heat or spike West African fish, meat, or vegetable dishes. They come as dry mixtures or as pastes.

Ingredients

A common hot sauce has chile peppers, onions, palm or peanut oil, and dried shrimp. There are other variations based on applications and preference. A popular dry, hot blend is made from black peppercorns, white peppercorns, green peppercorns, pink peppercorns, allspice berries, cubebs, ground ginger, grains of paradise, red cayenne pepper flakes, and ground dried habaneros. *Ata*, a Nigerian hot sauce, is made from onions, red chile peppers, red bell peppers, tomatoes, dried shrimp, and palm oil.

Applications

Hot pepper blends are used as rubs on meats or chicken before they are grilled. They are also used as a flavoring base for stews, soups, and sauces and as condiments for boiled, stewed, or fried root vegetables.

NORTH AMERICAN SPICE BLENDS

Prior to the arrival of Spanish, French, and English explorers, Native Americans grew chile peppers, sassafras, or allspice to season their wild rice, corn, venison, squash, beans, buffalo meat, and other foods. Early European immigrants brought cinnamon, nutmeg, clove, mustard, and molasses to the Americas. The African slaves brought groundnuts, benne (sesame seeds), cowpeas, and many West African ingredients. All of these influences created traditional American cooking. Poultry, old world, barbecue-style, Chesapeake Bay, pumpkin spice, and Old Bay seasonings are some traditional seasonings used for shrimp and crab boils, roast beef, steak, pork chops, pumpkin pie, and pickled vegetables. The later immigrants from northern, central, and southern Europe and East Asia contributed their own seasonings and foods. The Jews, Germans, Italians, Hungarians, Japanese, Chinese, and Koreans gave olive oil, sesame, sweet basil, ginger, soy sauce, garlic, pizzas, pastas, bagels, coleslaw, dumplings, cheese, chow mein, and fried rice. The latest migration of Latin Americans, eastern Europeans, South Asians, Southeast Asians, Caribbeans, and Middle Easterners have brought cilantro, lemongrass, curry powders, sumac, habaneros, lentils, adobos, and tropical fruits.

While North American cooking combined many different flavors, no uniform cooking style arose. Rather, many regional cuisines emerged that reflected the tastes of the local populations. In the Midwest, Scandinavians, Germans, and Eastern Europeans contributed dairy and baked products, hearty soups, sauerkraut, frankfurters, schnitzel, sausages, bratwurst, and cheeses. The Southern states had the most diverse flavors derived from Africans, Europeans, Native Americans, and Mexicans, such as gravy, chili con carne, barbecued ribs, gumbos, macaroni and cheese, and nachos.

The cuisine of the Pacific West has been influenced by many ethnic groups—Chinese, Koreans, Vietnamese, Mexicans, Spanish, Japanese, Hawaiians, and Native Americans. All of these influences are noted in the ingredients and preparation techniques found in this modern cuisine—stir-fries, olives, sushis, grills, pizzas, avocados, pork, figs, chile peppers, leafy greens, wine, taro, a variety of fruits, seafood, dumplings, and noodle soups.

Today, Cajun, Creole, Southern, Southwest, and other regional seasoning blends of North America are evolving into new, popular, trendy flavorings. Overseas, these blends are characterizing North America to the world.

POPULAR REGIONAL CUISINES OF NORTH AMERICA

Southern

Southern cooking is being revived as an important flavoring style today because many North Americans are going back to the basics. They are looking for foods they grew up with, the so-called comfort foods.

The South combines the home cooking of West Africans and Native Americans and the peasant cooking of southern Europe with influences from English and Spanish explorers. Of all of these various influences on Southern cooking, one of the greatest was from the West Africans who were brought here as slaves. Both before and after the Civil War, African-Americans contributed their unique soul food flavors that remain the heart of Southern cooking. Soul food has also influenced other regional cuisines as African-American cooks migrated from the South.

Economics have also influenced Southern cooking. Through wars and depressions and poverty, Southern cooks improvised by using less expensive cuts of meat, such as fatback, to season beans, peas, and vegetables. Pigs and chickens were less expensive to raise than beef, and foods were fried because lard was cheap and provided flavor.

Southern foods tend to be heavy and rich and generally not too spicy. Spices such as black pepper, cayenne pepper, onions, and vinegar-based sauces are typically used with cream, sugar, lard, salt, and butter. Tabasco sauce or other vinegar-based condiments provide heat. Mashed potatoes, macaroni and cheese, chicken-fried steak, roast turkey, corn muffins, biscuits, pickled pigs feet and ears, corn kernels, collard greens, black-eyed peas, relish, succotash, and barbecued ribs are many of the typical comfort foods from the South.

There is tremendous variety in traditional Southern cuisine. Virginia and Georgia are known for their seafood, ribs, corn dogs, biscuits and gravy, country ham, catfish stew, smothered pork chops, and fried turkey. The Carolinas offer seafood, fried chicken, pork, and rice. Northern Florida gives us seafood, sweet potatoes, citrus fruits, and pecans. In Memphis, we enjoy its barbecues, breaded okra, and mashed potatoes. The South also includes Louisiana, which is discussed separately because of its distinct spicy flavors.

Louisiana

Many ethnic groups have influenced the cooking of Louisiana—Arcadian French, Spanish, Native Americans, Haitians, Africans, and Italians. Two major styles of Louisiana cooking have emerged: Cajun and Creole.

There are differences between these types of Louisiana cooking. Cajun (from the Native American pronunciation of Arcadian, French settlers from Nova Scotia) food is so-called "unsophisticated" country cooking that is influenced by African, Native American, and Caribbean flavors. It is hot, very spicy, and uses a dark roux and a lot of animal fat. Cajun food has many one-pot meals, including jambalaya and etouffee. Creole food is so-called "refined" city food (influenced more by Italians, Spanish, and French) and uses a light roux with more cream, butter, garlic, and tomato sauce.

These cuisines have given us "spiced up" dishes such as gumbos, blackened fish, dirty rice, shrimp Creole, Jezebal sauce, shrimp bisque, red rice and beans, and crawfish etouffee. Roux is the basis of traditional Cajun-Creole cooking, whether a base for sauces, gumbos, or etouffees. Other important ingredients of these cuisines are garlic, parsley, onions, bell peppers, sassafras, tomatoes, green onions, celery, and fish flavorings.

Tex-Mex or Texan

Tex-Mex cooking incorporates many diverse flavors: Native American, Southern, Mexican, and European. This diverse mix has given us fajitas, chilis, burritos, nachos, tacos, rice and beans, corn biscuits, and many other familiar foods.

Texas is a large state, and its cuisine varies with neighboring influences. There are strong Mexican and Native American influences, especially in south and southwest Texas. Tomatillas, oregano, cumin, jalapenos, corn, black beans, avocados, and beef are staple ingredients in this region. Cowboy cooking and the flavors of northern Mexico predominate in the West, resulting in barbecues and fajitas. In the East, there is a strong African influence, with okra, groundnuts, cornmeal, and chile peppers. The Gulf Coast region is influenced by Louisiana, with rice, shellfish, oysters, jambalaya, Creole sauce, and gumbos.

Southwest

The new Southwest cuisine spearheaded the modern concept of fusion cooking in North America. This cuisine focuses on visual appeal and taste and combines preparation techniques and ingredients from Europeans, Mexicans, Native Americans, and, recently, from Asians.

Native Americans such as Hopi, Pueblo, Navajo, Apache, and Zuni provided the corn, squash, berries, nuts, game, and fish; Spaniards and Mexicans brought their chile peppers, tomatoes, garlic, beans, chocolate, avocado, and onions; while the

European missionaries added their wheat, fruits, and meats. Many varieties of chilies are used, including anchos, jalapenos, poblanos, chipotles, and green New Mexican chilies. Juniper berries, black pepper, garlic, onions, paprika, oregano, Mexican cilantro, cumin, coriander, and lime predominate their cooking. Recently, the Chinese and Japanese introduced newer seasonings, including wasabi, ginger, soy sauce, and mustard. All of these spices and ingredients are added to stews, sauces, chilis, soups, grilled meats, chicken, fish, and vegetables to create a Southwestern flair.

Today, the traditional regional North American cooking is evolving to new levels of visual appeal and creativity. Newer immigrants from Central and South America, the Caribbean, India, and Southeast Asia are adding their touches and giving rise to newer versions of the regional cuisines. New, nontraditional regional foods are appearing, such as Floribbean, which combines habaneros, mojos, salsas, yuca, mango, and marinated steaks from the Caribbean and Central America with key lime, fried chicken, tabasco pepper sauce, and lima beans from the South.

Other regions of the United States are also developing new flavors that reflect their current populations. For example, in New York, new fusion foods such as Italian-Thai, Puerto Rican-Chinese, or French-Indian will reflect its global diversity of Southeast Asians, Italians, Dominicans, Jamaicans, and Asian Indians. Thus, we can expect many traditional American comfort foods to be revived in new regional flavors. Chili, mashed potatoes, and Virginia ham will appear with many new twists such as chipotle chili, sofrito mashed potatoes, or Virginia curried ham.

Below is a description of several of the more popular U.S. regional blends.

POPULAR NORTH AMERICAN REGIONAL SPICE BLENDS

Cajun Blend

Description

Cajun blend is a classic Louisiana spice blend used with barbecued meats, fish, and rice and beans. There are many variations, depending upon its end usage—whether to flavor rice, as a dry rub to "blacken" fish and chicken, or as a cooking sauce. It is spicy, hot, and pungent. With blackened foods, it leaves a slightly bitter aftertaste.

Ingredients

Cajun blend combines sweet paprika, ground celery, garlic, black pepper, onion, cayenne red pepper, caraway, Greek oregano, cumin, dillseed, mace, cardamom, turmeric, basil, marjoram, thyme, rosemary, and bay leaf.

Applications

To prepare Cajun blend, all of the spices are ground, except for the leafy spices—bay, thyme, marjoram, oregano, and basil, which are crushed and added at the end.

Cajun blend is used as a dry rub on chicken, steak, or fish to create "blackened" finished products. The leafy spice particulates get burnt in the process, whereby adding the authentic blackened flavor and crunchy texture to the product. Cajun blend gives spicy, aromatic, and nutty notes to the finished applications. It is used with red beans, corn, and red and green bell peppers in "dirty rice," which is a popular flavored rice in the South.

Creole Sauce Blend

Description

Creole sauce blend is a smooth, spicy, slightly sweet seasoning that is used with Burgundy wine, tomatoes, bell peppers, celery, and shrimps. It can be considered the spicy ketchup of Louisiana.

Ingredients

Creole sauce blend is made from onions, garlic, seafood stock, bay leaf, thyme, parsley, ground black pepper, ground red cayenne pepper, and green onions. Other flavorings used are oregano and lemon juice.

Applications

To make Creole sauce blend, all spices are mixed well, except the leafy spices. Then bay, thyme, parsley, and green onions are added and blended gently. The sauce is simmered with chicken, fish, or shrimp and topped over white rice. This sauce can be used as a spicy accompaniment to grilled meats and fish or as a rub for barbecues. It can also be used as a dip for chips or fresh, cut vegetables.

Green Chile Blend

Description

Green chili blend is a Southwestern-style seasoning with a mix of European, Mexican, Native American, and, more recently, Asian spices.

Ingredients

The typical chilies used are anchos, poblanos, chipotles, and green New Mexican chilies. They are mixed with juniper berries, black pepper, garlic, onions, paprika, oregano, Mexican cilantro, cumin, coriander, and lime flavor. Wasabi, ginger, and mustard can also be added.

Applications

All of the spices are ground or pureed to make a smooth or coarsely textured sauce. Green chili blends enhance stews, chilis, soups, grilled meats, chicken, fish, or vegetables that require a Southwestern flair.

Chili Powder

Description

Chile powder was created by the poor Mexicans of Texas who added some spices and chile peppers to meat to make a hearty stew. Mexican cooks served this to the local cowboys who called this stew chili con carne because of the chile peppers that predominated the flavor. The seasoning blend that went into this stew finally came to be called chili powder.

Today, chili powder is a mildly pungent seasoning that is used commonly in Tex-Mex foods, Texan barbecues, chili con carne and other chilis. Mexicans use this spice blend for burritos, enchiladas, and tamale fillings. It has a red to burgundy color and a deep, rich flavor with very little heat. There are medium to hot chile powders for those who desire varying heat levels.

Ingredients

Chili powder is a blend of ground chile pepper with ground cumin, garlic, Mexican oregano, and salt. The ground chile pepper used is usually ancho chile pepper, which is a mild chile pepper. About 80% of the blend is ground chile pepper. The hotter versions have the ground red cayenne pepper.

Applications

To prepare chili powder, all of the ground ingredients are well mixed. This chili blend is then mixed with lime juice and water and used as a rub to prepare poultry, beef, or pork for grilling, broiling, or barbecuing. For chili con carne, chili powder is added to tomato, onion, coriander, red and green bell peppers, black pepper, allspice, cloves, and cilantro. Then this mixture is added to ground beef and cooked to a thick consistency. Red kidney beans, lamb, or mutton can be substituted for beef. Corn and other beans such as pinta, black and white, can be used to make a vegetarian chili with chili powder.

Chili blend can also be applied to curries, sauces, and condiments to create varying notes, to add zest, or to enhance existing flavors.

Barbecue Sauce Blends

Description

Barbecue is the icon of North American cooking and is meat, poultry, and seafood, whole or pieces, slowly cooked in a pit or on a spit using hot coals or flavored wood chips or hardwood. They are typically marinated or covered with barbecue sauces that are clear yellow to dark reddish brown with spicy, smoky, sweet, or mildly tart flavors. There are many types of North American barbecue sauces in Texas, South Carolina, Memphis, Kansas, Kentucky, and many other regions which all have their own distinctive, traditional-style flavors. Newer versions and concepts of barbecue sauces are emerging that incorporate many Latin, Asian, and Caribbean concepts, such as tandoori, hoisin, or chipotle flavors.

Ingredients

The basic barbecue sauce is made from tomatoes, mustard, onions, vinegar, garlic, and brown sugar with smoky or grilled notes. Beer, wine, molasses, and other spices are optional ingredients. Texas sauces are thick, spicy, and hot and consist of paprika, chili powder, black pepper, dark brown sugar, and vinegar, Worcestershire sauce, ketchup, dry mustard, and jalapenos. Carolina and Virginia sauce is tart and thin with cider vinegar, ketchup, mustard, crushed red pepper, and brown sugar. Memphis sauce is spicy and sweet, with paprika, garlic, brown sugar, onion, and black pepper. Kansas City type is a thick tomato based sauce with sweet and tangy notes. Today's BBQ contains olive oil, bell peppers, honey, corn syrup, orange juice, chipotle, serranos, ancho, thyme, cilantro soy sauce, tomato juice, and lemongrass.

Applications

Chicken, ribs, beef, pork, or seafood is basted with these sauces and slowly barbe-cued over charcoal, hickory wood smoke chips, or other flavored chips until the finished product develops a charred or dark brown color and an aromatic flavor.

GLOBAL SPICE BLENDS

Seasoning or spice blends are generally unique to particular regions, yet there are some spice blends that are found throughout the world. These have a common theme, but contain different spices because of their availability and regional and cultural preferences. Two primary examples are the green spice blends and chile blends. Every country has its own green blend and chile blend that is used as seasonings, whether in soups, sauces, stews, rices, pastas, salads, or as marinades.

POPULAR GLOBAL SPICE BLENDS

Green Blends

Description

Green blends are savory mixtures of green leafy spices and green chile peppers, with other spices and ingredients. To North Americans, Italian pesto is the most familiar green blend, but there are other green blends, such as *zhug*, salsa verde, *achari*, or *chat masala*. Almost every ethnic cuisine has its own version of a green blend, with varying flavors and textures based upon the ingredients, preparation styles, applications, and how it is presented in a dish. A green blend's color can range from a bright to dull green, depending upon the type and amount of green ingredients and spices, and whether it is served fresh or cooked.

Ingredients

Essential to any green blend are the ingredients that provide the characteristic color—green leafy spices or green chile peppers. Other ingredients such as nuts, garlic, ginger, tomatoes, or lemon are added to give the blend added character or flavor. The most commonly used green leafy spices are basil, cilantro, parsley, and mint. They are popular because their flavors combine well with a wide variety of ingredients—onions, cucumbers, coconut milk, green mango, lemongrass, coriander, and cheeses. Some green leafy spices, such as chives and parsley, provide the greenness but with little flavor, while basil, cilantro, sage, or mint provide the green blend's dominant flavor. Vegetables such as cucumber, celery, or bell peppers are sometimes chopped and added for a crunchy texture. Some of the well-known global green spice blends and their typical ingredients are listed in Table 36.

Applications

Green blends can be added as a major seasoning in a sauce or curry or to provide zest to a stew, salad, or soup. They are added fresh or cooked to flavor salsas, chutneys, mustards, dips, or salad dressings. Commercially, green blends come in dry or paste forms. They can be smooth, coarsely textured, or in chopped forms, depending on the end usage. Whether one desires a green blend's domineering notes or its enhancing profile, the result is a wonderful balance of flavors in the finished product.

Each ethnic green blend has its own favorite leafy spice that gives its distinctive flavor profile. For example, sweet basil characterizes Italian pesto, while the more pungent Thai basil is favored in Thai green curry, and cilantro is preferred in Mexican

TABLE 36
Global Green Spice Blends

Green Blends	Ingredients
Thai green curry paste	Thai basil, lemongrass, green cayenne chilies, cilantro, shallot, galangal, coriander seed, black pepper, shrimp paste, coconut milk, kaffir lime leaf
Italian pesto	Sweet basil, pine nut, parmesan and romano cheese, garlic, olive oil
Mexican salsa verde	Cilantro, green tomato, green jalapeno, onion, green bell pepper
Puerto Rican recaito	Cilantro, culantro, black pepper, garlic
Middle Eastern tabbouleh	Parsley, mint, bulgar wheat, lemon juice, tomato
Malaysian kurma	Cilantro, spearmint, green chili, coriander seed, star anise, cardamom, coconut milk
Indian green masala	Green cayenne chili, cilantro, coriander seed, ginger, dried mango
Argentinean chimmichurri	Parsley, cilantro, garlic, lime juice, black pepper
Kenyan achari	Green cayenne chile pepper, unripe mango, cilantro, shredded coconut

green salsas. Green cayenne, jalapeno, serrano, Cubanelle, or New Mexican chile peppers provide color, as well as heat, flavor, and texture to the blend. Other spices or flavorings such as lemongrass, culantro, sorrel, lime, avocado, or curry leaf are added in smaller amounts to enhance its characteristic ethnic profile.

With the increasing availability of new ethnic ingredients and the growing demand for variety and stronger flavors, adventurous consumers will seek other "new" pestos. Concepts for new pestos can be derived from the other green blends around the world. Lemongrass, sweet basil, cilantro, pine nuts, parmesan cheese, and green chilies can be combined to create a new version of the traditional Italian pesto, such as a spicier Thai-Italian pesto or Mexi-Italian pesto.

Chile Blends

Description

First cultivated in the Andes of South America, the chile pepper has traveled globally to become the dominant component of many spice blends. Chile blends now provide flavor and color to foods in countries in all corners of the world—India, Sri Lanka, Southeast Asia, North Africa, the Caribbean, Latin America, and the United States. Each of these regions has its own favorite chile peppers—Thais and Malaysians prefer cayennes and bird peppers; Caribbeans enjoy habaneros; Mexicans choose poblanos, jalapenos, and serranos; and Peruvians like ajis, rocotos, and pequins.

Chile blends are prized not only for their heat but also for their delicate flavors. Chile blends incorporate one or more types of chile peppers, as well as other ingredients, thus creating many different tastes. Chile blends can be a simple mixture of chilies, salt, vinegar, and sugar, such as *cili boh* or *sambal olek*, or a complex mixture with other added ingredients.

TABLE 37
Popular Global Chile Blends

Chile Blend	Ingredients
Tunisian harissa	Cayenne/New Mexican, garlic, cumin, cinnamon, coriander seed, caraway, olive oil
Malaysian sambal belacan	Cayenne, garlic, shallot, shrimp paste, lime juice
Bajan hot sauce	Habanero, mustard, papaya, sugar, chives, black pepper, vinegar
Kenyan bajia	Cayenne, tomato, coconut milk and fermented liquor of coconut
Argentinian chimichurri	Jalapeno, onion, parsley, oregano, black pepper, lemon juice, vinegar
South African blatjang	Cayenne/bird pepper, onion, almond paste, dried apricot, apricot puree, garlic, bay leaf, lime juice
North American hot sauce/ Tabasco sauce	Cayenne/Tabasco pepper, vinegar, salt

Chile peppers can be added fresh or dried, pureed, chopped, or ground to create spice blends. The particular form used depends on the end product and how it is used. Chile peppers can be roasted to modify their flavors and used seeded or deseeded to modify their heat.

Ingredients

Many ingredients are mixed with chile peppers to create unique flavor releases, to tone down their heat, or even to add color to the finished product. Garlic, shallots, lemongrass, coriander, cilantro, tomato, green mango, tamarind, brown sugar, and lime juice lend well to chile blends. Other spices, onions, nuts, seeds, chocolate, vegetables, fruits, fermented shrimp or fish, and coconut can also be used. The choices depend on the particular application, ethnic preferences, and availability. Chile blends lend well to pickling or preserving in brine or vinegar and with other seasonings and, if desired, fruits or vegetables. Chile blends come in dry, liquid, and paste forms.

Some global chile blends and their typical ingredients are shown in Table 37.

Applications

The number of applications that use chile blends is almost endless, and they are not limited to foods. Every ethnic group has its own chile blends that contain locally available and typically preferred ingredients—Mexicans with their moles and ceviches, Asian Indians with their curries and *achars*, Thais with their salads and kaengs, Malaysians with their *sambals* or *cili boh*, Mediterraneans with their rouille and harissa, Caribbeans with jerk blends or Indipep sauce, and Africans with their piri-piri *blatjang* and *berbere*.

Even certain Latin American drinks, such as *atole*, *tepache*, or *chilote* contain chile blends. Atole, which is a corn-chile-chocolate-based drink enjoyed by the Aztecs, continues to be a favorite Mexican breakfast item. Whether to provide heat, flavor, or simply to perk up a dish, chili blends can be added during cooking to provide zest to sauces, soups, or stews, used as marinades, or served as a dip or freshly made condiment/relish to a meal.

Chile blends go well with grilled, steamed, or roasted meats, chicken, or fish. Their flavors are complemented with certain spices and flavorings. They are given a creamy, sweet mouthfeel with coconut milk; sweet and sour notes with tomatoes and palm sugar; sharp and sour tastes with lime juice and tamarind; rich roasted brown and bitter notes with coffee or chocolate; and savory fermented flavors with shrimp paste and spices.

Certain profiles are added to chile blends to appeal specifically to different ethnic groups: a tartness for North Americans (Louisiana hot sauce); a fishy, pungent note for the Vietnamese (nuoc cham); a sweet and sour note for Chinese; a sweet garlicky note for the Thais (Sriracha sauce); a pungent and slightly bitter taste for the Oaxacans (moles); a sour and spicy green note for the Indians (achars); and a fruity, mustardy note for the Caribbeans.

8 Commercial Spice Blend and Seasoning Formulations

Americans are beginning to demand the flavor and consistency of freshly prepared foods. Many enjoy preparing their meals or being part of the preparation and cooking process. As explained in detail in Chapter 2, with growing frequency, Americans are seeking well-seasoned and balanced meals that provide variety and flavors from far-off lands. Many Americans want to experience spices and seasonings from their travels. The acculturated ethnic groups in the United States, whether Asian Americans or Latino Americans, desire the foods of their homelands, but with a difference—combining them with flavors that they have been exposed to at school, and at local supermarkets, restaurants, and friend's homes. As a consequence, both authentic spices and ingredients and "cross-over ethnic" or fusion foods, with their multidimensional flavor and texture profiles, continue to be the trend.

Consumers continue to look for flavor in meals, but at the same time, with quick and easy preparation. In short, they want simple guides and shortcuts, that also provide the flavor and variety in meals. Many ethnic meals take a long time to prepare, have complicated recipes for Americans, and have spices and other ingredients that are not easily found in local supermarkets. Or they must be purchased in volumes greater than required for the occasional ethnic meal.

Ready-made or commercial spice blends or seasonings solve this dilemma and meet consumer's requirements for flavor, variety and ease in preparation. Spice blends, savory and sweet, render food and beverages more palatable. Generally they are added to foods before they are cooked, but may also be added as toppings to dishes before they are served, including as garnishes for salads, soups, cooked rice, and noodles. Spice blends allow consumers to create meals that are authentic, but cut the long preparation times required for many ethnic meals. When spice blends are sold with simple recipes that explain how to use them, cooking with spices becomes a pleasurable activity.

Creating commercial batches of ethnic foods from Asia, Latin America, the Caribbean, or the Mediterraneans presents many creative and technical challenges to food developers. Commercial spice blends or seasonings can provide consistency in flavor, color, or texture required in food service and industrial kitchens. They consist of spices and other ingredients, including flavorings, salt, sugar, dextrose, corn syrup solids, starches, maltodextrin, yeast extracts or hydrolysates, hydrolysed plant protein (HPP), MSG, or nucleotides. These ingredients are added with spices to enhance or intensify the overall flavor of the application. However, there is a growing demand for spice blends that are natural, without chemical preservatives, MSG, or HPP. Many emerging spice blends or seasonings are being formulated to follow this market trend.

This chapter offers seven commercial spice blends and seasoning formulations kindly contributed by spice, flavor, and food companies, as examples of commercial products that are available for creating marinades, rubs, sauces, soups, dips, salad dressings, or condiments. They can be used as a guide for formulating your spice blends or seasonings. These formulas have dried spices and extractives as well as other ingredients that can enhance specific applications. The spices and other ingredients in the formula are based on a formula which totals 100%, and written in descending order of amount used in the formula (in accordance with regulations). When spice extractives are used in the formula, they are first plated on a carrier, such as salt, sugar, dextrose, or maltodextrin, before being mixed with other spices or ingredients. Anti-caking agents such as tri-calcium phosphate or silicon dioxide, are added (below 2% by weight of the total blend) to prevent caking. Spice particulates, including parsley or cilantro, and dehydrated or freeze-dried vegetables or fruits, are added after the blended ingredients are sifted so as not to disintegrate or break their form.

The seven commercial formulations provided in this chapter are:

1. Mexican Red Mole Seasoning for Chicken, Con Agra Food Ingredient Company, Cranberry, New Jersey
2. Jamaican Jerk Seasoning for Meats, Illes Seasonings & Flavors, Dallas, Texas
3. Brazilian Savory Pineapple Seasoning for Chicken, Meat or Fish, International Flavor & Fragrance, Dayton, New Jersey
4. Moroccan Seasoning for Lamb, T. Hasegawa USA, Northbrook, Illinois
5. Vietnamese Nuoc Cham (Fish Sauce Condiment) Seasoning, David Michael & Co., Inc., Philadelphia, Pennsylvania
6. Thai Red Curry Seasoning for Shrimp, Newly Wed Foods, Horn Lake, Missouri
7. South Indian Spice Blend for Lentil (or dal) Curry, Horizons Consulting LLC d/b/a Taste of Malacca, New Rochelle, New York

MEXICAN RED MOLE SEASONING FOR CHICKEN (CON AGRA FOOD INGREDIENT COMPANY)

Moles are traditional "royal" sauces from Mexico, with each region and even each family having their own style. Moles were originally created for rulers of pre-Hispanic Mexico. The red mole or mole rojo is popular in the regions around Mexico City and Oaxaca State, and is nutty and slightly sweet, with bitter roasted notes. The mole rojo sauce is cooked with or poured over roast pork, chicken, or turkey and eaten with tortillas or white rice. A traditional mole rojo contains ground almonds, peanuts or pumpkin seeds, onion, garlic, dried red chilies and cilantro, with wheat or corn flour as thickeners.

Red Mole Seasoning

Ingredients	%
Salt	25
Ancho chile pepper, ground	22
Garlic powder	9
Onion powder	8.65
Sugar	7.3
Tomato powder	6.97
Cocoa powder, 10/12 alkalized	4
Chicken flavor powder	3.4
Chipotle pepper	3
Cinnamon, ground	3
Red pepper, ground, 10,000 *SHU	2.88
Caramel color, powder	1
Oregano, ground	0.75
Allspice, ground	0.55
Yeast extract	0.5
Cloves, ground	0.4
Peanut flavor, artificial, liquid	0.35
Coriander, ground	0.3
Annatto extract, liquid	0.3
Citric acid, fine, granular	0.3
Silicon dioxide	0.2
Tween 80	0.15
	100

* SHU - Scoville Heat Units

Mix salt, sugar, and citric acid in a blender. Slowly add peanut flavor and annatto extract and blend well. Add the spices and blend well. Then add silicon dioxide and blend until uniform. Sift.

Application of Red Mole Seasoning (as a marinade) for chicken. For a 20% vacuum tumble marinade:

Chicken breast	100 lb
Cold water	16.2 lb
Red Mole Seasoning	3.5 lb
Phosphate	0.3 lb

Dissolve phosphate in cold water (maximum 45°F). Stir in Red Mole Seasoning until completely dissolved. Add chicken breasts to vacuum tumbler. Pour the Red Mole marinade on top. Pull a vacuum to approximately 25 psi. Tumble under vacuum for 20 minutes at 35 rpm. Remove chicken and refrigerate overnight (30°F–35°F) to stabilize for better flavor impact. Cook as desired.

JAMAICAN JERK SEASONING FOR MEATS (ILLES SEASONINGS & FLAVORS)

Jerk seasoning is the soul of Jamaica's food culture. It is a dry spice blend or a paste that is used as a marinade or rub for flavoring meats or fish before they are grilled or barbecued. The ingredients in the blend vary, depending on the cook. Jamaican jerk blend is generally a blend of Scotch bonnet peppers, garlic, scallions or onion, thyme, cinnamon, ginger, allspice, cloves, and or nutmeg. The blend is added with vinegar, sugar, or soy sauce (sometimes with dark rum), and is commonly applied as a rub or marinade for pork, chicken, and fish. In Jamaica, jerk pork ribs and chicken are the most popular items, sold at roadside food stalls. They are cooked slowly on half-cut steel drums placed over smoky open pit fires of allspice (or pimento) branches or wood.

Jamaican Jerk Seasoning

Ingredients	%
Salt	25
Sugar, white, granulated	18
Sugar, light brown	18
Onion, granulated	12
Chile pepper, ground	6.5
Allspice, ground	6.5
Black pepper, ground	3.0
Cinnamon, ground	1.6
Cocoa powder	0.6
Soybean oil (processing aid)	0.5
Thyme leaves, ground	0.3
Thyme leaves, whole	2.0
Red pepper, crushed	6.0
	100

Disperse oil on salt. Blend in sugar and cocoa powder. Add ground spices (except whole thyme leaves and crushed red pepper). Combine all ingredients and blend well. Sift. Then add whole thyme leaves and crushed red pepper and mix evenly.

Jamaican Jerk Seasoning is mixed with oil or water and made into a paste which is then rubbed onto meats (beef or pork), poultry, or seafood at 3% to 4% (and to soups, dips, stocks, or dressings at 4% to 6%). Let the seasoned meat, poultry, or seafood sit for about an hour or more. Grill, BBQ, or saute.

Jamaican jerk seasoning can also be used at 4% to 6% in a dip, dressing, soup, or stock liquid.

BRAZILIAN PINEAPPLE SEASONING FOR
PORK, CHICKEN, OR FISH
(INTERNATIONAL FLAVOR & FRAGRANCE)

Brazil, the largest country in South America, has a multiethnic cuisine, which reflects the flavors of its diverse population—Indians, Portuguese, Spanish, African, Middle Eastern, and Asian—as well as bordering countries. Seafood stews predominate in the North, while the South is the land of churrascos (grilled meats). Other predominant ingredients include rice, manioc, breads, potatoes, pastas, cheese, eggs, sausages, soy products, corn, bananas and other fruits, and beans. Tropical fruits like pineapple play a key role in flavoring Brazilian dishes. Pineapple is not only made into sweets, desserts, beverages, but also flavors many savory dishes. In the seasoning formula below, pineapple is combined with paprika, black pepper, oregano, vanilla, sugar and other ingredients to add a savory sweet seasoning for chicken, meats, or seafood.

Brazilian Pineapple Seasoning

Ingredient	%
Salt	38
Pineapple, freeze-dried	17.4
Sugar, fine	8.33
Dextrose	7.80
Red pepper 40 SHU	7.07
Paprika, ground	5.20
Onion, granulated	4.50
Black pepper, 40 mesh	3.70
Granulated garlic	1.50
Vanilla extract	1.10
Oregano, ground	0 .80
Thyme, ground	0.40
Basil, ground	0.40
Yeast extract autolysate	0.17
Chardex hickory (dry smoke)	0.17
Durkex oil	1.50
Citric acid	1.33
Silicon dioxide	0.63
	100

Disperse Durkey oil and vanilla extract on salt. Add remaining ingredients and blend. Add silicon dioxide and blend well. Sift.

Use at 2% to 3% on chicken, pork, fish, or vegetables, and prepare as desired.

MOROCCAN SEASONING FOR LAMB
(T. HASEGAWA USA)

With Arab, Berber, Turkish, and European influences, Moroccan cuisine has mild to moderately spiced dishes. Garlic, onion, cumin, mint, lemon, black pepper, parsley, paprika, and coriander are popular flavorings for salads, vegetable soups, stews, couscous, chickpeas, lamb, or fish. Every vendor or cook makes her own seasoning recipes, creating many versions of ras-el hannouth or chermoula or any other meat or chicken marinades used for kebabs, *keftas*, *merguez*, seafood, or tagines. Olive oil is a favorite in Moroccan cooking, in addition to clarified ghee.

Moroccan Seasoning

Ingredient	%
Olive oil, extra virgin	50.0
Lemon juice	17.0
Parlsey, chopped	10.1
Salt	8.50
Natural caramelized garlic flavor *WONF	2.16
Natural caramelized onion flavor WONF	2.16
Natural Black pepper flavor	2.02
Natural red chile pepper flavor	2.02
Natural capsicum flavor	1.50
Natural oregano flavor	1.37
Marjoram, chopped	1.00
Natural coriander flavor	0.94
Natural lemon WONF	0.87
Natural spearmint flavor WONF	0.36
	100

* WONF - with other natural flavors.

Combine all ingredients and flavors and mix well. Add the chopped parsley and marjoram. Blend well and pour over lamb. Coat lamb well in marinade. Let sit for few hours or overnight depending on the cut and size of lamb. Grill, saute, or oven-roast the lamb.

THAI RED CURRY SEASONING FOR SHRIMP
(NEWLY WED FOODS)

Thai red curry is a fiery, pungent, and coconut-laden curry that is a favorite with Thais and, today, Americans and others around the world. It possesses a hot, sweet, and sour flavor, very different from the Asian Indian curries. The commercial Thai red curry seasoning comes as a dry blend or a wet paste, with Thai red chilies, lemongrass, galangal, coriander root, and garlic. Ground spices (cumin and coriander), and leafy spices (cilantro, basil leaves, and kaffir lime leaves), add another dimension to its flavor. The seasoning is combined with coconut milk, brown sugar, fish sauce, and lemon juice. The sauce is then cooked with shrimp, chicken, fish, meats, and vegetables, such as red bell peppers, carrots, green beans, eggplant, and or bamboo shoots.

Thai Red Curry Seasoning

Ingredients	%
Thai red chilies, dried, ground	29.58
Sea salt	20.71
Shrimp, dried, ground	12.72
Cumin, ground	8.38
Garlic, dried, ground	7.20
Shallots, dried, ground	6.31
Lemongrass powder	5.03
Galangal powder	4.34
Coriander, ground	3.75
Kaffir lime leaf, dried	1.98
	100

Combine all ingredients, (except kaffir lime leaves), and blend well. Sift. Add kaffir lime leaves and blend.

Application: Use about 2% Thai Red Curry seasoning for shrimp red curry:

Raw shrimp, peeled, deveined	57.45%
Coconut milk	37.75%
Fish sauce	2.98%
Red curry seasoning	1.82%
	100%

Cook as desired. Add more salt if necessary. Garnish with fresh Thai basil, slivered bamboo shoots, onion, and green and red bell peppers.

VIETNAMESE FISH SAUCE SEASONING
FOR CONDIMENT
(DAVID MICHAEL & CO., INC.)

Fish sauce, or *nuoc mam*, is a widely used condiment in Vietnam, Thailand, Korea, and the Philippines. Fish sauce to Southeast Asians is as soy sauce to Chinese, or wine to the French. It enhances and perks up most dishes, whether roast chicken, grilled seafood, dips, or stir-fried vegetables. It is made from anchovies, salt, and water, fermented in large wooden vats in dark warehouses, anywhere from three months to a year. High quality fish sauce has a rich pungent flavor and a golden color. Lower quality fish sauces, with less fermentation, are less pungent but sweeter. During cooking, the pungent flavor mellows out and enhances the finished product.

It is commonly used for marinades, sauces, soups, stir-fries, or sausages. *Nuoc mam* is combined with sugar, vinegar, lime juice, shredded carrots, or spring onions for preparing the popular condiment *nuoc cham*. *Nuoc cham* is used as a dipping sauce for various types of fried or fresh spring rolls, cooked meats, and fish. It is also added to dressings for salads (prepared with papaya or lotus root), rice noodles, rice dishes, and steamed savory cakes. *Nuoc mam* and *nuoc cham* are essential items in every household kitchen in Vietnam or Thailand.

Fish Sauce Seasoning or *Nuoc Cham*

Ingredients	%
Fish Sauce* (Three Shrimps Brand)	25.55
Coconut Soda	49
Sugar	19.45
Lime Juice	6
	100

* (There are many brands of fish sauce in market but Three Shrimps Brands or Three Crab Brands are recommended)

Combine and mix all ingredients together and bring to a boil.Let simmer for 15 to 20 minutes. Allow to cool and regrigerate. Mix with shredded carrots, sliced chilies or chile sauce, and/orjulienned cucumbers.

Use as a dip or to perk up soups, salads, seafood, poultry, or meats.

SOUTH INDIAN SPICE BLEND FOR LENTIL (OR DAL) CURRY (HORIZONS CONSULTING LLC D/B/A TASTE OF MALACCA)

In India, curry powders vary with each dish or application, cook, and region. Numerous variations of curries abound depending on taste preferences and ingredients. In England curry has a touch of sweetness; in Jamaica, black pepper and turmeric predominate; the Thais add lemongrass, kaffir lime, and cilantro; Keralans enjoy curry leaves, green chilies, and coconut milk; while Malaysians savor star anise and fennel. But, the three essential spices in all curry powders are turmeric, coriander, and cumin. Chile powder or black pepper is optional if heat is desired. Then according to the type of application, other spices are layered on top, such as ginger, which is favored in fish, lentil, or vegetarian curries; fenugreek, which enhances fish dishes; while cinnamon, cardamom, clove, and nutmeg perk up meat dishes. Indians prepare lentils and beans in different styles, depending on the regions. Below is a simple curry blend for lentils (also can be used for vegetables) from southern India that meets the growing trend for all-natural labeling, without chemical preservatives, MSG, HPP, salt, or sugar.

South Indian Curry Blend

Ingredients	%
Turmeric, ground	30.00
Cumin, ground	25.44
Ginger, ground	19.05
Coriander, ground	14.59
Black pepper, ground, 35 mesh	3.81
Cinnamon, ground	2.54
Fenugreek, ground	2.54
Red pepper, 20,000 Scoville units	1.27
Clove, ground	0.76
	100

Preparation of lentil curry: Boil 1 cup lentils till somewhat soft. In a separate pan, heat oil, add 1 lb sliced ginger and half cup sliced onions, and saute till done. Blend in 1–1½ tbs South Indian curry blend and saute for a few seconds. Then add the sauteed spice blend-onion mixture to the cooked lentils, add some water to create sauce consistency and blend well. Add one medium sliced tomato and cook till done. Add salt to taste, blend and cook for another minute.

Optional: In a separate pan, heat oil and add 1 tsp mustard seeds till it "'pops." Toss in some fresh kari (or curry) leaves and add this to the lentil curry and blend well.

Serve with cooked rice or use as a dip for flatbreads such as naan, *paratha*, *roti canai*, pita bread, or *chappati*.

Bibliography

Achaya, K.T., 1994, *Indian Food: A Historical Companion*, Oxford University Press, New York.

Andrews, Jean S., 1984, *Peppers, Domesticated Capsicums*, University of Texas Press, Austin.

Ashurst, P.R., 1995, *Food Flavorings,* 2nd ed., Blackie Academic and Professional, imprint Chapman and Hall, New York.

ASTA Cleanliness Specifications for Spices Seeds and Herbs, Rev. April 28, 1999, ASTA, Inc., Engelwood Cliffs, NJ.

ASTA, 2001, Statistics report, American Spice Trade Association, Washington, DC.

Bremness, Lesley, 1991, *The Complete Book of Herbs*, Color Library Books Library, England.

Burdock, George A., 1994, *Fenarolis Handbook of Flavor Ingredients,* 3rd ed., CRC Press, Boca Raton, FL.

Chopra, Deepak, 1991, *Perfect Health,* Harmony Books, div. Crown Publishers, New York.

Code of Federal Regulations, Food and Drugs Administration: 21 CFR 101.22, 21CFR 101.3, 21CFR 102.3, 21CFR102.5; 21CFR 182.10, CFR 182.20; 21USCS 321, 21 USCS 342, 21 USCS 343.

Corn, Charles, 1998, *The Scents of Eden:A Narrative of the Spice Trade*, Kodansha International, Japan.

Czarnecki, Jack, 1995, *A Cook's Book of Mushrooms*, Artisan, div. Workman Publishing, New York.

Desai, Urmila, 1995, *The Ayurvedic Cookbook*, Lotus Press, Twin Lakes, WI.

DeWitt, Dave and Gerlach, Nancy, 1990, *The Whole Chile Pepper Book*, Little Brown & Company, Canada.

DeWitt, Dave, 1999, *The Chile Pepper Encyclopedia*, William Morrow,, imprint of Harper-Collins, London.

Disney, A.R., 1978, *Twilight of the Pepper Empire*, Harvard University Press, Cambridge,.

Farrell, Kenneth T., 1985, *Spices, Condiments and Seasonings,* Avi Publishing, Westport, CT.

Furth, Peter and Cox, Derryck, August, 2004, The Spice Market Expands, *Food Technology*, Vol. 58, No. 8.

Hadady, Letha, 1996, *Asian Health Secrets,* Crown Publishers, New York.

Heinerman, John, 1994, *Encyclopedia of Healing Juices*, Prentice Hall, New York.

Johari, Harish, 1994, *The Healing Cuisine*, Healing Arts Press, Rochester, VT.

Katzer, Gernot, 1999, *Spice Pages*, retrieved from http://www.uni-graz.at/~katzer/engl/index.html.

Kenji, Hirasa and Mitsuo, Takemasa, 1998, *Spice Science and Technology,*. Marcel Dekker, New York, pp. 163–198.

Khan, M.T., *Spices in Indian Economy*, Academic Foundation, Delhi, India.

Lawless, Julia, 1995, *The Illustrated Encyclopedia of Essential Oils*, Barnes and Noble, London.

Lust, John B., 1974, *The Herb Book,* 2nd ptg., Benedict Lust Publications, Simi Valley, CA.

Mahindru, S.N., 1982, *Spices in Indian Life,* 1st ed., Sultan Chand & Sons, New Delhi, India.

Mazza, G. and Oomah, B.D., Eds., 2000, *Herbs, Botanicals and Teas,* Technomic,, Lancaster, PA.

Mindell, Earl, 1994, *Food as Medicine*, Simon & Schuster, New York.

Naj, Amal, 1993, *Peppers: A Story of Hot Pursuits*, Vintage Books, div. Random House, New York.

Norman, Jill, 1991, *The Complete Book of Spices*, Viking Studio Books, div. Penguin Books, New York.

Onstad, Dianne, 1996, *Whole Foods Companion*, Chelsea Green Publishing Company, VT.

Ortiz, Elisabeth L., 1994, *Encyclopedia of Herbs, Spices and Flavorings*, Carroll and Brown, London.

Parry, John W., 1953, *Story of Spices*, Chemical Publishing, New York.

Pennington, Jean A.T., 1998, *Food Values of Portions Commonly Used*, Lippincott-Raven Publishers, Philadelphia, PA.

Pruthi, J.S., 1976, *Spices and Condiments*, National Book Trust, New Delhi, India.

Purseglove, J.W., Brown, E.G., Green, C.L., and Robbins, S.R.J., 1981, *Spices, Volume 1*, Longman Group, Harlow, England.

Purseglove, J.W., Brown, E.G., Green, C.L., and Robbins, S.R.J., 1987–88, *Spices, Volume 2*, Longman Group, Harlow, England.

Raghavan, Susheela, retrieved from www.susheelaconsulting.com or www.tasteofmalacca.com.

Raghavan, Susheela, May, 2000, Foodservice Trends: Preparation Methods To Create Flavor, *Food Product Design*.

Raghavan, Susheela, May, 2000, Spices: Tools for Alternative or Complementary Medicine, *Food Technology*.

Raghavan, Susheela, May, 2000, Thai Dining, *Food Product Design*.

Raghavan, Susheela, January, 2000, Latin American cuisine, *Food Product Design*.

Raghavan, Susheela, January, 2001, North African cuisine, *Food Product Design*.

Raghavan, Susheela, December, 2003, Brazilian cuisine, *Food Product Design*.

Raghavan, Susheela, August, 2004, Developing Ethnic Foods and Ethnic Flair with Spices, *Food Technology*, Vol. 58, No. 8.

Raghavan, Susheela, December, 2005, Touring South American Cuisine, *Food Product Design*.

Reineccius, Gary, 1994, *Source Book of Flavors,* Chapman & Hall, New York, pp. 186–199, 255–360.

Ridley, Henry N., 1912, *Spices*, Macmillan, London.

Rosengarten, Frederic, Jr., 1979, *The Book of Spices*, Pyramid Communications, New York.

Schivelbusch, Wolfgang, 1993, *Taste of Paradise,* Vintage Books, div. Random House, New York, pp. 3–14.

Stobart, Tom, 1982, *Herbs, Spices and Flavorings*, Milton Peter Corp. with The Overlook Press, Woodstock, NY.

Stuckey, Maggie, 1997, *The Complete Spice Book*, St. Martin's Press, New York.

Tiwari, Maya Ayurveda, 1995, *A Life of Balance*, Healing Arts Press, VT.

Toussaint-Samat, Maguelonne, Reprinted 1993, *A History of Foods,* Blackwell Publishers, Cambridge, MA.

Uhl, Susheela R., September, 1995, Making Ethnic Authentic, *Food Product Design*.

Uhl, Susheela R., April, 1996, Designing for Hispanic Marketplace, *Food Product Design*.

Uhl, Susheela R., July, 1996, Ingredients: The Building Blocks for Developing New Ethnic Foods, *Food Technology*.

Uhl, Susheela R., November, 1996, Legumes, *Food Product Design*.

Uhl, Susheela R., November, 1997, Ethnic Side Dishes, *Food Product Design*.

Uhl, Susheela R., May, 1997, Spices & Spice Extractives, *Food Product Design*.

Uhl, Susheela R., September, 1998, 4 Steps to Create Authentic Ethnic Foods, Special Edition, *Food Product Design*.

Uhl, Susheela R., June, 1998, Asian Ingredients in American Cooking, *Food Forum.*
Uhl, Susheela R., March, 1998, Cilantro: Flavorings, Fine Cooking, Taunton Press, Newton, CT.
Uhl, Susheela R., June, 1998, Ethnic Entrees, *Food Product Design.*
Uhl, Susheela R., April, 1998, Hot & Spicy Foods, *Food Product Design.*
Uhl, Susheela R., July, 1998, New Ethnic Entrees, *Food Product Design.*
Uhl, Susheela R., May, 1999, Creatively Carribean, *Food Product Design.*
Uhl, Susheela R., December, 1999, Exploring Asian Ethnic, *Food Product Design.*
Uhl, Susheela R., April, 1999, Mediterranean Medley, *Food Product Design.*
Uhl, Susheela, R., January, 2000, Ethnic Extravaganza, *Refrigerated and Frozen Foods.*
Uhl, Susheela, R., May, 2000, Spice Translator, retrieved from www.SusheelaConsulting.com.

COOKBOOKS

AFRICAN

Cusick, Heidi Haughy, 1995, *Soul and Spice*, Chronicle Books, San Francisco.
Hachten, Harva, 1998, *Best of Regional African Cooking*, Hippocrene Books, New York.
Hafner, Dorinda, 1993, *A Taste of Africa,* Ten Speed Press, Berkley, CA.
Harris, Jessica B., 1998, *The Africa Cookbook: Taste of a Continent*, Simon & Schuster, New York.

ASIAN

Alejandro, Reynaldo G., 1998, *The Food of Philippines,* Periplus Edns. (HK) Ltd.
Bhumichitr, Vatcharin, 1994, *Vatch's Thai Cookbook,* Pavilion Books, Great Britain.
Brennen, Jennifer, 1984, *Encyclopedia of Chinese and Oriental Cooking*, St. Martins Press, New York.
Choi, Trieu Thi and Isaak, Marcel, 1997, *The Food of Vietnam,* Periplus Edns.
DeWitt, Dave and Pais, Arthur, 1994, *A World Of Curries*, Little, Brown & Company, Boston.
Kosaki, Takawaki and Wagner, Walter, 1996, *The Food of Japan,* Periplus Edns.
Krauss, Sven, Ganguillet, Laurent, and Vira Sanguanwong, 1996, *The Food of Thailand,* Periplus Edns.
Madhur, Jaffrey, 1995, *Flavors of India,* BBC Books, London.
Millon, Marc and Kim, 1991, *Flavors of Korea*, Andre Deutsch, London.
Pham, Mai, 1996, *The Best of Vietnamese and Thai Cooking*, Prima Publishing, Roclin, CA.
Sahni, J., 1996, *Savoring Spices and Herbs*, William Morrow, New York.
Sri, Owen, 1994, *Indonesian Regional Cooking*, Doubleday, Great Britain.
Tettoni, Luca Invernizzi and Hutton, Wendy, 1995, *The Food of Malaysia,* Perplus Edns .
Tettoni, Luca Invernizzi and Hutton, Wendy, 1996, *The Food of China,* Periplus Edns.von Holzen, Heinz and Arsana, Lother, 1995, *The Food of Indonesia,* Periplus Edns.
Yan, Martin, 1995, *Culinary Journey through China*, Publishers Group West, San Francisco.

CARIBBEAN

Dedeaux, Devra, 1991, *The Sugar Reef Caribbean Cookbook*, Dell, New York.
Hafner, Dorinda, 1996, *Taste of the Caribbean*, Ten Speed Press, Berkley, CA.

EUROPEAN

Child, Julia, 1998, *The French Chef Cookbook*, Ballantine, New York.
Ducasse, Alain, 1998, *The Flavors of France,* Artisan, div. Workman, New York.
Root, Waverly, 1992, *The Food of France*, Vintage Books, div. Random House, New York.

LATIN AMERICAN

Botafogo, Dolores, 1993, *The Art of Brazilian Cooking*, Hippocrene Books, New York.
Green, Linette, 1994, *A Taste of Cuba*, Penguin, New York.
Karoff, Barbara, 1990, *South American Cooking*, Aris Books, Editorial Offices and Test Kitchen, Berkley, CA.
Ortiz, Elisabeth Lambert, 1994, *The Book of Latin American Cooking*, Ecco Press, Hopewell, NJ.
Ortiz, Yvonne, 1997, *A Taste Of Puerto Rico*, Penguin, New York.
Quintana, Patricia, 1993, *The Taste of Mexico*, Stewart, Tabori and Chang, New York.
Romano, Dora, 1993, *Rice and Beans and Tasty Things: A Puerto Rican Cookbook*, Ramallo, Puerto Rico.
Zarela, Martínez, 1992, *Foods from My Heart,* MacMillan, New York.
Zaslavsky, Nancy, 1995, *A Cooks Tour of Mexico*, St. Martin's Press, New York.

MEDITERRANEAN

Chapman, Pat, 1989, *Homestyle Middle Eastern Cooking,* The Crossing, Freedom, CA.
Hazan, Giuliano, 1993, *The Classic Pasta Cookbook*, Dorling Kindersley, New York.
Hazan, Marcella, 1984, *More Classic Italian Cooking*, Ballantine Books, div. Random House, New York.
Hazan, Marcella, 1997, *Marcella Cucina*, HarperCollins, New York.
Hess, Reinhardt and Sabine, Salzer, 1999, *Regional Italian Cuisine*, Barron's Educational Series, Hauppauge, NY.
Roden, Claudia, 1974, *A Book of Middle-Eastern Food*, Random House, New York.
Roden, Claudia, 1987, *Mediterranean Cookbook*, Alfred A. Knopf, New York.
Root, Waverly, 1992, *The Food of Italy*, Vintage Books, div. Random House, New York.
Wolfert, Paula, 1994, *Mediterranean Cooking*, HarperCollins, New York.

NORTH AMERICAN

Pyles, Stephan, 1993, *The New Texas Cuisine*, Doubleday, div. of Bantam Doubleday, see Publishing Group Inc., New York.
Raichlen, Steven, 1993, *Miami Spice: The New Florida Cuisine*, Workman , New York.
Stromquist, Joan, 1993, *Sante Fe: Hot and Spicy Recipe*, Tierra Publications, Sante Fe, NM.

Index

D

J

Jackfruit, 206
Jaggery, 209
Jamaican jerk seasoning. *See also* Jerk seasoning
   commercial preparation, 298
Jamaican sorrel, 173, 193
Japanese cooking, 246
   characteristics of, 15
   wrappers in, 196
Japanese seven-spice blend, 249
Jasmine, 193
Jerk seasoning, 274, 279
   commercial preparation, 298
Juniper, 121–122

K

Kaffir lime, 122–123, 206
Kala masala, 253
Kalamansi, 205
   sour, 206
Kampuchean cuisine, 246
   regional curries, 258
Kapha dosha, 44
   spices to balance/reduce, 44
Kari leaf, 123–124
   in etymology of curry, 251
Kari podis, 251
Karnataka, 3
Kashmiri masala, 253
Kerala, 3
Kernels, 25
Kimchis, 21
King Solomon, 5
Kokum fruit, 207
Kola nuts, 217
Korma, 253
Krill, in Malaysian cuisine, 199

L

Labeling, 55–56
   of spice extractives, 33
Languedoc regional cuisine, 263
Laos, regional curries, 258
Latin America
   Caribbean regional cuisines, 231–232
   Central American regional cuisines, 232
   ethnic spices and seasonings of, 18
   Mexican regional cuisines, 230–231
   regional cuisines of, 230–234
   South American regional cuisines, 232–234

   spice blends of, 229–230, 234–240
   wrappers in, 196–197
La tin American spice blends, 234
   adobos, 234–2345
   mojo blends, 237
   mole blends, 235–236
   recado blends, 238–239
   salsa blends, 237–238
sofrito blends, 239–240
Latin Americans, 11
Latino cuisines, 12, 15
Laurel leaf, 73–75
   Boldo leaf, 76–77
   Indian bay leaf, 75
   Indonesian bay leaf, 76
   West Indian bay leaf, 76
Lavender, 124–126, 193
Leafy plant parts, xvii, 25
Leaves
   emerging global spices/flavorings, 19
   as wrappers, 194
Leeks, Aryan disdain for, 3
Legumes
   as emerging global spices/flavorings, 19, 213
   in ethnic cuisines, 213–215, 215
   soybeans, 214–215
Lemon, in ethnic cuisines, 205–206
Lemon balm, 126
Lemon basil, 71
Lemon verbena, 127
Lemongrass, 22, 127–129
Lentils, 214
   as dals in South Indian vegetarian cuisine, 254
Licorice
   in ethnic cuisines, 189–190
   reported therapeutic effects, 47
Light, effects of exposure to, 61
Lily buds, 192
Lime, in ethnic cuisines, 205–206
Liquid soluble spices, 32
   uniform dispersion of color and flavor with, 33
Liver, benefits from curcumin, 47
Local dialects, spice names in, xix
Long pepper, 3, 152–158
Lotus flower, 193
Lotus leaf, as wrappers, 195
Lotus root, in ethnic cuisines, 190
Louisiana cuisine, 286–287
Lovage, 129

M

Mace, 130–131
   antioxidant properties, 41

Z

T - #0364 - 071024 - C8 - 234/156/16 - PB - 9780367390099 - Gloss Lamination